Molecular Magnets

Molecular Magnets

Special Issue Editors

Maria Bałanda
Magdalena Fitta

MDPI • Basel • Beijing • Wuhan • Barcelona • Belgrade

MDPI

Special Issue Editors

Maria Bałanda
Polish Academy of Sciences
Poland

Magdalena Fitta
Polish Academy of Sciences
Poland

Editorial Office
MDPI
St. Alban-Anlage 66
4052 Basel, Switzerland

This is a reprint of articles from the Special Issue published online in the open access journal *Crystals* (ISSN 2073-4352) in 2019 (available at: https://www.mdpi.com/journal/crystals/special_issues/Molecular_Magnets)

For citation purposes, cite each article independently as indicated on the article page online and as indicated below:

LastName, A.A.; LastName, B.B.; LastName, C.C. Article Title. *Journal Name* **Year**, *Article Number*, Page Range.

ISBN 978-3-03897-710-0 (Pbk)
ISBN 978-3-03897-711-7 (PDF)

Contents

About the Special Issue Editors

Maria Bałanda, Professor, senior researcher at the H. Niewodniczański Institute of Nuclear Physics, Polish Academy of Sciences, Kraków, Poland. Graduated in physics from the Jagiellonian University, (Kraków), received a Ph.D. in the field of HTc superconductivity. Her habilitation, obtained in 2007 from her institution, was devoted to low dimensional molecular magnets. She has performed experiments with AC susceptometry/DC magnetometry, X-ray and neutron diffraction, as well as muon spin rotation techniques. She has published ca. 120 articles in peer-reviewed journals on magnetic oxides, intermetallics, superconductors and molecular systems and is the author or co-author of chapters in the books "Relaxation Phenomena" Ed. W. Haase & S. Wróbel, Springer; "Molecular Magnetic Materials" Ed. B. Sieklucka & D. Pinkowicz, Wiley; "Investigations of Phase Transitions" Ed. E. A. Mikuli, UJ (in Polish). Her fields of interest include magnetic and superconducting materials, phase transitions, magnetic relaxation, and low-dimensional and functional molecular magnets.

Magdalena Fitta, Ph.D., researcher at the H. Niewodniczański Institute of Nuclear Physics, Polish Academy of Sciences, Kraków, Poland. She graduated in physics from the AGH University of Science and Technology, (Kraków), and received her Ph.D. (2011) from the Institute of Nuclear Physics, Polish Academy of Sciences with research on molecular magnets. She was awarded a pre-doctoral Santiago Grisolia scholarship by the Generalitat Valenciana (2008) and the START scholarship for the young researchers by the Foundation for Polish Science (2012). She has performed experiments with AC susceptometry/DC magnetometry, X-ray diffraction, as well as UV-Vis and IR spectroscopy. She has published ca. 50 peer-reviewed journal articles on topics related to intermetallic compounds, thin films of molecular magnets and nanoparticles. Her current research interests focus on functional molecular systems with tunable magnetic properties, low dimensional molecular magnets as well as composite materials.

crystals

MDPI

Editorial

Molecular Magnets

Maria Bałanda * and **Magdalena Fitta**

Institute of Nuclear Physics Polish Academy of Sciences, Radzikowskiego 152, 31-342 Krakow, Poland; magdalena.fitta@ifj.edu.pl
* Correspondence: maria.balanda@ifj.edu.pl

Received: 28 February 2019; Accepted: 4 March 2019; Published: 6 March 2019

Molecular magnetism is an interdisciplinary research area, which deals with design, synthesis and physical characterization as well as the theoretical modeling of molecular materials showing acquired properties. The features that distinguish molecular magnets from traditional magnetic materials are: low-density, transparency to electromagnetic radiation, and sensitivity to external stimuli such as light, pressure, temperature, chemical modification or magnetic/electric field. Furthermore, molecular magnetism offers an exceptional collection of materials of various magnetic dimensionality: from 0D single-molecule magnets and 1D single-chain magnets, regarded as molecular nanomagnets due to slow relaxation and bistability at low temperatures, through 2D molecular layers, to 3D coordination polymers showing the collective ordering of magnetic moments below the critical temperature T_c. Research into molecule-based materials, both theoretical and experimental became more intense at the end of 20th century and has concentrated on (i) low dimensional materials, motivated by their potential applicability in high-density magnetic storage or nanoscale devices and (ii) on "functional" materials, strongly responding to change in external parameters, that may be used in sensors of a different type. This Special Issue shows the rich palette of the properties of magnetic molecular materials and presents current work on this interesting and important topic. The issue contains four review articles and also includes the results by the authors, as well as original contributed papers.

Molecular magnets involve well-localized magnetic moments, which make them the perfect playground for the investigation of intriguing phenomena and testing theoretical models. The interplay of spatial anisotropy of the exchange coupling and the intrinsic or magnetic-field induced spin anisotropy were discussed in the two-dimensional magnetic models by Orendáčová et al. [1]. In this excellent review, authors provide a concise introduction of the development of the theory and numerical approaches aimed at the description of low-dimensional magnetism. The main body of this paper presents a thorough study of the ground state and finite-temperature properties of the S = ½ models interpolating between the quantum Heisenberg antiferromagnetic chain and the rectangular spin lattice. The effect of the possible inter-layer exchange coupling resulting in the stabilization of the 3D long-range order is also discussed. Some physical consequences following from the characteristics of the underlying model are evidenced by reporting consistent experimental data on magnetic and calorimetric properties of low-dimensional Cu(II) based metal-organic magnets.

A complete understanding of the mechanisms of the magnetic interaction in molecular magnets requires advanced calculations such as proposed by Brumfield and Haraldsen in [2]. Based on the Heisenberg spin-spin exchange model, the general relationships for the quantum energy levels, the spin states for isosceles spin trimmers of spins equal ½ up to 5/2 are provided. Dependence on heat capacity, magnetic susceptibility on temperature, and the corresponding inelastic neutron scattering structure factors have been determined. As stated by the authors, results of the calculations could help with the general analysis and characterization of magnetic molecule-based systems.

The influence of external pressures on the structural and magnetic properties of molecular magnets has been reviewed by Zentkova and Mihalik [3]. The underlying mechanisms of the effect are presented on the example of $Cr(CN)_6$-based Prussian blue analogues (PBs)—TM-$Cr(CN)_6$ and

K-M-Cr(CN)$_6$, where TM = Mn, Ni, Co, Cr, and M = Mn, Ni. After reviewing the sensitivity of PBs to external stimuli and the high potential of PBs for applications, the authors describe the super-exchange interaction between the metallic ions and give an analysis of the pressure effect in case of predominated ferromagnetic, ferrimagnetic as well as mixed ferro–ferromagnetic interactions. Results of structural, magnetic and Raman spectra measurements at pressures of up to GPa for different analogues are shown. The paper clearly explains the reason of various pressure-induced T$_c$ change observed for subsequent samples under study. The pressure effect on magnetic properties was also checked for hexacyanometalate-based nanoparticles and core–shell heterostructures.

Similar to other magnetic materials, molecular magnets also exhibit a magnetocaloric effect (MCE). The review of the magneto-thermal properties of octacyanometallate-based and molecular magnets showing the different types of crystal architecture is presented by Fitta et al. [4] MCE study was performed for 3D and 2D coordination polymers and the high spin cluster built with [Nb^{IV}(CN)$_8$] or [W^{V}(CN)$_8$] molecular blocks. The results were obtained by means of two experimental methods, i.e., calorimetry and magnetometry. Moreover, the study of a new effect-rotating magnetocaloric effect (RMCE) in two 2D molecular magnets was presented. Dependence of MCE on temperature and on the value of the magnetic field was tested. It was found that the Ni$_9$[W(CN)$_8$]$_6$ cluster compound has a potential for cryogenic magnetic cooling. Conclusions related to MCE scaling and critical behavior in some systems under study were also included.

Two new oxime-based cationic [Mn$_6$]$^{2+}$ complexes, synthesized and characterized structurally and magnetically, are presented by Rojas-Dotti et al. [5]. A ground state spin value of both systems is $S = 12$. In these compounds the slow relaxation of magnetization occurs, which is consistent with the single-molecule magnet (SMM) behavior. Moreover, the energy barrier for the relaxation of magnetization determined for one of the considered compounds is the highest reported so far for cationic oxime-based [Mn$_6$]$^{2+}$ systems. As mentioned by the authors, such a type of the cationic SMM can be used as precursors of new multifunctional magnetic materials through the incorporation of anionic species that bring for instance conductivity or luminescence to the final material.

Example of the interesting solvatomagnetic compound based on [Cu(cyclam)]$^{2+}$ and [W(CN)$_8$]$^{3-}$ building blocks is presented by Nowicka et al. in [6]. The removal of water molecules from the ladder-chain crystal structure of the compound during the dehydration process leads to the modification of the geometry of the bonds, and finally to the single-crystal-to-single-crystal structural transformation. The noticeable change of magnetic properties was observed and reflected in switching the predominant intra-chain interactions from ferromagnetic in a hydrated compound to antiferromagnetic in an anhydrous sample. The dehydration process is not fully reversible, probably due to the formation of intra-chain hydrogen bonds.

An important group of molecular magnets are purely organic molecular materials, as they show electric conductivity and non-trivial magnetic properties New examples of organic magnetic materials are presented by Pinkowicz et al. [7]. The paper reports the synthesis, crystal structures and magnetic characterizations of a series of six dioxothiadiazole-based radical compounds. Structurally, the presented compounds are formed by alternating cation–anion layers or chains of π-conjugated molecules. Magnetic data reveal weak antiferromagnetic interactions between the radical anions. Magnetic interaction pathways between pairs of radical anions are justified by the presence of C-H N hydrogen bonds. Authors discuss the influence of the structural differences on the magnetic properties of the radical salts under the study.

Finally, research on the π-d interacting magnetic molecular superconductor κ-(BETS)$_2$FeX$_4$ (X = Cl, Br) studied by means of an angle-resolved heat capacity is reviewed by Fukuoka et al. [8]. The π-d interacting systems consisting of organic donor molecules and counter anions containing magnetic ions, are of interest due to the cooperative phenomena between conducting electrons and localized spins, leading to unique magnetic and transport properties. The experimental method used enables high accuracy investigations of the anisotropy of the magnetic heat capacity against the in-plane magnetic field. Instrument applicable to tiny single crystals of molecular magnetic materials was constructed by the authors of the article. Uncommon crossover from a 3D magnetic ordering to a 1D magnet was observed at the field parallel to the *a* axis, while the superconducting transition temperature also showed a remarkable anisotropy against the in-plane magnetic field. These valuable results point to the influence of the 3d electron spins on the superconducting state of the π electron system.

Acknowledgments: Guest editors appreciate all the authors of the articles for sending their works. They hope the content of this special issue presents a comprehensive report on the current work on molecular magnets and will be interesting for the readers.

References

1. Orendáčová, A.; Tarasenko, R.; Tkáč, V.; Čižmár, E.; Orendáč, M.; Feher, A. Interplay of Spin and Spatial Anisotropy in Low-Dimensional Quantum Magnets with Spin 1/2. *Crystals* **2019**, *9*, 6. [CrossRef]
2. Brumfield, A.; Haraldsen, J.T. Thermodynamics and Magnetic Excitations in Quantum Spin Trimers: Applications for the Understanding of Molecular Magnets. *Crystals* **2019**, *9*, 93. [CrossRef]
3. Zentkova, M.; Mihalik, M. The Effect of Pressure on Magnetic Properties of Prussian Blue Analogues. *Crystals* **2019**, *9*, 112. [CrossRef]
4. Fitta, M.; Pełka, R.; Konieczny, P.; Bałanda, M. Multifunctional Molecular Magnets: Magnetocaloric Effect in Octacyanometallates. *Crystals* **2019**, *9*, 9. [CrossRef]
5. Rojas-Dotti, C.; Moliner, N.; Lloret, F.; Martínez-Lillo, J. Ferromagnetic Oxime-Based Manganese(III) Single-Molecule Magnets with Dimethylformamide and Pyridine as Terminal Ligands. *Crystals* **2019**, *9*, 23. [CrossRef]
6. Pacanowska, A.; Reczyński, M.; Nowicka, B. Modification of Structure and Magnetic Properties in Coordination Assemblies Based on $[Cu(cyclam)]^{2+}$ and $[W(CN)_8]^{3-}$. *Crystals* **2019**, *9*, 45. [CrossRef]
7. Pakulski, P.; Arczyński, M.; Pinkowicz, D. Bis(triphenylphosphine)iminium Salts of Dioxothiadiazole Radical Anions: Preparation, Crystal Structures, and Magnetic Properties. *Crystals* **2019**, *9*, 30. [CrossRef]
8. Fukuoka, S.; Fukuchi, S.; Akutsu, H.; Kawamoto, A.; Nakazawa, Y. Magnetic and Electronic Properties of π-d Interacting Molecular Magnetic Superconductor κ-$(BETS)_2FeX_4$ (X = Cl, Br) Studied by Angle-Resolved Heat Capacity Measurements. *Crystals* **2019**, *9*, 66. [CrossRef]

crystals

MDPI

Review

Interplay of Spin and Spatial Anisotropy in Low-Dimensional Quantum Magnets with Spin 1/2

Alžbeta Orendáčová *, Róbert Tarasenko, Vladimír Tkáč, Erik Čižmár, Martin Orendáč and Alexander Feher

Institute of Physics, Faculty of Science, P.J. Šafárik University, Park Angelinum 9, 041 54 Košice, Slovakia; robert.tarasenko@upjs.sk (R.T.); tkac.vladimir@upjs.sk (T.V.); erik.cizmar@upjs.sk (E.Č.); martin.orendac@upjs.sk (M.O.); alexander.feher@upjs.sk (A.F.)
* Correspondence: alzbeta.orendacova@upjs.sk; Tel.: +421552342280

Received: 26 November 2018; Accepted: 19 December 2018; Published: 21 December 2018

Abstract: Quantum Heisenberg chain and square lattices are important paradigms of a low-dimensional magnetism. Their ground states are determined by the strength of quantum fluctuations. Correspondingly, the ground state of a rectangular lattice interpolates between the spin liquid and the ordered collinear Néel state with the partially reduced order parameter. The diversity of additional exchange interactions offers variety of quantum models derived from the aforementioned paradigms. Besides the spatial anisotropy of the exchange coupling, controlling the lattice dimensionality and ground-state properties, the spin anisotropy (intrinsic or induced by the magnetic field) represents another important effect disturbing a rotational symmetry of the spin system. The $S = 1/2$ easy-axis and easy-plane XXZ models on the square lattice even for extremely weak spin anisotropies undergo Heisenberg-Ising and Heisenberg-XY crossovers, respectively, acting as precursors to the onset of the finite-temperature phase transitions within the two-dimensional Ising universality class (for the easy axis anisotropy) and a topological Berezinskii–Kosterlitz–Thouless phase transition (for the easy-plane anisotropy). Experimental realizations of the $S = 1/2$ two-dimensional XXZ models in bulk quantum magnets appeared only recently. Partial solutions of the problems associated with their experimental identifications are discussed and some possibilities of future investigations in quantum magnets on the square and rectangular lattice are outlined.

Keywords: Heisenberg; $S = 1/2$ XXZ model; spin anisotropy; square lattice; chain; rectangular lattice; Berezinskii-Kosterlitz-Thouless phase transition; phase diagram; quantum magnet

1. Introduction

The history of low-dimensional magnetism started in 1925 by the theoretical work of Ising who found an exact solution of the hypothetic system of spins arranged into a chain and oriented in one direction [1]. Famous Onsager's solution of the two-dimensional Ising model on a square lattice [2] showed the existence of long-range order (appearance of nonzero spontaneous magnetization) at a finite temperature, T_C. Intensive theoretical studies of one-dimensional (1D) and two-dimensional (2D) spin systems were stimulated by the effort to understand the properties of three-dimensional (3D) phase transitions and critical phenomena [3].

In 1966, Mermin and Wagner theoretically proved the absence of conventional long-range order (LRO) at a finite temperature in 1D and 2D Heisenberg (isotropic) and easy-plane (XY) magnets [4]. Nevertheless, Stanley and Kaplan conjectured a possibility of an unusual phase transition for 2D Heisenberg ferromagnet on triangular, square and honeycomb lattice arguing that the absence of a spontaneous magnetization at finite temperatures does not imply the absence of any phase transition [5]. Subsequent work indicated that the evidence for this was stronger for XY than for

Heisenberg models [6–8]. Berezinskii revealed that eigenstates of the Hamiltonian describing the 2D lattice of planar rotators, 2D Bose liquid, and 2D XY magnet can be sorted into two classes; localized *"vortices"* characterized by a nonzero circulation along a minimum closed contour of the square lattice and displaced harmonic oscillations—*spin waves* with zero circulation [9,10]. Below a critical temperature related to some phase transition, the vortices form configurations with a total zero circulation. Kosterlitz and Thouless introduced a definition of a *topological* long-range order adopted from the dislocation theory of melting [11]. In a 2D crystal, authors showed that at low temperatures, dislocations with Burgers vector of the magnitude b tend to form closely bound dipole pairs with resulting $b = 0$. Above some critical temperature, the pairs start to dissociate and the dislocations will appear spontaneously. The same type of argument can be applied for the 2D XY model and 2D neutral superfluid. While in the 2D XY model, a logarithmically large energy barrier, $V(r) \sim -\ln(r)$, stabilizes a topological order formed by the bound pairs of vortices, in the case of the 2D Heisenberg model, there is no topological order, since energy barriers separating different configurations are small, allowing continuous changes between individual configurations [11].

Many theoretical studies showed that phase transitions in the 2D systems such as granular superconducting films, superfluid films, 2D Coulomb gas etc., with continuous symmetry of the order parameter, may be described by the 2D XY model [12]. These topological phase transitions are related to the dissociation of the pairs of topological excitations (vortices in superconducting or superfluid films, dislocations in 2D crystals, dipole pairs of oppositely charged particles in 2D plasma, etc.), and belong to the same universality class. In later literature, these transitions were named as Berezinskii–Kosterlitz–Thouless (BKT) transition, occurring at a finite critical temperature, T_{BKT}.

Unlike the theory, the move of low-dimensional magnets from the abstract to real world started much later, during the period of seventies and eighties, when the first real materials appeared, resembling the behavior theoretically predicted for the 1D and 2D magnetic models [13].

The physics of low-dimensional systems is interesting in its own right and some phenomena have no parallel in three-dimensional physics. Besides the absence of the aforementioned conventional LRO at finite temperatures, the main feature of low-dimensional systems is a failure of the classical spin-wave theory which is not able to describe the complexity of the non-linear spin dynamics [14]. It was realized that localized solitary excitations, large-amplitude waves propagating with a permanent profile, are possible in classical magnetic chains [15]. The classical theory of solitons has been remarkably successful in describing the properties of real magnetic chains as $(CH_3)_4NMnCl_3$ (TMMC) with spin $S = 5/2$ and even $(C_6H_{11}NH_3)CuBr_3$ (CHAB) with $S = 1/2$ [16]. Various theoretical approaches tried to find a quantum analog to the classical solitary excitations leading to the concept of quantum solitons which proved useful for the description of the ground-state properties of an $S = 1$ Heisenberg antiferromagnetic (HAF) chain in the famous Haldane's conjecture [17,18]. Within the semi-classical approximation Haldane showed that the ground state of the HAF chain with integer spin is characterized by the presence of topological solitons. Consequently, the ground state is disordered with $S = 0$, separated from the excited $S = 1$ magnon state by the Haldane gap, arising from the presence of strong quantum fluctuations preventing the onset of the Néel order even at $T = 0$. Experimentally, the Haldane phase was most comprehensively studied in the $S = 1$ chain material $Ni(C_2H_8N_2)_2NO_2ClO_4$ (NENP), confirming the theoretical predictions [19,20].

The presence of strong magnetic fluctuations in the low-dimensional magnetic subsystems of high-T_C superconductors triggered renewed intensive theoretical and experimental interest in the 2D quantum magnets [21,22]. In this context, the frustration effects became widely studied to understand the pairing processes. Sophisticated mathematical and computational methods enabled theorists to solve the variety of more complex low-dimensional quantum frustrated lattices including Shustry-Shuterland [23], Kagomé lattice [24], Kitaev honeycomb model [25] and others, having exotic properties and many of them still waiting for their discovery in the real world [26].

Besides the study of rather exotic 1D and 2D magnetic models, the theorists try to incorporate the effect of inter-chain/inter-layer coupling, crystal field, spin anisotropy, dilution and other effects present in real compounds. While a close collaboration of chemists, experimental and theoretical physicists working in the area of low-dimensional magnetism has long-lasting tradition, the participation of quantum chemists in this field appeared only recently. The community realized the large importance of quantum-mechanical calculations based on the first principles which can become crucial for the identification of the studied material. This cooperation is also stimulated by the practical needs—the search for variety of materials involving aforementioned low-dimensional and strongly frustrated systems which can be potentially used in nanotechnologies [27], quantum computing [28–30], refrigerating due to enhanced magneto-caloric effect [31–33], etc. Besides the aforementioned practical applications and solutions of fundamental problems, current low-dimensional magnetism is also characterized by the search for the analogies in different fields of physics (quantum tunneling [34], Bose-Einstein condensation [35], quantum phase transitions [36,37], thermal Hall effect [38], etc.).

This brief introduction to the history and current state of the low-dimensional magnetism tried to point out that this field covers extremely wide area of research comprising directions which seem to run independently until they mix together forming new qualities. In this context, this modern, widely developing area changes to a field with a strong interdisciplinary character based on the close interaction of theory and experiment, accompanied by the intensive collaboration of physicists, chemists and material engineers.

In this review, we will restrict to the ground-state and finite-temperature properties of the $S = 1/2$ models interpolating between the quantum Heisenberg antiferromagnetic chain and the square lattice. Both models represent important paradigms of the low-dimensional magnetism. Their ground states are completely different, since they are determined by different strength of quantum fluctuations. The involving of additional exchange interactions results in the variety of the quantum models which can be derived from the aforementioned paradigms. Besides the spatial anisotropy of the exchange coupling which controls the lattice dimensionality, the spin anisotropy (intrinsic or induced by magnetic field) plays an important role in the symmetry of the order parameter. Despite the fact, that some of the models including the square lattice were theoretically studied many years ago, their experimental realizations appeared only recently, mostly in the bulk Cu(II) based metal-organic magnets. The problems associated with their experimental identifications are also discussed.

2. Spatial Anisotropy of the Exchange Coupling: From the Chain to the Square Lattice

2.1. The S = 1/2 Heisenberg Antiferromagnetic Chain

Let us consider a pair of the $S = 1/2$ spins coupled by the isotropic Heisenberg interaction

$$H = JS_1 \cdot S_2 \tag{1}$$

where $J > 0$ corresponds to an antiferromagnetic (AF) interaction. The ground state of the coupled spins is represented by the singlet state

$$|s\rangle = \frac{1}{\sqrt{2}}(|\uparrow\downarrow\rangle - |\downarrow\uparrow\rangle) \tag{2}$$

with the energy $-3J/4$. The quantum-mechanical state represented by the Equation (2) is known as a *valence bond* and can be regarded as the ultimate expression of quantum fluctuations which govern the formation of more exotic magnetic states depending on the lattice symmetry, spin value etc.

$$H = J\sum_i (S_i^x S_{i+1}^x + S_i^y S_{i+1}^y + S_i^z S_{i+1}^z) \tag{3}$$

The $S = 1/2$ Heisenberg antiferromagnetic (HAF) spin chain with the nearest-neighbor (*nn*) interactions has a unique ground state with power-law correlations and the gapless excitation spectrum [39]. Using a concept of the valence bonds, this ground state has a character of a resonating-valence-bond (RVB) state, in which the quantum fluctuations restore the translational symmetry and mix in the bonds of greater length than those connecting only the *nn* spins [40]. An alternative approach is the mapping of the quantum spin 1/2 HAF chain to a chain of interacting spinless fermions [26]. The absence of the fermion (hole) at a site *i* means a spin state $\left|S_i^z = -\frac{1}{2}\right\rangle$, whereas the presence of the fermion (particle) means $\left|S_i^z = \frac{1}{2}\right\rangle$. In this language, the model becomes a realization of the 1D Tomonaga–Luttinger liquid [41]. A corresponding ground state has correlations decreasing as a power law with a distance and elementary excitations form a particle-hole continuum. The excitation of a single hole carrying spin 1/2 is called a spinon and can be created by a turning half of the chain upside down which results in the two neighboring sites with the spin up. The *x-y* part of the Hamiltonian in the Equation (3) moves the spinon by two lattice sites (Figure 1).

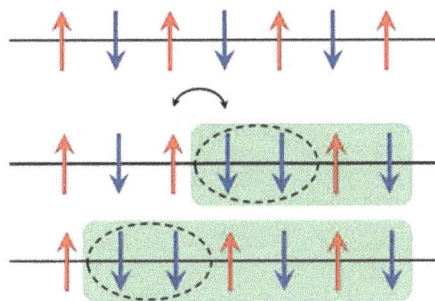

Figure 1. Schematic representation of a spinon excitation by reversing all spins beyond a certain lattice site. The spinon (domain wall) propagates along the chain moving by two lattice sites.

On the other hand, a standard magnon excitation carrying spin 1, is created by a flipping single spin at a lattice site. It corresponds to the two holes in the spectrum. As a consequence of the Fermion nature of the half-integer spins, the spin 1 carried by the spin wave excitation is decomposed into the pair of the $S = 1/2$ quantum solitons, known as spinons [41]. The pairs of spinons form a gapless continuum of excitations experimentally observed in various quasi-low dimensional magnets. One example of such material is $KCuF_3$ with strong AF exchange coupling propagating along the *c*-direction, while much weaker ferromagnetic coupling along the *a* and *b* directions is responsible for the onset of the long-range order at $T_N = 39$ K. The inelastic neutron scattering measurements in zero magnetic field probed the energy spectrum of $KCuF_3$ below and above T_N. In comparison with low temperatures, the measurements made at 50 K indicated only little change in the scattering cross-section. The calculated spectrum of the $S = 1/2$ HAF chain is in good agreement with the experimental data [42] (Figure 2).

While in $KCuF_3$ the inter-chain coupling, J', is rather weak ($J' \approx 0.01J$), in Cs_2CuCl_4, the J' achieves nearly 20% of the intra-chain value, $J/k_B \approx 4$ K, and the onset of LRO occurs at $T_N = 0.62$ K. The low J-value guarantees rather easy achievement of the critical magnetic field, B_c, necessary for decoupling the spin chains as well as the saturation field, B_{sat}, above which the ground state achieves a full ferromagnetic polarization. Theoretical studies of the $S = 1/2$ HAF chain in the magnetic field [14] showed that the field splits the triplet excitation continuum into the separate continua, the positions of which alter with increasing field. Above the saturation field, the excitations have a character of well-defined magnon dispersion. Correspondingly, in Cs_2CuCl_4, the 1D regime was expected to set at the fields lower than $B_{sat} \approx 6$ T. The magnetic excitations were studied as a function of the field by measuring the inelastic scattering at low temperatures, $T = 0.06$ K [43]. It was found that the intensity of the magnetic excitation decreases with increasing field and the line shape changes above $B_c = 1.66$ T, where the 1D regime occurs. Corresponding spectra were found to be in good agreement with the predictions for the $S = 1/2$ HAF chain in the magnetic field [43].

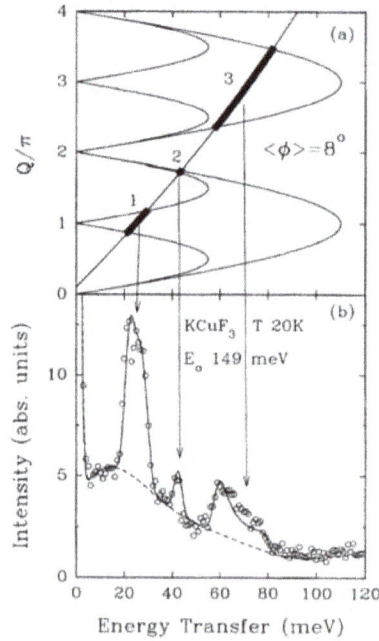

Figure 2. (a) Spectrum of the $S = 1/2$ HAF chain in zero magnetic field. The thin solid line represents a scattering trajectory for a detector at the scattering angle of 8′ and an incident energy of $E_0 = 149$ meV in KCuF$_3$. The scattering occurs when the trajectory intersects with the continuum (bold line); (b) Scattering measured in the low-angle detector banks at $T = 20$ K. (Reproduced with permission from reference [42]).

The existence of the two-spinon continuum was experimentally confirmed in the organometallic compound Cu(C$_4$H$_4$N$_2$)(NO$_3$)$_2$, which proved to be an excellent realization of the $S = 1/2$ HAF chain with $J = 0.9$ meV and a negligible inter-chain coupling ($J'/J < 10^{-4}$) [44].

These rather demanding neutron scattering studies are usually preceded by more accessible experimental techniques providing information on the finite-temperature macroscopic properties as specific heat, susceptibility and magnetization which serve as an important tool for the reliable identification of the magnetic system. For that purpose, many theoretical studies of the $S = 1/2$ HAF chain based on different methods were performed to yield theoretical predictions usable for the analysis of experimental results [45–48]. The studies of real compounds approximating the model of the $S = 1/2$ HAF chain point at the importance of additional exchange couplings, responsible for the deviations from the ideal chain behavior [13,42–44,49].

2.2. The S = 1/2 Heisenberg Antiferromagnet on the Spatially Anisotropic Square Lattice

The absence of the Néel order in the ground state is believed to be a general feature of one-dimensional isotropic antiferromagnets [17,18,41]. Inter-chain coupling can change the ground-state properties and introduce dimensional crossover phenomena. A two-dimensional array of the spin chains coupled by the inter-chain interaction as depicted in Figure 3, forms a spatially anisotropic square lattice, often called as a rectangular lattice, which can be described by the Hamiltonian

$$H = J \left[\sum_{\langle i,j \rangle_J} S_{i,j} \cdot S_{i+1,j} + R \sum_{\langle i,j \rangle_{J'}} S_{i,j} S_{i,j+1} \right]. \tag{4}$$

The parameter $R = J'/J$ represents the ratio of the AF inter-chain to AF intra-chain coupling and $\langle i, j \rangle_{J,J'}$ denotes the nearest neighbors along the chain and perpendicular to the chain direction (Figure 3). For $R = 0$, the 2D model (Equation (4)) simply reduces to the isolated $S = 1/2$ HAF chains (Equation (3)) with the zero value of the order parameter (staggered magnetization), $m = 0$, reflecting the absence of the LRO in the ground state. For $R = 1$, Equation (4) represents the model of HAF on the *spatially isotropic square lattice* (or simply *square lattice*). According to Mermin-Wagner theorem [4] thermal fluctuations are strong enough to destroy the Néel LRO at finite temperatures. However, it was not clear, whether also quantum fluctuations can destroy the Néel LRO at zero temperature.

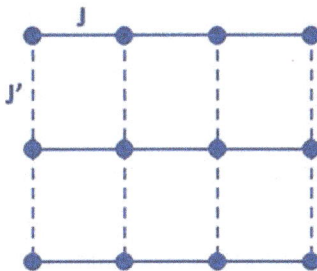

Figure 3. Cartoon of the spatially anisotropic square ($\equiv$ rectangular) lattice.

After intensive studies of the ground-state properties of the spin $1/2$ HAF square lattice, over many decades a final agreement was achieved about the semi-classical Néel LRO at zero temperature [26]. The quantum fluctuations lead to the significant reduction of the order parameter, $m \approx 0.3$, which amounts about 60% of the classical value, $m_0 = 1/2$. The quantum fluctuations are strong enough to preserve spin-rotation symmetry such as the RVB state which may be relevant at high energies. While low-energy excitations are gapless magnons, recent experimental and theoretical studies showed that at higher energies, the existence of pairs of fractional $S = 1/2$ quasiparticles, 2D analogs of 1D spinons was established [50]. Using various theoretical approaches, excitation spectra and finite-temperature properties of the square lattice were investigated including specific heat, uniform and staggered susceptibility, correlation length etc. to provide useful tools for the identification of the model realization in the real world [51–55]. In these studies, the compound La_2CuO_4, with the exchange coupling $J/k_B \approx 1500$ K proved to be a model system for the testing of the theories, especially those investigating ground-state properties. Naturally, the huge intra-layer coupling prevented any studies of the compound at moderate temperatures, $T \approx J/k_B$. Later, other 2D magnetic systems, mostly copper(II) coordination complexes, were identified as excellent realizations of the spin $1/2$ HAF square lattice with much lower exchange coupling. The analysis of the specific heat of $Cu(en)_2Ni(CN)_4$ (en = $C_2H_8N_2$) and $Cu(bmen)_2Pd(CN)_4$ (bmen = N, N'-dimethyl-1,2-diaminoethane), revealed excellent agreement with the theoretical prediction for the spin $1/2$ HAF square lattice with $J/k_B = 0.36$ K and 0.48 K, respectively [56,57]. Unlike La_2CuO_4, small coupling and corresponding saturation field $B_{sat} \approx 1$ T, enable comfortable studies in the wide region of temperatures and magnetic fields. In these octahedral Cu(II) complexes comprising of weakly bound electroneutral covalent chains, the exchange coupling between Cu(II) ions is mediated predominantly through hydrogen bonds. On the other hand, in the tetragonal compound $[Cu(pz)_2(NO_3)][PF_6]$ (pz = pyrazine) and monoclinic $Cu(pz)_2(ClO_4)_2$, the copper sites are connected within the square layers by bridging pyrazine molecules with the exchange coupling $J/k_B = 10.5$ K and 17.5 K, respectively [58,59]. Even larger exchange coupling was indicated in $Cu(HCOO)_2 \cdot 4H_2O$ with $J/k_B = 72$ K [60]. Magneto-structural investigations [61–63] of monoclinic compounds $(5MAP)_2CuBr_4$ and $(5BAP)_2CuBr_4$ (5MAP = 5-methyl-2-aminopyridinium, 5BAP = 5-bromo-2-aminopyridinium) revealed that the magnetic interaction occurs between Cu(II) sites with four equivalent nearest neighbors through Br $\cdots$ Br contacts forming 2D square layers with exchange coupling $J/k_B \approx 7$ K. The layers of $CuBr_4$ tetrahedrons are separated by the bulk of organic cations which stabilize 3D structure. Systematic study of the compounds from the series A_2CuX_4 [A = 5MAP, 5BAP, 5-chloro-2-aminopyridinium $\equiv$ 5CAP, 5-cyano-2-aminopyridinium $\equiv$ 5CNAP, etc., X = Br, Cl] found that the increasing size of the A cation improves the isolation of individual magnetic square layers but at the same time it reduces the strength of the intra-layer exchange coupling [62–64]. Apparently, the chemical modification of the structure can control magnetic properties demonstrating the flexibility of molecular magnetism.

Intensive theoretical studies of the ground-state and finite-temperature properties of the square lattice ($R = 1$) provided reach variety of theoretical predictions allowing reliable identification of real candidates. Besides the materials manifesting the model realization of the $S = 1/2$ HAF square lattice, there appeared two-dimensional quantum systems with some amount of the spatial anisotropy of the intra-layer exchange coupling which could not be well determined due to the lack of proper theoretical predictions [65–70]. Using various approaches, the ground-state properties of the spatially anisotropic square lattice (Equation (4)) were investigated in the vicinity of the isotropic model ($R = 1$) and 1D chain ($R = 0$). Many theoretical studies tried to solve the question about the existence of a critical R_c value above which a long-range order is established in the ground state ($m > 0$). Conventional spin-wave theories as well as various numerical techniques predicted a final value, $R_c \approx 0.1$–0.2, below which a 2D spin-liquid state with $m = 0$ can be stabilized [71,72]. For small R, a single-chain mean-field theory [73] predicted $R_c = 0$ and a gradual increase of m, proportional to $\sqrt{R}$. Multi-chain mean-field calculations complemented by large-scale Monte Carlo simulations of the 2D Hamiltonian (Equation (4)), confirmed that $R_c = 0$ and showed that for $R \to 0$, m vanishes slower than $\sqrt{R}$ due to a logarithmic correction to this form [74]. Applying various techniques [72], the order-disorder ground-state transition was indicated for $R \approx 0.2$ (Figure 4a). The sharp change of m for $R < 0.2$ was interpreted as a crossover in the magnetic behavior of the $S = 1/2$ HAF rectangular lattice, accompanied by a sharp change in the spatial dependence of spin correlations; with decreasing R, the rising quantum fluctuations gradually reduce the size of the order parameter and for small R, the system approaches 1D behavior with algebraic decrease of correlation functions [72]. This conclusion was further supported by quantum Monte Carlo studies [75] of finite-temperature properties of the rectangular lattice (Equation (4)). For small R, the temperature dependence of the uniform susceptibility, χ, follows that of a single chain, while deviations appear below temperatures $k_BT/J \approx 5R$. For larger R, the χ values lie between those of the chain and square lattice. The 1D - 2D dimensional crossover is evident also in the behavior of the correlation length, ξ, depicted in Figure 4b. For $R > 0$ and low temperatures, the quantity is well described by the relation [75]

$$\xi = \frac{A\exp[2\pi\rho_s(R)/(k_BT)]}{1 + 0.5T/[2\pi\rho_s(R)]} \tag{5}$$

where the spin stiffness, ρ_s, depends on the spatial anisotropy R. As can be seen in Figure 4b, the intra-chain correlation length gradually approaches 1D behavior when decreasing R. Alike susceptibility, for small R, the ξ values merge with those for a single chain at sufficiently high temperatures. Using these theoretical predictions, authors of reference [75] identified the quantum magnet Sr_2CuO_3 as a realization of the $S = 1/2$ HAF on the rectangular lattice with the intra-chain coupling $J/k_B = 2200$ K, $R = 0.002$ and the staggered magnetization, $m \approx 0.03$. In comparison with the $m \approx 0.3$ derived for the square lattice, the significant reduction results from the strong enhancement of quantum fluctuations.

While the ground-state properties of the rectangular lattice were already understood quite well, corresponding finite-temperature studies appeared only recently [76,77]. Quantum Monte Carlo simulations of the susceptibility and magnetization enabled to identify the realizations of the $S = 1/2$ HAF rectangular lattice, namely $Cu(pz)Cl_2$ ($J/k_B = 28$ K, $R = 0.3$), $Cu(pz)(N_3)_2$ ($J/k_B = 15$ K, $R = 0.46$) and $Cu(2\text{-apm})Cl_2$ (2-apm = 2-amino-pyrimidine) with $J/k_B = 116.3$ K and $R = 0.084$ [77].

It should be noted that a reliable identification of a real compound requires rather complex and careful analysis of bulk properties. It proved that making final conclusion on the basis of even several experimental techniques cannot be sufficient since various models can provide the same theoretical prediction. In that case, first-principle calculations can provide valuable information how to proceed in further analysis as was demonstrated in the analysis of the susceptibility of Cu(NCS)$_2$ where the first-principle calculations enabled to identify the material as the realization of the rectangular lattice with $J/k_B \approx 170$ K and $R \approx 0.08$ [78].

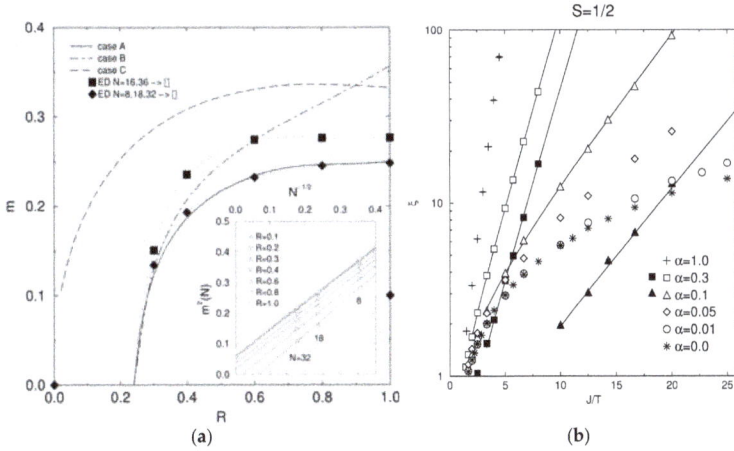

(a)   (b)

Figure 4. (a) Ground-state staggered magnetization m of the rectangular lattice for different R values calculated by various approaches; the short-dashed lines are fits to the exact diagonalization (ED) data. Cases *A, B, C* correspond to different choice of parameters in Green's function approach. The inset demonstrates the deviation of the ED results from the N-1/2 scaling law at $R < 0.3$ (solid lines are least-squares fits). (Reproduced with permission from reference [72]); (b) The correlation length for the $S = 1/2$ HAF on the rectangular lattice as a function of the inverse temperature for various R values (α stands for R). Open (filled) symbols denote the correlation length along (perpendicular to) the chain. For small α ($\alpha < 0.05$), the values of inter-chain correlation length are smaller than one lattice constant and are not shown in the figure. Solid lines are fits to Equation (5), showing the exponential dependence of ξ on $1/T$. (Reproduced with permission from reference [75]).

Similarly, it was shown in reference [79], that the susceptibility and isothermal magnetization of Cu(PM)(EA)$_2$ (PM = [C$_6$H$_2$(COO)$_4$]$^{4-}$, EA = [C$_2$H$_5$NH$_3$]$^+$) can be described by the model (Equation (4)) with various R in a wide range of temperatures and magnetic fields (Figure 5).

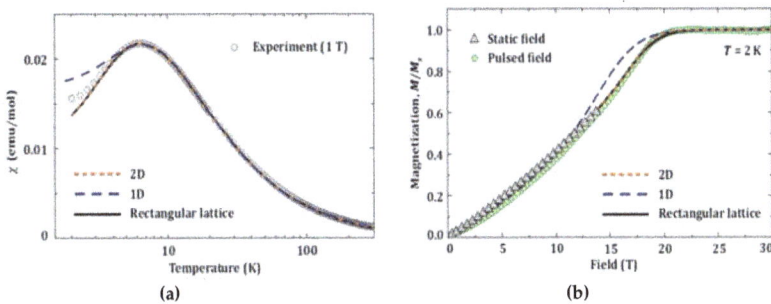

(a)   (b)

Figure 5. (a) Fits of the magnetic susceptibility of Cu(PM)(EA)$_2$ with the model (Equation (4)) for $R = 0$ (1D) with $J/k_B = 10.2$ K, $g = 2$; $R = 1$ (2D) with $J/k_B = 6.8$ K, $g = 2.05$; $R = 0.7$ (rectangular lattice) with $J/k_B = 8.0$ K, $g = 2.07$; (b) Field dependence of the magnetization of Cu(PM)(EA)$_2$. Lines show the simulations with the same parameters as those used for the susceptibility. (Reproduced with permission from reference [79]).

Apparently, using proper parameters, various theoretical predictions yield identical behavior for both, susceptibility and magnetization [79]. Thus, making conclusions on the basis of magnetic data only, can be very tricky, since the height of the maximum, χ_{max}, as well as its position, T_{max}, depend on the J and g-factor values. Correspondingly, the proper choice of g and J parameters can yield an excellent agreement with the data (Figure 5). Unlike the susceptibility, another bulk property, specific heat, can provide more valuable information since the maximum, C_{max}, depends on the used model (Figure 6). The best agreement was found for $R = 0.7$ as predicted by the first-principle calculations [79].

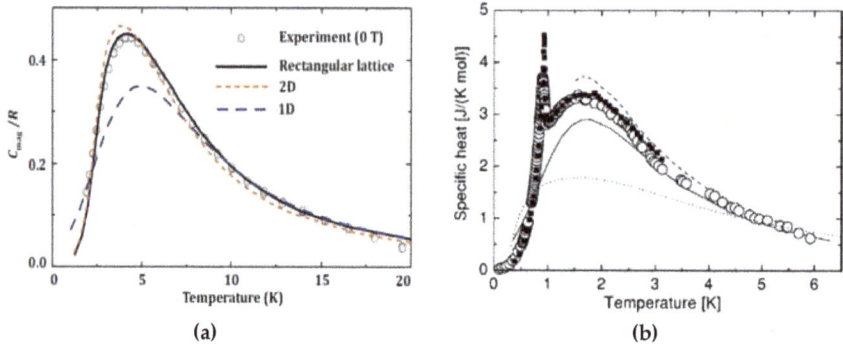

Figure 6. (a) Fits of the magnetic specific heat of Cu(PM)(EA)$_2$ in $B = 0$ with the model (Equation (4)) for $R = 0$ (1D), $R = 1$ (2D) and $R = 0.7$ (rectangular lattice) with the same parameters as in Figure 5. (Reproduced with permission from reference [79]); (b) Fits of the magnetic specific heat of Cu(en)(H$_2$O)$_2$SO$_4$ in $B = 0$ (open and full circles correspond to powder and single crystal, respectively) with the model (Equation (4)) for $R = 0$ with $J/k_B = 3.6$ K, (solid line), $R = 1$ with $J/k_B = 2.8$ K (dashed line) and the model of the $S = 1/2$ HAF on the triangular lattice with $J/k_B = 4.4$ K (dotted line). (Reproduced with permission from reference [65]).

Such calculations proved to be very important in the identification of the 2D quantum magnet Cu(en)(H$_2$O)$_2$SO$_4$. Previous analysis of powder thermodynamic data [65] identified the material as a potential realization of the partially frustrated $S = 1/2$ HAF on the spatially anisotropic triangular lattice (SATL). The lack of proper theoretical predictions enabled the analysis only in the frame of the limiting models of the SATL, i.e., the square lattice, chain and triangular lattice and could not provide a reliable information whether the spin system approaches the properties of the chain or the square lattice (Figure 6b). Considering $d_{x^2-y^2}$ ground state of the Cu(II) ion, it was assumed that potential exchange pathways form SATL with a dominant exchange coupling creating the square lattice, while weaker interactions were expected to occur along one of the diagonals of the square plaquettes [65]. The analysis of single-crystal electron paramagnetic resonance spectra [80] indicated the need to revisit the concept of SATL in Cu(en)(H$_2$O)$_2$SO$_4$, which triggered first-principle calculations of exchange couplings [81]. The calculations revealed the formation of a spatially anisotropic zig-zag square lattice (Figure 7a) comprised of 2D array of weakly coupled zig-zag chains with $R = J'/J \approx 0.15$. Corresponding quantum Monte Carlo (QMC) calculations of finite-temperature properties of the $S = 1/2$ HAF on the spatially anisotropic zig-zag square lattice (SAZZSL) including specific heat, susceptibility and magnetization, provided theoretical predictions in a wide range of temperatures and magnetic fields [82]. Subsequent analysis of single-crystal Cu(en)(H$_2$O)$_2$SO$_4$ experimental data within the SAZZSL model found the excellent agreement for $J/k_B = 3.4$ K and $R = 0.35$ (Figure 8).

The comparison of the finite-temperature properties of the $S = 1/2$ HAF on the rectangular and zig-zag square lattice surprisingly revealed the *identical* behavior in the whole range of the spatial anisotropy R [77,82]. Thus, to decide which model is appropriate for the description of the real compound, first-principle studies are very important.

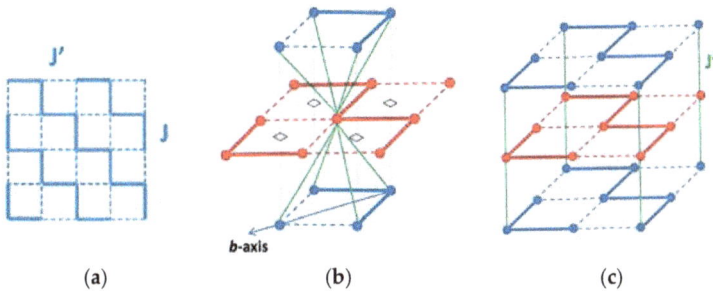

Figure 7. (a) Spatially anisotropic zig-zag square lattice interpolates between the chain ($R = J'/J = 0$) and the isotropic square lattice ($R = 1$); (b) Position of adjacent layers (blue) in $Cu(en)(H_2O)_2SO_4$ with respect to the central 2D layer (red). Diamonds denote projections of the positions of blue circles into the central layer; (c) Position of adjacent layers (blue) in the verdazyl radical with respect to the central 2D layer (red). In (b) and (c), the inter-layer exchange couplings J'' are depicted by green lines (see Section 2.3).

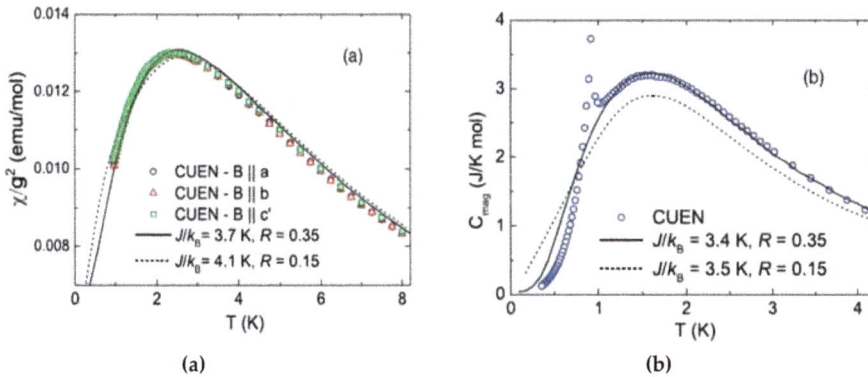

Figure 8. (a) Temperature dependence of normalized single-crystal $Cu(en)(H_2O)_2SO_4$ (CUEN) susceptibility in the field 10 mT. Solid and dashed lines represent corresponding QMC calculations ($L = 128$) for the $S = 1/2$ HAF SAZZSL model with $R = 0.35$ and 0.15, respectively; (b) Temperature dependence of magnetic specific heat of CUEN single crystal in zero magnetic field. Solid and dashed lines have the same meaning as in (a). (Reproduced with permission from reference [82]).

As was already mentioned, the compound $Cu(pz)_2(ClO_4)_2$ was considered as a model system for the realization of the isotropic square lattice [59]. Recent density functional theory (DFT) simulations using structural $Cu(pz)_2(ClO_4)_2$ data at 10 K revealed the formation of SAZZSL with $R = 0.7$ [83] while other authors working with 100 K structural data arrived to the rectangular lattice with $R \approx 0.93$ [84]. While it is hard to discriminate between two different results on the basis of magnetic bulk data, as was done in refs. [83,84], specific heat analysis would be subservient to make the decision. As was shown in refs. [79,82], the difference in the maximum specific heat values for $R = 1$ and 0.7 is easily distinguishable. This approach was applied also in the determination of the 2D magnetic lattice in the verdazyl radical α-2,3,5-Cl_3-V (3-(2,3,5-trichlorophenyl)-1,5-diphenylverdazyl] (Figure 7) where DFT calculations revealed the formation of SAZZSL with $R = 0.56$ [85].

All these results indicate, that rather simple combination of bulk properties and first-principle calculations can provide a reliable identification of magnetic lattice in the real compounds. However, unlike the ideal 2D models, in real compounds, the interlayer interactions are always present and depending on their strength, they can affect bulk properties, usually leading to the onset of the 3D long-range order.

2.3. The Crossover from 2D to 3D

The need to understand the effect of the inter-layer coupling, J'', on the properties of two-dimensional magnets stimulated theoretical studies of the 3D array of coupled layers with the square-lattice motif described by the Equation (4), forming the $S = 1/2$ HAF on the spatially anisotropic simple cubic lattice [51,86–88]. The studies showed that the coupling between the magnetic layers can induce a phase transition to 3D LRO at a finite temperature [51]

$$k_B T_N \approx J'' \left(\frac{m}{m_0} \right)^2 \left(\frac{\xi}{a} \right)^2 \tag{6}$$

where a is a lattice constant, m/m_0 is the reduced staggered magnetization at zero temperature, and correlation length is defined by the Equation (5). For the isotropic square lattice ($R = 1$), $(m/m_0)^2 \approx 0.3$ and the spin stiffness in the Equation (5) is $\rho_s \approx 0.18\ J$. Apparently, even a minute amount of J'' is capable to induce a 3D LRO at a finite temperature. Large-scale QMC studies of the $S = 1/2$ HAF on the spatially anisotropic simple cubic lattice (with $R = 1$ within layers) provided an empirical formula for T_N which enables the estimation of the inter-layer coupling [88]

$$T_N = 4\pi\rho_s / [b - \ln(J''/J)] \tag{7}$$

with $\rho_s = 0.183\ J$ and $b = 2.43$.

As was already mentioned, in the absence of J'' the ground-state staggered magnetization vanishes with $R \rightarrow 0$ (Figure 4a). On the other hand, the increase of J'' leads to the enhancement of m in the whole range of R [86]. Thus, the inclusion of the un-frustrated interlayer coupling stabilizes a collinear Néel order already for very small values of J''. The competition of the inter-layer coupling and the intra-layer spatial anisotropy R projects also to the finite-temperature properties [86]; for $R \rightarrow 0$ within the layer, enhanced quantum fluctuations lead to reducing T_N. On the other hand, nonzero J'' reduces the strength of the quantum fluctuations, which results in the enhancement of T_N. In highly anisotropic 2D systems ($J'' \ll J, J'$), most of the entropy is removed above T_N and the effective number of degrees of freedom associated with the 3D LRO is significantly reduced. In such extreme conditions, a λ-like anomaly in the specific heat associated with the onset of 3D LRO completely vanishes. Quantum Monte Carlo studies of the $S = 1/2$ HAF on the spatially anisotropic simple cubic lattice ($R = 1$ within layers) showed, that a sharp peak dominates the specific heat behavior for the strong inter-layer coupling, while for $J'' < 0.05\ J$ a clear separation of two peaks occurs [87]. Finally, for $J'' < 0.015\ J$, the sharp peak completely vanishes and despite the onset of 3D LRO, the specific heat follows the behavior of the 2D system (Figure 9a).

Such extreme spatial two-dimensionality accompanied by the absence of the λ-like anomaly in the specific heat in zero magnetic field was already experimentally observed in Cu(tn)Cl$_2$ (tn = C$_3$H$_{10}$N$_2$) [89] (Figure 9b) and Cu(pz)$_2$(pyO)$_2$(PF$_6$)$_2$ (pyO = pyridine-N-oxide) [90]. Thus, one has to be careful in the declaration of the absence of the 3D LRO on the basis of the specific heat only and other experiments are necessary to confirm the assumption. Besides demanding neutron diffraction experiments, much simpler susceptibility measurements or electron paramagnetic—antiferromagnetic resonance can provide reliable information about the onset of the 3D LRO. Neglecting any other phenomena but J'', the absence or the presence of the λ-like anomaly can be affected by the geometry of inter-layer exchange pathways in real compounds. The comparison of magnetic specific heats of Cu(tn)Cl$_2$ and Cu(en)(H$_2$O)$_2$SO$_4$ suggests that the 2D correlations responsible for the appearance of a round maximum have the same character while the manifestation of the 3D correlations is completely different (Figure 9b). The severe weakening of their effect in Cu(tn)Cl$_2$ was ascribed to the combined effect of a geometrical frustration and the large distances (about 10 Å) between Cu(II) ions in the adjacent layers [89].

On the other hand, the strength of the inter-layer coupling in Cu(en)(H$_2$O)$_2$SO$_4$ was estimated to be much lower than 0.015 J [82] thus, according to reference [87], no phase transition should be visible in the specific heat data. In Cu(en)(H$_2$O)$_2$SO$_4$, the adjacent layers are shifted along the b-axis by $b/2$. As a consequence, a central spin from the magnetic layer has a high number of nearest neighbors ($z = 8$) from the adjacent layers (Figure 7). It should be noted, that QMC simulations [87] were performed for the geometry of the simple cubic lattice, thus only $z = 2$ was considered for the central spins in the layer.

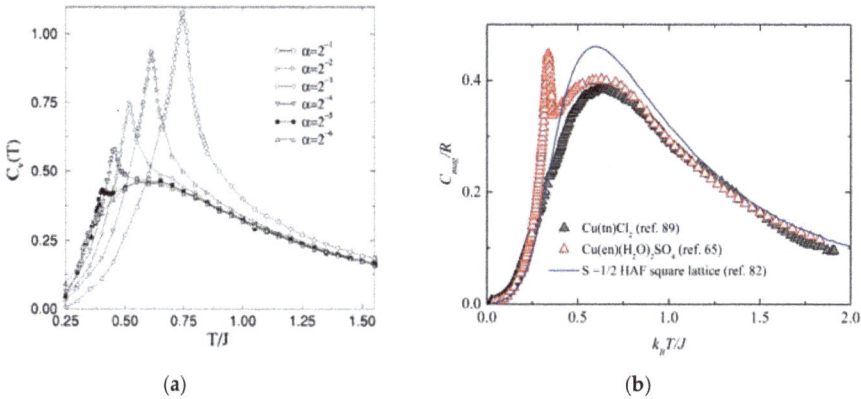

(a)

(b)

Figure 9. (a) The specific heat of the $S = 1/2$ HAF model on the spatially anisotropic simple cubic lattice with $R = 1$ within the layers. Quantum Monte Carlo calculations are performed over a wide temperature range for several different spatial anisotropies ($\alpha = J''/J$). The system size is 48 × 48 × 12. The separation of the 3D ordering peak from the broad maximum arising from the 2D physics is clearly visible for $\alpha < 2^{-3}$. (Reproduced with permission from reference [87]); (b) Temperature dependence of Cu(en)(H$_2$O)$_2$SO$_4$ ($J/k_B = 2.8$ K) and Cu(tn)Cl$_2$ ($J/k_B = 3.1$ K) magnetic specific heat compared with the theoretical prediction for the $S = 1/2$ HAF on the square lattice ($R = 1$).

The important role of the number of neighbors in adjacent planes was demonstrated also in the verdazyl radical α-2,3,5-Cl$_3$-V with a large inter-layer coupling, $J'' \approx 0.24 J$. This is another extreme, since such a high value of J'' should completely suppress the two-dimensional character of the magnetic system in the verdazyl radical. However, thermodynamic properties clearly demonstrate a high measure of magnetic two-dimensionality, comparable to Cu(en)(H$_2$O)$_2$SO$_4$. This contradiction can be explained by very low number of interacting nearest neighbors from the adjacent planes. As can be seen in Figure 7, in verdazyl radical, effectively only $z = 0.5$ per central spin can be considered, thus reducing the effect of rather strong inter-layer coupling can be expected [85].

Generally, if only isotropic inter-layer coupling is responsible for the onset of the 3D LRO, the finite-temperature phase transition in such spatially anisotropic systems remains in the universality class of the 3D classical Heisenberg model [87]. The universality can be changed when other mechanisms are present as the magnetic field and spin anisotropy [91].

3. The Effect of the Spin Anisotropy and Magnetic Field in the $S = 1/2$ HAF on the Square Lattice

3.1. $B = 0$

The influence of the spin-orbit (LS) coupling projects into the anisotropy of local as well as global magnetic properties. In the case of isolated ions with the electronic spin 1/2 (e.g., Cu(II) ion with the $3d^9$ open shell), the interplay of the LS coupling and the crystal field produced by the local surrounding generates the anisotropy of the magnetic moment, observed in the form of the anisotropic g-tensor [92]. In the presence of magnetic interactions between the paramagnetic ions, the exchange anisotropy (symmetric and/or antisymmetric) appears with the relative strength comparable with the order of the g-factor anisotropy [93]. Accordingly, experimental studies of quasi-2D spin 1/2 magnets revealed that the spin anisotropy is very weak, ranging from 10^{-4} to 10^{-2} times the intra-layer exchange coupling J [59,80,94].

In case of the symmetric exchange anisotropy, theoretical studies of the $S = 1/2$ antiferromagnetic XXZ model on the square lattice ($R = 1$) revealed, that even extremely weak spin anisotropies can influence thermodynamic properties at low temperatures, inducing a crossover between the isotropic Heisenberg and the anisotropic regime [95–97]. The $S = 1/2$ AF XXZ model (i.e., anisotropic Heisenberg antiferromagnet) on the square lattice can be defined by the Hamiltonian [97]

$$H = \frac{J}{2}\sum_{i,d}\left[(1 - \Delta_\mu)\left(S_i^x S_{i+d}^x + S_i^y S_{i+d}^y\right) + (1 - \Delta_\lambda)S_i^z S_{i+d}^z\right] \tag{8}$$

where $i = (i_1, i_2)$ runs over the sites of the square lattice, d connects the i-th site to the nearest neighbors, $J > 0$ is antiferromagnetic coupling, Δ_μ and Δ_λ are the easy-axis and easy-plane anisotropy parameters, respectively. For $\Delta_\mu = \Delta_\lambda = 0$, the Equation (8) reduces to the Equation (4) with $R = 1$ describing the isotropic HAF on the square lattice. The parameters $\Delta_\lambda = 0$, $0 < \Delta_\mu \leq 1$ define the easy-axis anisotropy, while the easy-plane anisotropy corresponds to $\Delta_\mu = 0$, $0 < \Delta_\lambda \leq 1$. In the case of the weak easy-axis anisotropy $\Delta_\mu = 10^{-2}$ and 10^{-3}, quantum Monte Carlo studies of finite-temperature properties revealed the existence of a phase transition in the 2D Ising universality class occurring at finite temperatures, $k_B T_I \approx 0.28\,J$ and $0.22\,J$, respectively [97]. In the specific heat, the onset of the 2D LRO was indicated as a small sharp peak superimposed on the left side of a round maximum. As the anisotropy decreased, the sharp peak diminished, moving to low temperatures, while the round maximum converged to that of the HAF on the square lattice. Similarly, the uniform susceptibility follows the prediction for the HAF on the square lattice down to $k_B T \approx 0.4\,J$ for $\Delta_\mu = 10^{-2}$. At lower temperatures, the transverse susceptibility, χ^{xx}, and longitudinal, χ^{zz}, separate from the isotropic Heisenberg curve well above the phase transition; at the transition temperature T_I, the χ^{xx} displays a minimum, while χ^{zz} monotonically decreases to zero. Apparently, the susceptibility measurement in two different orientations of magnetic field provides a tool for a reliable identification of the phase transition. Similar features of the susceptibility and specific heat were observed in the XXZ model with the easy-plane anisotropy for $\Delta_\lambda = 2 \times 10^{-2}$ and 10^{-3}. A crossover temperature, T_{CO}, from the isotropic Heisenberg to the easy-plane (XY) behavior was estimated [96]

$$\frac{k_B T_{CO}}{J} \simeq \frac{4\pi\rho_s/J}{\ln(C/\Delta_\lambda)} \tag{9}$$

with the parameter $C = 160$ and $\rho_s = 0.214\,J$. The onset of the XY regime below T_{CO} is accompanied with the formation of the pairs of vortices and antivortices. Concerning the specific heat, a position of a tiny peak superimposed on the left side of a round maximum, corresponds to the maximum of the temperature derivative of the vortex density, while a phase transition of Berezinskii–Kosterlitz–Thouless type is set at lower temperature. For $\Delta_\lambda = 2 \times 10^{-2}$, the uniform susceptibility follows the behavior of the HAF on the square lattice down to $0.4J$. At lower temperatures, the transverse and longitudinal susceptibility separate from the isotropic Heisenberg curve well above the transition temperature, T_{BKT}; at T_{CO}, the χ^{zz} component displays a minimum, while χ^{xx} decreases faster than the isotropic Heisenberg curve, achieving some nonzero value at $T = 0$. As authors showed, in the experiments with the real quasi-2D quantum magnets, the measurements of a single-crystal uniform susceptibility can help to determine the onset of a phase transition as well as the type of the spin anisotropy. If a minimum in the χ^{zz} component is observed *above* T_N (i.e., the temperature of a phase transition to the 3D LRO), this is a signature of the easy-plane anisotropy, while the occurrence of the minimum *at* the transition temperature suggests the easy-axis anisotropy (Figure 10a).

The QMC studies [97] revealed that the critical temperatures $T_{I,BKT}$ remain finite for any finite easy-plane and easy-axis anisotropy (Figure 10b). Thus, unlike the isotropic Heisenberg model on the square lattice, for the XXZ analogue, the quantum and thermal fluctuations are not able to destroy the phase transitions at finite temperatures.

The compound $Sr_2CuO_2Cl_2$ was identified as the first experimental $S = 1/2$ XXZ square-lattice antiferromagnet with a huge intra-layer coupling $J/k_B = 1450$ K, extremely weak inter-layer coupling $J''/J \cong 10^{-5}$ and extremely weak easy-plane anisotropy $\Delta_\lambda \approx 10^{-3}$ [96].

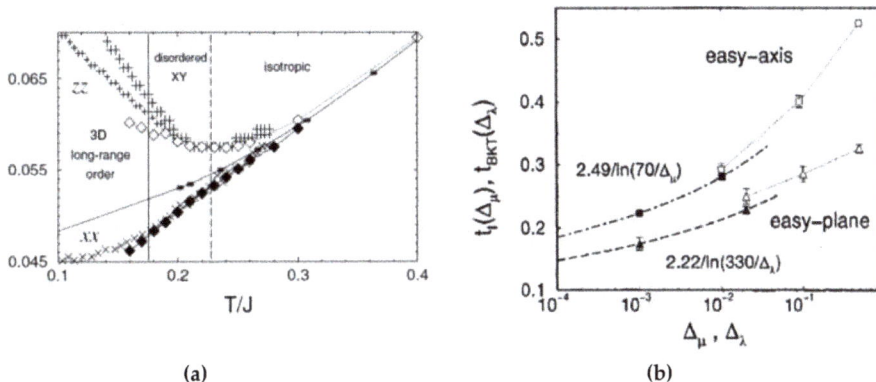

(a)

(b)

Figure 10. (a) Uniform susceptibility of $Sr_2CuO_2Cl_2$ (thin and thick pluses represent χ^{zz}, crosses - χ^{xx}) compared to the theoretical predictions for the $S = 1/2$ easy-plane XXZ model on the square lattice with $\Delta_\lambda = 10^{-3}$ (χ^{zz}—open diamonds, χ^{xx}—full diamonds). Stars represent the susceptibility of the isotropic HAF on the square lattice. (Reproduced with permission from reference [96]); (b) Phase diagram of the $S = 1/2$ XXZ model on the square lattice for weak anisotropies. The curves in the low-anisotropy region, i.e., $\Delta_{\mu,\lambda} < 10^{-1}$, represent the behavior of the reduced critical temperatures $t_{I,BKT} = k_B T_{I,BKT}/J$, described by the corresponding expressions. (Reproduced with permission from reference [97]).

A careful analysis of the low-temperature single-crystal susceptibility and magnetization of other Cu(II) based quantum magnets, Cu(pz)$_2$(NO$_3$)][PF$_6$], Cu(pz)$_2$(ClO$_4$)$_2$, and Cu(pz)$_2$(BF$_4$)$_2$, allowed to determine the presence of the weak easy-plane anisotropy $\Delta_\lambda \approx 10^{-3}$ [59]. In these quasi-2D $S = 1/2$ XXZ square-lattice magnets, the intra-layer coupling is two orders of magnitude lower than in $Sr_2CuO_2Cl_2$. The authors found that for the compounds with the high degree of the lattice two-dimensionality (i.e., $J''/J \leq 10^{-3}$), the spin anisotropy correlates well with the ratio of the anisotropy field, B_A, and the saturation field, $\Delta_\lambda \approx B_A/B_{sat}$ while a strong inter-layer coupling disturbs this coincidence [59].

3.2. $B \neq 0$

Extensive theoretical studies of the antiferromagnetic XXZ model on the square lattice in the external magnetic field revealed, that depending on the field orientation, a crossover between different spin symmetries can be induced. The studies of the classical as well as quantum version of the *easy-plane* XXZ model (Equation (8) with $\Delta_\mu = 0, 0 < \Delta_\lambda \leq 1$) showed the reinforcement of the XY anisotropy when the field is applied along the hard axis z [98,99]. On the other hand, at least in the classical version of the XXZ model with the strong easy-plane anisotropy, the application of the magnetic field within the easy plane ($B \perp z$) breaks the XY symmetry, introducing the symmetry of Ising type and a transition to the 2D LRO can be expected [98]. However, for the extremely small easy-plane anisotropy, the behavior of the AF XXZ model on the square lattice in $B \perp z$ is unclear.

In the case of the *easy-axis* XXZ model (Equation (8) with $\Delta_\lambda = 0, 0 < \Delta_\mu \leq 1$) in the magnetic field perpendicular to the easy z axis, $B \perp z$, the system retains a weak Ising anisotropy at all fields up to the saturation value. The ordered AF phase is separated from the paramagnetic phase by a line of the second-order transitions within the 2D Ising universality class [100]. The magnetic field applied along the easy axis ($B\|z$) competes with the spin anisotropy which tends to align the spins along the easy axis. Classical Monte Carlo studies of this 2D model showed the persistence of the collinear Néel order in low magnetic fields. The ordered antiferromagnetic phase is separated from the paramagnetic phase by a line of second-order phase transitions within the 2D Ising universality class [101–104]. The magnetic field acts as the effective easy-plane anisotropy, thus further increase leads to the spin-flop (SF) transition at the field, B_{SF}, accompanied with the reorientation of spins to be orthogonal to the field and gradually canting in its direction. The spin-flop phase is separated from the paramagnetic one by a critical line of BKT transitions [101–104]. In the limit $\Delta_\mu \to 0$, the ordered AF phase gradually vanishes, i.e., $B_{SF} \to 0$ and in the isotropic HAF limit, only the spin flop phase remains, separated from the paramagnetic state by the critical line of BKT transitions [101,105].

Quantum Monte Carlo studies of the $S = 1/2$ HAF on the square lattice in the external magnetic field described by the Hamiltonian

$$H = H_0 - g\mu_B B \sum_i S_i^z \tag{10}$$

(H_0 is the Hamiltonian described by the Equation (8) for $\Delta_\mu = \Delta_\lambda = 0$ and $i = (i_1, i_2)$ runs over the sites of the square lattice), were performed in a wide range of magnetic fields $h = g\mu_B B/(SJ)$ from zero to the saturation field $h_{sat} = 8$ [106,107]. It was found, that alike in the classical counterpart [101,105], the infinitesimal uniform field induces a BKT transition at a finite temperature. The magnetic phase diagram is characterized by a non-monotonous behavior of the critical temperatures (Figure 11a). In the weak field, the Hamiltonian (Equation (10)) can be mapped on the easy-plane XXZ model in zero magnetic field (Equation (8)) with $\Delta_\lambda \approx 0.1\, h^2$ and the transition temperature

$$\frac{k_B T_{BKT}}{J} \cong \frac{4\pi \rho_s / J}{\ln(C/h^2)} \tag{11}$$

In higher fields, two competing effects appear; a suppression of the fluctuations of the S^z component, resulting in the enhancement of the effective easy-plane anisotropy, which tends to increase T_{BKT}.

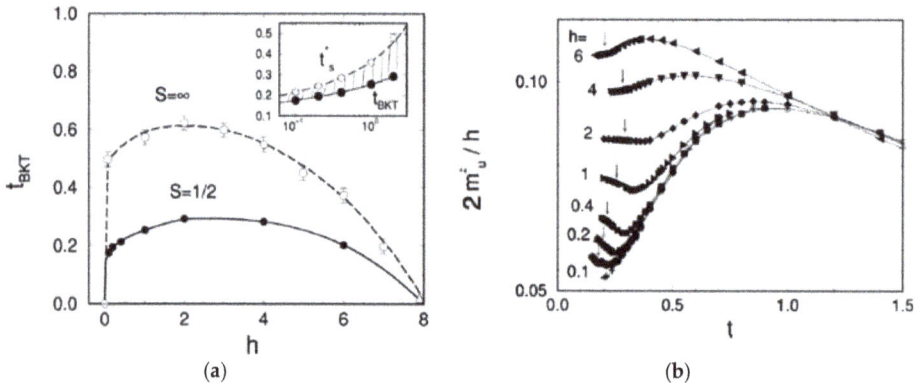

Figure 11. (a) Magnetic phase diagram of the $S = 1/2$ HAF on the square lattice (full symbols) in reduced coordinates $t_{BKT} = k_B T_{BKT}/J$ and $h = g\mu_B B/(SJ)$. Open symbols refer to the classical limit of the model. Inset: t_{BKT} vs. h for weak fields, t_s^* represents a crossover temperature from the isotropic to XY behavior. Shaded area marks the region of disordered XY behavior. (Reproduced with permission from reference [106]); (b) Field-induced uniform magnetization vs reduced temperature for different field values. The stars represent the zero-field uniform susceptibility of the $S = 1/2$ HAF on the square lattice. The arrows indicate the onset of BKT transition. (Reproduced with permission from reference [106]).

On the other hand, the average projection of the spins in the *xy* plane decreases, the system behaves as a planar rotator with a reduced rotator length, which leads to the reduction of T_{BKT}. The effective rotator length goes to zero at the saturation field and corresponding BKT critical temperature vanishes. Apparently, the interplay between the two effects is responsible for the non-monotonous dependence of T_{BKT}, which for the $S = 1/2$ achieves maximum values at higher fields than in the classical counterpart (Figure 11a). This shift was ascribed to the effect of quantum fluctuations [106].

The QMC simulations of the finite-temperature properties of the $S = 1/2$ HAF on the square lattice (Equation (10)) including the specific heat and uniform magnetization revealed typical features associated with the onset of the XY regime above the BKT transition. In the specific heat, the onset of the BKT transition demonstrates at the temperatures about 20–30% below a field-induced narrow peak superimposed on the round maximum. The peak is associated with a massive unbinding pairs of vortices (V) and antivortices (AV). A comparative study of the entropy derivatives in $h = 0$ and $h > 0$ revealed, that for $T < T_{BKT}$, the entropy in $h > 0$ grows slower due to the presence of the quasi-LRO established in the magnetic field after binding V + AV pairs, while above T_{BKT}, the growth is much faster due to unbinding V + AV pairs. Finally, at high temperatures, the entropy growth does not depend on the field in fully disordered systems [106]. A uniform magnetization is another bulk quantity, which can be experimentally measured; the QMC calculations showed that for $h < 1$, the temperature dependence of the uniform magnetization coincides with the uniform susceptibility of the $S = 1/2$ HAF on the square lattice in $h = 0$. Deviations appear at low temperatures $k_B T < J$, displaying a minimum, which indicates a field-induced crossover from the isotropic to the XY regime (Figure 11b).

In the real quasi-two-dimensional magnets, the strength of the interlayer coupling determines to which extent the aforementioned features characterizing the ideal 2D systems can be preserved. Depending on the type of the spin anisotropy and orientation of magnetic field, a corresponding 3D critical behavior can change from the already mentioned 3D Heisenberg to the 3D XY or 3D Ising universality class.

Recent experimental studies of real $S = 1/2$ quantum magnets with the extremely weak inter-layer coupling revealed a non-monotonous behavior of the 3D transition temperature in the magnetic field analogic to the *B-T* diagram in Figure 11a. Quantum Monte Carlo studies of the $S = 1/2$ HAF on the spatially anisotropic simple cubic lattice (with $R = 1$ within layers) in the magnetic field [108] simulated the experimental specific heat of the quasi-two-dimensional $S = 1/2$ quantum antiferromagnet $[Cu(HF_2)(pz)_2]BF_4$ with $J/k_B = 5.9$ K and $J''/J \approx 3 \times 10^{-2}$. In zero magnetic field, the compound undergoes a phase transition to the 3D LRO at $T_N = 1.6$ K. The application of the magnetic field up to 8 T led to the enhancement of the transition temperature, and a further field increase resulted in the conventional reduction of $T_N(B)$. While in the presence of the nonzero J'' the finite-temperature critical 2D properties are lost, the QMC studies found that in the quasi-2D magnets with extremely weak inter-layer coupling, this is just the nonmonotonic behavior of $T_N(B)$, which preserves also in the real systems (Figure 12a). On the other hand, the strong J'' will smear even this feature characterizing the ideal 2D magnets and a conventional decrease of the transition temperature will be observed in all magnetic fields. Such behavior was observed in $(5CAP)_2CuCl_4$, the quasi-2D $S = 1/2$ HAF on the square lattice with $J''/J \approx 0.25$ [64].

While extremely weak inter-layer coupling in $Cu(tn)Cl_2$ did not allow the formation of a sharp specific heat λ-like anomaly in the zero magnetic field, the application of the magnetic field of 0.75 T was capable to induce a weak anomaly at about 0.7 K. A further increase of the field enhanced the anomaly, shifting its position towards higher temperatures. In the fields above 2 T, the amplitude gradually decreased and the anomaly shifted to lower temperatures [89].

A typical increase of $T_N(B)$ was also observed in the $Cu(pz)_2(ClO_4)_2$ with $J''/J < 10^{-3}$. Since the saturation field is very large, the *B-T* phase diagram was recorded only for the fields lower than $B_{sat}/4$. This value corresponds to the fields below which, the BKT temperature in the ideal 2D case of the $S = 1/2$ HAF on the square lattice grows with the magnetic field [109]. The fact, that the phase diagrams of the $Cu(pz)_2(ClO_4)_2$ measured in the fields parallel and perpendicular to magnetic layers were found to be identical, was ascribed to a very weak intrinsic spin anisotropy.

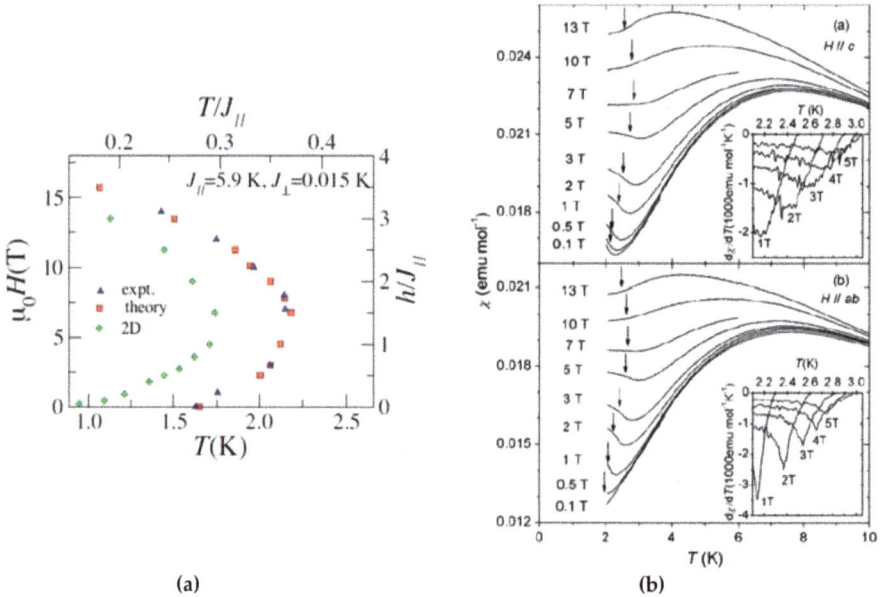

Figure 12. (a) Magnetic phase diagram of $[Cu(HF_2)(pz)_2]BF_4$ compared with QMC simulations of the $S = 1/2$ HAF on the spatially anisotropic simple cubic lattice ($R = 1$ within layers) with the system parameters $(J_{||} \equiv J, J_{\perp} \equiv J'', h = g\mu_B\mu_0 H)$. For comparison, the results for the pure 2D case ($J'' = 0$) are also shown. (Reproduced with permission from reference [108]); (b) $\chi(T)$ of $[Cu(pz)_2(pyO)_2](PF_6)_2$ for the magnetic field applied perpendicular to the easy plane ab ($\mu_0 H || c$) and within the easy plane ab ($\mu_0 H || ab$). The arrows mark the positions of the specific heat peaks. The insets show $\partial\chi(T)/\partial T$ below 5 T. (Reproduced with permission from reference [90]).

As for the case of $Cu(tn)Cl_2$, no sharp specific heat anomaly was observed down to 0.5 K in $[Cu(pz)_2(pyO)_2](PF_6)_2$, (pyO = pyridine-N-oxide), a quasi-2D $S = 1/2$ HAF on the square lattice with $J/k_B = 8.2$ K, $J''/J \approx 2 \times 10^{-4}$ and the weak easy-plane anisotropy $\Delta_\lambda \approx 0.007$ [90]. The application of the magnetic field of 1 T induced a weak anomaly at about 2 K, gradually growing and shifting towards higher temperatures. In relatively low fields, the shape of the anomaly depended on the orientation of the magnetic field. When the field was applied perpendicular to the easy plane, in accord with the theory [99], the field plays a role of an effective easy-plane anisotropy, thus fingerprints of the BKT transition in the whole field region up to B_{sat} can be expected. Considering the field applied within the easy plane, authors expected a field-induced easy-axis anisotropy, established in the fields up to 5 T [90]. In these fields, the formation of tiny λ-like anomalies, typical for the Ising transitions was observed, while at the fields above 5 T, the anomalies evolved to a broad anomaly of the expected BKT type and the system behaved as the isotropic 2D HAF in the field. Authors studied also a temperature dependence of the susceptibility in constant magnetic fields. In both orientations, the susceptibility is characterized by a broad maximum typical for 2D magnets with an upturn below 3 K (Figure 12b). In the field perpendicular to the easy plane, a minimum occurs always above the transition temperature to the 3D LRO, a feature typical for the onset of the XY regime, while in the field applied within the easy plane, such behavior occurs only above 1 T. Experimental studies of other quasi-2D $S = 1/2$ antiferromagnets with a weak easy-plane anisotropy in the magnetic field applied within the easy plane found, that for the fields lower than the anisotropy field B_A, the temperature dependence of the susceptibility has no upturn and its qualitative behavior follows χ^{xx} in Figure 10a. On the other hand, in the fields exceeding B_A, the upturn gradually develops and a characteristic minimum forms at temperatures above $T_N(B)$ as a typical sign of the onset of the field-induced XY regime (see Figure 8 of reference [59]). Thus, considering an excellent spatial two-dimensionality of [Cu(pz)$_2$(pyO)$_2$](PF$_6$)$_2$, the application of the aforementioned relation $\Delta_\lambda \approx B_A/B_{sat}$ provides $B_A \approx 0.2$ T. Taking into account a combined effect of the inter-layer coupling and the spin anisotropy, some higher external field should compensate both effects, to set the 2D XY regime. The experimental susceptibility data in the fields applied within the easy plane in Figure 12b suggest, that this condition has already been fulfilled at least for $B = 1$ T.

It should be noted, that at present, there are no theoretical studies of the $S = 1/2$ XXZ model with the extremely weak easy-plane anisotropy in the magnetic field applied within the easy plane which would provide reliable information about the field-induced spin crossover between the Ising and the XY regime. Despite the absence of the theoretical work, experimental studies of the magnetic phase diagrams of the spin 1/2 quasi-2D magnets were performed, to investigate the field-induced spin crossover in detail. Previous studies of Cu(pz)$_2$(ClO$_4$)$_2$, the 2D square-lattice magnet with $J''/J \approx 8.8 \times 10^{-4}$, $T_N = 4.2$ K and $B_{sat} \approx 49$ T, identified the presence of the easy-plane anisotropy $\Delta_\lambda \approx 4.6 \times 10^{-3}$ [59]. Subsequent antiferromagnetic resonance experiments and magnetization measurements in the ordered phase refined the character of the spin anisotropy comprising of the out-of-plane (easy-plane) anisotropy $\Delta_\lambda \approx 3.1 \times 10^{-3}$ and the in-plane anisotropy $\Delta_{in} \approx 3.1 \times 10^{-4}$, the latter breaking a continuous symmetry within the easy xy plane [94,110]. Thus, the description of Cu(pz)$_2$(ClO$_4$)$_2$ within the XYZ model with extremely weak anisotropy is more realistic.

The interplay of the in-plane anisotropy and magnetic field was already theoretically investigated in the AF chains [16], thus some qualitative conclusions can be extrapolated also to the 2D systems. For the field applied along the easy axis x, the magnetic field opposes the effect of the in-plane anisotropy. The resulting effective anisotropy $\Delta_{eff} = \Delta_{in} - a(g\mu_B B/J)^2$ controls the ground-state symmetry [16]. For the small magnetic field, the intrinsic anisotropy dominates, stabilizing the collinear AF Néel order along the x axis. With the increasing field, the influence of Δ_{in} gradually weakens and finally, a spin-flop transition from the easy (x) to the middle axis (y) occurs within the easy plane [16].

In Cu(pz)$_2$(ClO$_4$)$_2$, for the field applied along the easy axis, the AF-SF critical line ended in a bicritical point at 4 K and the field about 0.73 T (Figure 13). For the constant fields $B < 0.5$ T, the temperature dependence of the corresponding normalized uniform magnetization $M(T)/B$ is characterized by a sharp change of the slope in the vicinity of the transition temperature, separating the collinear AF Néel phase and the paramagnetic (PM) phase. For the fields, $0.5 < B < 0.73$ T, the $M(T)/B$ curves cross the SF-AF and AF-PM critical lines. The former crossing is accompanied with a pronounced step in the curves (Figure 13a). For higher fields, the spin-flop phase is stabilized with the XY regime, manifesting by the upturn in the $M(T)/B$ curves with a minimum, ascribed to the 2D Heisenberg-XY crossover. The kinks in the curves were associated with the onset of the 3D LRO. The application of the field along the hard axis z, leads to the behavior of $M(T)/B$ typical for the XY regime, since the magnetic field enforces the effect of the intrinsic easy-plane anisotropy, resulting in the effective easy-plane anisotropy $\Delta_{eff} = \Delta_\lambda + a(g\mu_B B/J)^2$ [94]. Concerning the middle axis y, authors of reference [16] expected enforcing of the in-plane easy-axis anisotropy, $\Delta_{eff} = \Delta_{in} + a(g\mu_B B/J)^2$. However, as was shown in Cu(pz)$_2$(ClO$_4$)$_2$, in this orientation, the fields above 2 T introduced the XY regime with the effective anisotropy $\Delta_{eff} \approx a(g\mu_B B/J)^2$ (Figure 16 in reference [94]). Apparently, alike in the case of the aforementioned easy-plane XXZ model, the theoretical studies of the $S = 1/2$ 2D XYZ model on the square lattice with a weak spin anisotropy are necessary, to verify the persistence of the Ising-like ground state as well as the Ising-XY crossover, induced by the field applied along the middle axis.

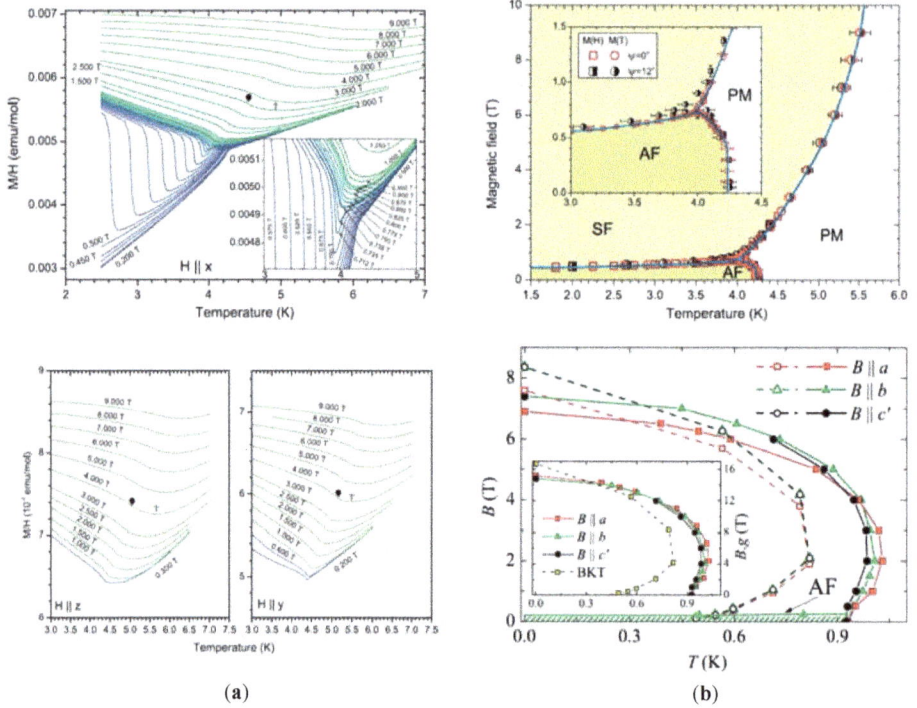

Figure 13. (**a**) (top, bottom) Normalized magnetization $M(T)/B$ of Cu(pz)$_2$(ClO$_4$)$_2$ for various magnetic fields $B \equiv \mu_0 H$ applied along the easy (x), middle (y) and hard (z) axis. In all three cases, the cross and droplet symbols mark the minimum and kink, respectively. (Reproduced with permission from reference [94]); (**b**) (top) Magnetic phase diagram of Cu(pz)$_2$(ClO$_4$)$_2$ in the field applied along the easy axis. AF, SF and PM denote the collinear antiferromagnetic phase, the spin-flop phase and the paramagnetic phase, respectively. The phase boundaries are constructed from the anomalies in the $M(T)$ and $M(H)$ curves, red and black points correspond to the orientation of the field ($\psi = 0°$ and 12°) with respect to the easy axis. Lines are a guide to the eye. In the inset, an expanded region around the bicritical point is shown. (Reproduced with permission from reference [94]); (**b**) (bottom) Magnetic phase diagram of Cu(en)(H$_2$O)$_2$SO$_4$. The a, b, c' axes correspond to the hard, easy and middle axis, respectively. Open symbols represent the theoretical predictions [97] for a field-induced BKT transition in the $S = 1/2$ HAF square lattice ($R = 1$, $J/k_B \equiv J_{eff}/k_B = 2.8$ K, $g_a = 2.200$, $g_b = 2.005$ and $g_{c'} = 2.000$). Inset: the same diagram, B is replaced by the product $B \cdot g$. (Reproduced with permission from reference [82]).

The analysis of thermodynamic properties showed that also Cu(en)(H$_2$O)$_2$SO$_4$ can be treated as the realization of the $S = 1/2$ 2D XYZ model [82]. In this respect, both, Cu(pz)$_2$(ClO$_4$)$_2$ and Cu(en)(H$_2$O)$_2$SO$_4$ should represent the same model, i.e., the $S = 1/2$ XYZ model on the spatially anisotropic zig-zag square lattice, differing in the amount of the spatial and the spin anisotropy. Within the effective $S = 1/2$ HAF square lattice model, the rescaling $B_{sat}/7$ and $J_{eff}/(7k_B)$ parameters of Cu(pz)$_2$(ClO$_4$)$_2$ will provide $B^*_{sat} \approx 7$ T and $J^*_{eff}/k_B \approx 2.6$ K, the values which are reported for Cu(en)(H$_2$O)$_2$SO$_4$ [65]. Concerning the spin anisotropy, applying the relation for the anisotropy-induced energy gaps in the excitation spectrum [94]

$$E_g^{in,out} \cong 1.2\sqrt{2J(\Delta J^{in,out})} \tag{12}$$

for the rescaled J_{eff}^*/k_B and the gap value $E_g^{in} \approx 0.3$ K reported for Cu(en)(H$_2$O)$_2$SO$_4$ [82], the relation provides the in-plane anisotropy $\Delta J^{in} \approx 12$ mK, which is about two times larger than $\Delta J^{in} \approx 5.3$ mK reported for Cu(pz)$_2$(ClO$_4$)$_2$ [94]. This result correlates well with the recent antiferromagnetic resonance experiments in Cu(en)(H$_2$O)$_2$SO$_4$ [111]. In addition, they determined the ratio of the spin gaps $E_g^{out}/E_g^{in} \approx 2.3$, correspondingly $\Delta J^{out}/\Delta J^{in} \approx 5$. In Cu(pz)$_2$(ClO$_4$)$_2$, the ratio of the spin gaps and the spin anisotropies is about 3.3 and 10, respectively, which is much closer to the easy-plane XXZ model. Apparently, the size as well as the relative strength of the in-plane anisotropy in Cu(en)(H$_2$O)$_2$SO$_4$ is two times larger than in Cu(pz)$_2$(ClO$_4$)$_2$. It seems, that this difference correlates well with the difference in the spatial anisotropies; $R = 0.7$ in Cu(pz)$_2$(ClO$_4$)$_2$ is two times lower than $R = 0.35$ in Cu(en)(H$_2$O)$_2$SO$_4$.

Previous studies of the electron paramagnetic resonance spectra in Cu(en)(H$_2$O)$_2$SO$_4$ already revealed the aforementioned strong rhombic character of the spin anisotropies [80]. It was assumed, that a dipolar coupling between nearest neighbors could provide the main source of the spin anisotropy in Cu(en)(H$_2$O)$_2$SO$_4$, where the exchange pathways are very long, resulting in weak exchange couplings. Consequently, the spin anisotropy could reflect the spatial distribution of Cu(II) magnetic moments. The dipolar coupling was considered as the only origin of the extremely weak easy-axis anisotropy $\Delta_\mu \approx 10^{-4}$, observed in the verdazyl radical [85]. As for Cu(pz)$_2$(ClO$_4$)$_2$, the exchange coupling is much stronger than in Cu(en)(H$_2$O)$_2$SO$_4$, but the strength of the in-plane anisotropy is weaker than in Cu(en)(H$_2$O)$_2$SO$_4$. The origin of the in-plane spin anisotropy in Cu(pz)$_2$(ClO$_4$)$_2$ was ascribed to the weak rhombic distortion of the crystal lattice, being of the same order as the spin anisotropy [94]. Similar symmetry arguments were used to explain the extremely weak spin anisotropy in [Cu(HF$_2$)(pz)$_2$]SbF$_6$ with a tetragonal structural symmetry in the ordered phase, while stronger anisotropies observed in the Cu(II) magnets with lower crystal symmetry were ascribed to their reduced structural symmetry [84].

Experimental studies of the Cu(II) compounds showed that sufficiently large magnetic fields, $g\mu_B B > E_g^{in,out}$, can overcome the effect of the intrinsic spin anisotropies and the systems behave as a Heisenberg antiferromagnet in the magnetic field, which induces the XY regime [82,94]. As was demonstrated in Cu(en)(H$_2$O)$_2$SO$_4$, after rescaling magnetic field with the high-temperature g-factors (to remove the effects of the local surrounding of Cu(II) ions introduced via a spin-orbit coupling), the magnetic fields exceeding the B_{SF} values, produce in all orientations the same behavior of the transition temperatures (Figure 13b). A comparison of the rescaled magnetic diagram with that for the $S = 1/2$ 2D HAF model on the square lattice ($R = 1$) revealed large discrepancies for $B \approx 0$, due to the presence of the inter-layer coupling in Cu(en)(H$_2$O)$_2$SO$_4$. On the other hand, the differences in the vicinity of the saturation field result from the higher measure of the quantum fluctuations in Cu(en)(H$_2$O)$_2$SO$_4$, introduced by the spatial anisotropy of the intra-layer exchange coupling.

4. Summary and Concluding Remarks

The effects of the spatial anisotropy of the exchange coupling as well as the intrinsic or magnetic-field induced spin anisotropy were discussed in the two-dimensional magnetic models and their experimental counterparts realized by the low-dimensional Cu(II) based metal-organic magnets.

The ground-state and finite-temperature properties of the $S = 1/2$ HAF chain model and HAF on the square lattice with the nearest neighbors were presented separately as important paradigms of the low-dimensional magnetism. Then, the attention was focused on the $S = 1/2$ HAF model on the rectangular lattice, which interpolates between the aforementioned 1D and 2D models, enabling the tuning the strength of the quantum fluctuations, which destabilize the ordered collinear Néel ground state when approaching the 1D limit. The problems with the distinction between the square, rectangular, and zig-zag square lattice in the analysis of the experimental data are discussed and the importance of the specific heat data and first-principle calculations is demonstrated on individual examples.

The impact of the inter-layer exchange coupling on the stabilization of the 3D long-range order in the quasi-2D magnets was presented. An ambiguous indication of the phase transition in the experimental data can occur when the inter-layer coupling is too weak. Besides neutron scattering, other experimental methods, as antiferromagnetic resonance or magnetic susceptibility can provide a proper information on the onset of the 3D LRO.

The effect of the extremely weak spin anisotropy in the 2D systems was presented by the $S = 1/2$ easy-axis and easy-plane XXZ models on the square lattice in zero and nonzero magnetic field. In $B = 0$, theoretical studies predict the onset of the 2D LRO, induced by the easy-axis anisotropy, while Berezinskii–Kosterlitz–Thouless phase transition appears in the presence of the intrinsic easy-plane anisotropy. It was shown, that under some conditions, the external magnetic field can act as a source of the effective easy-plane anisotropy. The fingerprints of the BKT transition in the real quasi-2D Cu(II) based molecular magnets were discussed. Finally, the interplay of the intrinsic spin anisotropy, magnetic field, and the inter-layer coupling in real quasi-2D Cu(II) based molecular magnets was demonstrated as specific features of magnetic properties and the corresponding magnetic phase diagrams.

The aforementioned discussion of the 2D theoretical models was restricted to the rectangular lattice with the nearest neighbors. Apparently, even this simple un-frustrated model can provide a rich variety of the phenomena, some of which still waiting for the theoretical and experimental investigations. While the excitation spectra of the $S = 1/2$ HAF chain and the square lattice were studied in detail theoretically and experimentally, analogical studies for $R < 1$ are still missing. Concerning the excitation spectrum of the HAF on the square lattice, the formation of a field-induced low-energy roton-like structure with the softening of the roton gap near B_{sat}, was theoretically predicted along with the suggestions, how to observe this feature experimentally [112]. Theoretical studies also predict the existence of the roton minimum in the high-energy part of the excitation spectrum of the same model in zero magnetic field [113].

As was shown in reference [72,74], in the $S = 1/2$ HAF on the rectangular lattice the amount of quantum fluctuations increases when approaching 1D limit. Corresponding theoretical studies of the $S = 1/2$ XXZ model on the rectangular lattice would be desirable to find, to which extent the interplay of the intrinsic/effective spin anisotropy and the enhanced quantum fluctuations can change the physical picture already derived for the XXZ model on the square lattice. Experimental investigations of the spin anisotropy in various quasi-2D quantum magnets found correlations between the character of the spin anisotropy and the structural symmetry. To achieve a 2D Heisenberg limit by minimizing the intra-layer spin anisotropies, the synthesis of molecule-based materials with a large inter-layer spacing and the tetragonal crystal symmetry in the ordered magnetic phase is required [84]. The engineering of proper experimental systems would enable to control the third lattice dimension in the square-lattice antiferromagnets as well as the intra-layer spin anisotropy, to obtain experimental model systems. Besides the chemical synthesis, the application of the pressure represents another tool for the structure modifications of molecular-based magnets. In these materials, hydrogen bonds and other electrostatic contacts play an important role in the packing to the 3D crystal structure with rather soft mechanical properties, characterized by low Debye temperatures and low-energy vibrational modes [114–116]. Correspondingly, unlike inorganic materials, the application of pressure to such soft structures could introduce new properties at relatively low pressures [117,118].

Besides bulk materials, the modification of their properties by a deposition of molecular-based magnets on substrates [119,120], growing in the confined geometry [121], etc. opens new possibilities for the experimental and theoretical investigations. Such studies have already strongly multidisciplinary character, requiring the cooperation of the experts from different fields and besides fundamental research, the results can have potential practical applications e.g., in the information technologies.

Author Contributions: Conceptualization, A.O.; Investigation, E.Č. and R.T.; Writing—Original Draft Preparation, A.O.; Writing-Review & Editing, R.T.; Visualization, V.T.; Funding Acquisition, M.O. and A.F.

Funding: The work was supported by VEGA Grant No. 1/0269/17 of the Scientific Grant Agency of the Ministry of Education, Science, Research and Sport of the Slovak Republic and the Slovak Academy of Sciences, the Slovak Research and Development Agency Projects No. APVV-14-0078 and No. APVV-14-0073, and the Project No. ITMS 26220120047 of European Regional Development Fund.

Conflicts of Interest: The authors declare no conflict of interest. The funders had no role in the design of the study; in the collection, analyses, or interpretation of data; in the writing of the manuscript, and in the decision to publish the results.

References

1.   Ising, E. Beitrag zur Theorie des Ferromagnetismus. *Z. Phys.* **1925**, *31*, 253–258. [CrossRef]
2.   Onsager, L. Crystal Statistics. I. A Two-Dimensional Model with an Order-Disorder Transition. *Phys. Rev.* **1944**, *65*, 117–149. [CrossRef]
3.   Domb, C. On the Theory of Cooperative Phenomena in Crystals. *Adv. Phys.* **1960**, *9*, 149–361. [CrossRef]
4.   Mermin, N.D.; Wagner, H. Absence of Ferromagnetism or Antiferromagnetism in One- or Two-Dimensional Isotropic Heisenberg Models. *Phys. Rev. Lett.* **1966**, *17*, 1133–1136. [CrossRef]
5.   Stanley, H.E.; Kaplan, T.A. Possibility of a Phase Transition for the Two-Dimensional Heisenberg Model. *Phys. Rev. Lett.* **1966**, *17*, 913–915. [CrossRef]
6.   Wegner, F. Spin-Ordering in a Planar Classical Heisenberg Model. *Z. Phys.* **1967**, *206*, 465–470. [CrossRef]
7.   Stanley, H.E. Dependence of Critical Properties on Dimensionality of Spins. *Phys. Rev. Lett.* **1968**, *20*, 589–592. [CrossRef]
8.   Moore, M.A. Additional Evidence for a Phase Transition in the Plane-Rotator and Classical Heisenberg Models for Two-Dimensional Lattices. *Phys. Rev. Lett.* **1968**, *23*, 861–863. [CrossRef]
9.   Berezinskii, V.L. Destruction of Long-range Order in One-dimensional and Two-dimensional Systems having a Continuous Symmetry Group I. Classical Systems. *Sov. Phys. JETP* **1971**, *32*, 493–500.
10. Berezinskii, V.L. Destruction of Long-range Order in One-dimensional and Two-dimensional Systems Possessing a Continuous Symmetry Group. II. Quantum Systems. *Sov. Phys. JETP* **1971**, *34*, 610–616.
11. Kosterlitz, M.J.; Thouless, D.J. Ordering, Metastability and Phase Transitions in Two-Dimensional Systems. *J. Phys. C* **1973**, *6*, 1181–1203. [CrossRef]
12. Minnhagen, P. The Two-Dimensional Coulomb Gas, Vortex Unbinding, and Superfluid-Superconducting Films. *Rev. Modern Phys.* **1987**, *59*, 1001–1066. [CrossRef]
13. de Jongh, L.J.; Miedema, A.R. Experiments on Simple Magnetic Model Systems. *Adv. Phys.* **1974**, *23*, 1–260. [CrossRef]
14. Müller, G.; Thomas, H.; Beck, H.; Bonner, J.C. Quantum Spin Dynamics of the Antiferromagnetic Linear Chain in Zero and Nonzero Magnetic Field. *Phys. Rev. B* **1981**, *24*, 1429–1467. [CrossRef]
15. Mikeska, H.J. Solitons in a One-Dimensional Magnet with an Easy Plane. *J. Phys. C* **1978**, *11*, L29–L32. [CrossRef]
16. Mikeska, H.J.; Steiner, M. Solitary Excitations in One-Dimensional Magnets. *Adv. Phys.* **1991**, *40*, 191–356. [CrossRef]
17. Haldane, F.D.M. Continuum Dynamics of the 1-D Heisenberg Antiferromagnet: Identification with the O(3) Nonlinear Sigma Model. *Phys. Lett. A* **1983**, *93*, 464–468. [CrossRef]
18. Haldane, F.D.M. Nonlinear Field Theory of Large-Spin Heisenberg Antiferromagnets: Semiclassically Quantized Solitons of the One-Dimensional Easy-Axis Néel State. *Phys. Rev. Lett.* **1983**, *50*, 1153–1156. [CrossRef]
19. Renard, J.P.; Verdaguer, M.; Regnault, L.P.; Erkelens, W.A.C.; Rossat-Mignod, J.; Stirling, W.G. Presumption for a Quantum Energy Gap in the Quasi-One-Dimensional $S = 1$ Heisenberg Antiferromagnet $Ni(C_2H_8N_2)_2NO_2(ClO_4)$. *Europhys. Lett.* **1987**, *3*, 945–951. [CrossRef]
20. Ma, S.; Broholm, C.; Reich, D.H.; Sternlieb, B.J.; Erwin, R.W. Dominance of Long-Lived Excitations in the Antiferromagnetic Spin-1 Chain NENP. *Phys. Rev. Lett.* **1992**, *69*, 3571–3574. [CrossRef]
21. Dagotto, E. Correlated Electrons in High-Temperature Superconductors. *Rev. Mod. Phys.* **1994**, *66*, 763–840. [CrossRef]
22. Huse, D.A.; Elser, V. Simple Variational Wave Functions for Two-Dimensional Heisenberg Spin-$\frac{1}{2}$ Antiferromagnets. *Phys. Rev. Lett.* **1988**, *60*, 2531–2534. [CrossRef] [PubMed]

23. Shastry, B.S.; Sutherland, B. Exact Ground-State of a Quantum-Mechanical Antiferromagnet. *Physica B & C* **1981**, *108*, 1069–1070. [CrossRef]
24. Bernhard, B.H.; Canals, B.; Lacroix, C. Green's Function Approach to the Magnetic Properties of the Kagomé Antiferromagnet. *Phys. Rev. B* **2002**, *66*, 104424. [CrossRef]
25. Savary, L.; Balents, L. Quantum Spin Liquids: A Review. *Rep. Prog. Phys.* **2017**, *80*, 016502. [CrossRef] [PubMed]
26. Scholwöck, U.; Richter, J.; Farnel, D.J.J.; Bishop, R.F. *Quantum Magnetism*, 1st ed.; Springer: Berlin/Heidelberg, Germany, 2004; ISBN 978-3-540-40066-0.
27. Blundel, S.J.; Pratt, F.L. Organic and Molecular Magnets. *J. Phys. Condens. Matter* **2004**, *16*, R771–R828. [CrossRef]
28. Meier, F.; Levy, J.; Loss, D. Quantum Computing with Spin Cluster Qubits. *Phys. Rev. Lett.* **2003**, *90*, 047901. [CrossRef]
29. Jenkins, M.D.; Duan, Y.; Diosdado, B.; García-Ripoll, J.J.; Gaita-Ariño, A.; Giménez-Saiz, C.; Alonso, P.J.; Coronado, E.; Luis, F. Coherent Manipulation of Three-Qubit States in a Molecular Single-Ion Magnet. *Phys. Rev. B* **2017**, *95*, 064423. [CrossRef]
30. Gaudenzi, R.; Burzurí, E.; Maegawa, S.; van der Zant, H.S.J.; Luis, F. Quantum Landauer Erasure with a Molecular Nanomagnet. *Nat. Phys.* **2018**, *14*, 565–568. [CrossRef]
31. Zhang, X.X.; Wei, H.L.; Zhang, Z.Q.; Zhang, L. Anisotropic Magnetocaloric Effect in Nanostructured Magnetic Clusters. *Phys. Rev. Lett.* **2001**, *87*, 157203. [CrossRef]
32. Zhitomirski, M.E. Enhanced Magnetocaloric effect in Frustrated Magnets. *Phys. Rev. B* **2003**, *67*, 104421. [CrossRef]
33. Trippe, C.; Honecker, A.; Klümper, A.; Ohanyan, V. Exact Calculation of the Magnetocaloric Effect in the spin-1/2 XXZ Chain. *Phys. Rev. B* **2010**, *81*, 054402. [CrossRef]
34. Thomas, L.; Lionti, F.; Ballou, R.; Gatteschi, D.; Sessoli, R.; Barbara, B. Macroscopic Quantum Tunnelling of Magnetization in a Single Crystal of Nanomagnets. *Nature* **1996**, *383*, 145–147. [CrossRef]
35. Radu, T.; Wilhelm, H.; Yushankhai, V.; Kovrizhin, D.; Coldea, R.; Tylczynski, Z.; Lühmann, T.; Steglich, F. Bose-Einstein Condensation of Magnons in Cs_2CuCl_4. *Phys. Rev. Lett.* **2005**, *95*, 127202. [CrossRef] [PubMed]
36. Zurek, W.H.; Dorner, U.; Zoller, P. Dynamics of a Quantum Phase Transition. *Phys. Rev. Lett.* **2005**, *95*, 105701. [CrossRef] [PubMed]
37. Blanc, N.; Trinh, J.; Dong, L.; Bai, X.; Aczel, A.A.; Mourigal, M.; Balents, L.; Siegrist, T.; Ramirez, A.P. Quantum Criticality among Entangled Spin Chains. *Nat. Phys.* **2018**, *14*, 273–276. [CrossRef]
38. Kasahara, Y.; Sugii, K.; Ohnishi, T.; Shimozawa, M.; Yamashita, M.; Kurita, N.; Tanaka, H.; Nasu, J.; Motome, Y.; Shibauchi, T.; et al. Unusual Thermal Hall Effect in a Kitaev Spin Liquid Candidate α-$RuCl_3$. *Phys. Rev. Lett.* **2018**, *120*, 217205. [CrossRef]
39. Affleck, I.; Lieb, E. A Proof of Part of Haldane's Conjecture on Spin Chains. *Lett. Math. Phys.* **1986**, *12*, 57–69. [CrossRef]
40. Affleck, I.; Kennedy, T.; Lieb, E.H.; Tasaki, H. Rigorous Results on Valence-Bond Ground States in Antiferromagnets. *Phys. Rev. Lett.* **1987**, *59*, 799–802. [CrossRef]
41. Giamarchi, T. *Quantum Physics in One Dimension*, 1st ed.; Clarendon Press: Oxford, UK, 2003; ISBN 9780198525004.
42. Tennant, D.H.; Cowley, R.A.; Nagler, S.E.; Tsvelik, A.M. Measurement of the Spin-Excitation Continuum in One-Dimensional $KCuF_3$ using Neutron Scattering. *Phys. Rev. B* **1995**, *52*, 13368–13380. [CrossRef]
43. Coldea, R.; Tennant, D.A.; Cowley, R.A.; McMorrow, D.F.; Dorner, B.; Tylczynski, Z. Quasi-1D S = 1/2 Antiferromagnet Cs_2CuCl_4 in a Magnetic Field. *Phys. Rev. Lett.* **1997**, *79*, 151–154. [CrossRef]
44. Stone, M.B.; Reich, D.H.; Broholm, C.; Lefmann, K.; Rischel, C.; Landee, C.P.; Turnbull, M.M. Extended Quantum Critical Phase in a Magnetized Spin-1/2 Antiferromagnetic Chain. *Phys. Rev. Lett.* **2003**, *91*, 037205. [CrossRef] [PubMed]
45. Bonner, J.C.; Fisher, M.E. Linear Magnetic Chains with Anisotropic Coupling. *Phys. Rev.* **1964**, *135*, A640–A658. [CrossRef]
46. Blöte, H.W.J. The Specific Heat of Magnetic Linear Chains. *Physica B & C* **1975**, *79*, 427–466. [CrossRef]
47. Klümper, A.; Johnston, D.C. Thermodynamics of the Spin-1/2 Antiferromagnetic Uniform Heisenberg Chain. *Phys. Rev. Lett.* **2000**, *84*, 4701–4704. [CrossRef] [PubMed]

48. Johnston, D.C.; Kremer, R.K.; Troyer, M.; Wang, X.; Klümper, A.; Bud'ko, S.L.; Panchula, A.F.; Canfield, P.C. Thermodynamics of Spin S = 1/2 Antiferromagnetic Uniform and Alternating-Exchange Heisenberg Chains. *Phys. Rev. B* **2000**, *61*, 9558–9606. [CrossRef]

49. Tarasenko, R.; Čižmár, E.; Orendáčová, A.; Kuchár, J.; Černák, J.; Prokleška, J.; Sechovský, V.; Orendáč, M. S = 1/2 Heisenberg Antiferromagnetic Spin Chain [Cu(dmbpy)(H₂O)₂SO₄] (dmbpy = 4,4'-dimethyl-2,2'-bipyridine): Synthesis, Crystal Structure and Enhanced Magnetocaloric Effect. *Solid State Sci.* **2014**, *28*, 14–19. [CrossRef]

50. Piazza, B.D.; Mourigal, M.; Christensen, N.B.; Nilsen, G.J.; Tregenna-Piggott, P.; Perring, T.G.; Enderle, M.; McMorrow, D.F.; Ivanov, D.A.; Ronnow, H.M. Fractional Excitations in the Square-Lattice Quantum Antiferromagnet. *Nat. Phys.* **2015**, *11*, 62–68. [CrossRef]

51. Chakravarty, S.; Halperin, B.I.; Nelson, D.R. Low-Temperature Behavior of Two-Dimensional Quantum Antiferromagnets. *Phys. Rev. Lett.* **1988**, *60*, 1057–1060. [CrossRef]

52. Chakravarty, S.; Halperin, B.I.; Nelson, D.R. Two-Dimensional Quantum Heisenberg Antiferromagnet at Low Temperatures. *Phys. Rev. B* **1989**, *39*, 2344–2371. [CrossRef]

53. Makivić, M.S.; Ding, H.-Q. Two-Dimensional Spin-1/2 Heisenberg Antiferromagnet: A Quantum Monte Carlo Study. *Phys. Rev. B* **1991**, *43*, 3562–3574. [CrossRef]

54. Cuccoli, A.; Tognetti, V.; Vaia, R.; Verrucchi, P. Two-Dimensional Quantum Heisenberg Antiferromagnet: Effective-Hamiltonian Approach to the Thermodynamics. *Phys. Rev. B* **1997**, *56*, 14456–14468. [CrossRef]

55. Kim, J.-K.; Troyer, M. Low Temperature Behavior and Crossovers of the Square Lattice Quantum Heisenberg Antiferromagnet. *Phys. Rev. Lett.* **1998**, *80*, 2705–2708. [CrossRef]

56. Orendáč, M.; Orendáčová, A.; Černák, J.; Feher, A. Magnetic Specific Heat Analysis of Cu(C₂H₈N₂)₂Ni(CN)₄: A Quasi-Two-Dimensional Heisenberg Antiferromagnet. *Solid State Commun.* **1995**, *94*, 833–835. [CrossRef]

57. Hanko, J.; Orendáč, M.; Kuchár, J.; Žák, Z.; Černák, J.; Orendáčová, A.; Feher, A. Hydrogen Bonds Mediated Magnetism in Cu(bmen)₂Pd(CN)₄. *Solid State Commun.* **2007**, *142*, 128–131. [CrossRef]

58. Turnbull, M.M.; Albrecht, A.S.; Jameson, G.B.; Landee, C.P. High-Field Magnetization Studies of Two-Dimensional Copper Antiferromagnets. *Mol. Cryst. Liq. Cryst.* **1999**, *335*, 245–252. [CrossRef]

59. Xiao, F.; Woodward, F.M.; Landee, C.P.; Turnbull, M.M.; Mielke, C.; Harrison, N.; Lancaster, T.; Blundell, S.J.; Baker, P.J.; Babkevich, P.; et al. Two-Dimensional XY Behavior Observed in Quasi-Two-Dimensional Quantum Heisenberg Antiferromagnets. *Phys. Rev. B* **2009**, *79*, 134412. [CrossRef]

60. Clarke, S.J.; Harrison, A.; Mason, T.E.; Visser, D. Characterisation of Spin-Waves in Copper(II) Deuteroformate Tetradeuterate: A Square S = 1/2 Heisenberg Antiferromagnet. *Solid State Commun.* **1999**, *112*, 561–564. [CrossRef]

61. Zhou, P.; Drumheller, J.E.; Rubenacker, G.V.; Halvorson, K.; Willett, R.D. Novel Low-Dimensional Spin 1/2 Antiferromagnets: Two-Halide Exchange Pathways in A₂CuBr₄ Salts. *J. Appl. Phys.* **1991**, *69*, 5804. [CrossRef]

62. Woodward, F.M.; Landee, C.P.; Giantsidis, J.; Turnbull, M.M.; Richardson, C. Structure and Magnetic Properties of (5BAP)₂CuBr₄: Magneto-Structural Correlations of Layered S = 1/2 Heisenberg Antiferromagnets. *Inorg. Chim. Acta* **2001**, *324*, 324–330. [CrossRef]

63. Woodward, F.M.; Albrecht, A.S.; Wynn, C.M.; Landee, C.P.; Turnbull, M.M. Two-Dimensional S = 1/2 Heisenberg Antiferromagnets: Synthesis, Structure, and Magnetic Properties. *Phys. Rev. B* **2002**, *65*, 144412. [CrossRef]

64. Coomer, F.C.; Bondah-Jagalu, V.; Grant, K.J.; Harrison, A.; McIntyre, G.J.; Rønnow, H.M.; Feyerherm, R.; Wand, T.; Meißner, M.; Visser, D.; et al. Neutron Diffraction Studies of Nuclear and Magnetic Structures in the S = 1/2 Square Heisenberg Antiferromagnets (d₆-5 CAP)₂CuX₄ (X = Br and Cl). *Phys. Rev. B* **2007**, *75*, 094424. [CrossRef]

65. Kajňaková, M.; Orendáč, M.; Orendáčová, A.; Vlček, A.; Černák, J.; Kravchyna, O.V.; Anders, A.G.; Bałanda, M.; Park, J.-H.; Feher, A.; et al. Cu(H₂O)₂(C₂H₈N₂)SO₄: A Quasi-Two-Dimensional S = 1/2 Heisenberg Antiferromagnet. *Phys. Rev. B* **2005**, *71*, 014435. [CrossRef]

66. Potočňák, I.; Vavra, M.; Čižmár, E.; Tibenská, K.; Orendáčová, A.; Steinbornc, D.; Wagner, C.; Dušek, M.; Fejfarová, K.; Schmidt, H.; et al. Low-Dimensional Compounds Containing Cyano Groups. XIV. Crystal Structure, Spectroscopic, Thermal and Magnetic Properties of [CuL₂][Pt(CN)₄] Complexes (L=ethylenediamine or N,N-dimethylethylenediamine). *J. Solid State Chem.* **2006**, *179*, 1965–1976. [CrossRef]

67. Zeleňák, V.; Orendáčová, A.; Císařová, I.; Černák, J.; Kravchyna, O.V.; Park, J.-H.; Orendáč, M.; Anders, A.G.; Feher, A.; Meisel, M.W. Magneto-Structural Correlations in Cu(tn)Cl$_2$ (tn = 1,3-Diaminopropane): Two-Dimensional Spatially Anisotropic Triangular Magnet Formed by Hydrogen Bonds. *Inorg. Chem.* **2006**, *45*, 1774–1782. [CrossRef] [PubMed]

68. Vlček, A.; Orendáč, M.; Orendáčová, A.; Kajňaková, M.; Papageorgiou, T.; Chomič, J.; Černák, J.; Massa, W.; Feher, A. Magneto-Structural Correlation in Cu(NH$_3$)$_2$Ag$_2$(CN)$_4$. Crystal Structure, Magnetic and Thermodynamic Properties of an $S = 1/2$ Low-Dimensional Heisenberg Antiferromagnet. *Solid State Sci.* **2007**, *9*, 116–125. [CrossRef]

69. Sedláková, L.; Tarasenko, R.; Potočňák, I.; Orendáčová, A.; Kajňaková, M.; Orendáč, M.; Starodub, V.A.; Anders, A.G.; Kravchyna, O.; Feher, A. Magnetic Properties of S = 1/2 Two-Dimensional Quantum Antiferromagnet Cu(D$_2$O)$_2$(C$_2$H$_6$D$_2$N$_2$)SO$_4$. *Solid State Commun.* **2008**, *147*, 239–241. [CrossRef]

70. Orendáč, M.; Čižmár, E.; Orendáčová, A.; Tkáčová, J.; Kuchár, J.; Černák, J. Enhanced Magnetocaloric Effect in Quasi-One-Dimensional $S = 1/2$ Heisenberg Antiferromagnet [Cu(dmen)$_2$(H$_2$O)]SiF$_6$. *J. Alloys Compd.* **2014**, *586*, 34–38. [CrossRef]

71. Parola, A.; Sorella, S.; Zhong, Q.F. Realization of a Spin Liquid in a Two Dimensional Quantum Antiferromagnet. *Phys. Rev. Lett.* **1993**, *71*, 4393–4396. [CrossRef]

72. Ihle, D.; Schindelin, C.; Weiße, A.; Fehske, H. Magnetic Order-Disorder Transition in the Two-Dimensional Spatially Anisotropic Heisenberg Model at Zero Temperature. *Phys. Rev. B* **1999**, *60*, 9240–9243. [CrossRef]

73. Affleck, I.; Halperin, B. On a Renormalization Group Approach to Dimensional Crossover. *J. Phys. A Math. Gen.* **1996**, *29*, 2627–2631. [CrossRef]

74. Sandvik, A.W. Multichain Mean-Field Theory of Quasi-One-Dimensional Quantum Spin Systems. *Phys. Rev. Lett.* **1999**, *83*, 3069–3072. [CrossRef]

75. Kim, Y.J.; Birgeneau, R.J. Monte Carlo Study of the S = 1/2 and S = 1 Heisenberg Antiferromagnet on a Spatially Anisotropic Square Lattice. *Phys. Rev. B* **2000**, *62*, 6378–6384. [CrossRef]

76. Jiang, F.-J.; Kämpfer, F.; Nyfeler, M. Monte Carlo Determination of the Low-Energy Constants of a Spin-1/2 Heisenberg Model with Spatial Anisotropy. *Phys. Rev. B* **2009**, *80*, 033104. [CrossRef]

77. Keith, B.C.; Landee, C.P.; Valleau, T.; Turnbull, M.M.; Harrison, N. Two-Dimensional Spin-1/2 Rectangular Heisenberg antiferromagnets: Simulation and Experiment. *Phys. Rev. B* **2011**, *84*, 104442; Erratum *Phys. Rev. B* **2011**, *84*, 229901, doi:10.1103/PhysRevB.84.229901. [CrossRef]

78. Cliffe, M.J.; Lee, J.; Paddison, J.A.M.; Schott, S.; Mukherjee, P.; Gaultois, M.W.; Manuel, P.; Sirringhaus, H.; Dutton, S.E.; Grey, C.P. Low-Dimensional Quantum Magnetism in Cu(NCS)$_2$: A Molecular Framework Material. *Phys. Rev. B* **2018**, *97*, 144421. [CrossRef]

79. Nath, R.; Padmanabhan, M.; Baby, S.; Thirumurugan, A.; Ehlers, D.; Hemmida, M.; Krug von Nidda, H.-A.; Tsirlin, A.A. Quasi-Two-Dimensional S = 1/2 Magnetism of Cu[C$_6$H$_2$(COO)$_4$][C$_2$H$_5$NH$_3$]$_2$. *Phys. Rev. B* **2015**, *91*, 054409. [CrossRef]

80. Tarasenko, R.; Orendáčová, A.; Čižmár, E.; Maťaš, S.; Orendáč, M.; Potočňák, I.; Siemensmeyer, K.; Zvyagin, S.; Wosnitza, J.; Feher, A. Spin Anisotropy in Cu(en)(H$_2$O)$_2$SO$_4$: A Quasi-Two-Dimensional S = 1/2 Spatially Anisotropic Triangular-Lattice Antiferromagnet. *Phys. Rev. B* **2013**, *87*, 174401. [CrossRef]

81. Sýkora, R.; Legut, D. Magnetic Interactions in a Quasi-One-Dimensional Antiferromagnet Cu(H$_2$O)$_2$(en)SO$_4$. *J. Appl. Phys.* **2014**, *115*, 17B305. [CrossRef]

82. Lederová, L.; Orendáčová, A.; Chovan, J.; Strečka, J.; Verkholyak, T.; Tarasenko, R.; Legut, D.; Sýkora, R.; Čižmár, E.; Tkáč, V.; et al. Realization of a Spin-1/2 Spatially Anisotropic Square Lattice in a Quasi-Two-Dimensional Quantum Antiferromagnet Cu(en)(H$_2$O)$_2$SO$_4$. *Phys. Rev. B* **2017**, *95*, 054436. [CrossRef]

83. Barbero, N.; Shiroka, T.; Landee, C.P.; Pikulski, M.; Ott, H.-R.; Mesot, J. Pressure and Magnetic Field Effects on a Quasi-Two-Dimensional Spin-1/2 Heisenberg Antiferromagnet. *Phys. Rev. B* **2016**, *93*, 054425. [CrossRef]

84. Goddard, P.A.; Singleton, J.; Franke, I.; Möller, J.S.; Lancaster, T.; Steele, A.J.; Topping, C.V.; Blundell, S.J.; Pratt, F.L.; Baines, C.; et al. Control of the Third Dimension in Copper-Based Square-Lattice Antiferromagnets. *Phys. Rev. B* **2016**, *93*, 094430. [CrossRef]

85. Yamaguchi, H.; Tamekuni, Y.; Iwasaki, Y.; Otsuka, R.; Hosokoshi, Y.; Kida, T.; Hagiwara, M. Magnetic Properties of a Quasi-Two-Dimensional S = 1/2 Heisenberg Antiferromagnet with Distorted Square Lattice. *Phys. Rev. B* **2017**, *95*, 235135. [CrossRef]

86. Siurakshina, L.; Ihle, D.; Hayn, R. Theory of Magnetic Order in the Three-Dimensional Spatially Anisotropic Heisenberg Model. *Phys. Rev. B* **2000**, *61*, 14601–14606. [CrossRef]
87. Sengupta, P.; Sandvik, A.W.; Singh, R.R.P. Specific Heat of Quasi-Two-Dimensional Antiferromagnetic Heisenberg Models with Varying Interplanar Couplings. *Phys. Rev. B* **2003**, *68*, 094423. [CrossRef]
88. Yasuda, C.; Todo, S.; Hukushima, K.; Alet, F.; Keller, M.; Troyer, M.; Takayama, H. Néel Temperature of Quasi-Low-Dimensional Heisenberg Antiferromagnets. *Phys. Rev. Lett.* **2005**, *94*, 217201. [CrossRef] [PubMed]
89. Orendáčová, A.; Čižmár, E.; Sedláková, L.; Hanko, J.; Kajňaková, M.; Orendáč, M.; Feher, A.; Xia, J.S.; Yin, L.; Pajerowski, D.M.; et al. Interplay of Frustration and Magnetic Field in the Two-Dimensional Quantum Antiferromagnet Cu(tn)Cl$_2$. *Phys. Rev. B* **2009**, *80*, 144418. [CrossRef]
90. Kohama, Y.; Jaime, M.; Ayala-Valenzuela, O.E.; McDonald, R.D.; Mun, E.D.; Corbey, J.F.; Manson, J.L. Field-Induced XY and Ising Ground States in a Quasi-Two-Dimensional S = 1/2 Heisenberg Antiferromagnet. *Phys. Rev. B* **2011**, *84*, 184402. [CrossRef]
91. Irkhin, V.Y.; Katanin, A.A. Thermodynamics of Isotropic and Anisotropic Layered Magnets: Renormalization-Group Approach and 1/N Expansion. *Phys. Rev. B* **1998**, *57*, 379–391. [CrossRef]
92. Hathaway, B.J.; Billing, D.E. The Electronic Properties and Stereochemistry of Mono-Nuclear Complexes of the Copper(II) Ion. *Coord. Chem. Rev.* **1970**, *5*, 143–207. [CrossRef]
93. Moriya, T. Anisotropic Superexchange Interaction and Weak Ferromagnetism. *Phys. Rev.* **1960**, *120*, 91–98. [CrossRef]
94. Povarov, K.Y.; Smirnov, A.I.; Landee, C.P. Switching of Anisotropy and Phase Diagram of the Heisenberg Square-Lattice S = 1/2 Antiferromagnet Cu(pz)$_2$(ClO$_4$)$_2$. *Phys. Rev. B* **2013**, *87*, 214402. [CrossRef]
95. Ihle, D.; Schindelin, C.; Fehske, H. Magnetic Order in the Quasi-Two-Dimensional Easy-Plane XXZ Model. *Phys. Rev. B* **2001**, *64*, 054419. [CrossRef]
96. Cuccoli, A.; Roscilde, T.; Vaia, R.; Verrucchi, P. Detection of XY Behavior in Weakly Anisotropic Quantum Antiferromagnets on the Square Lattice. *Phys. Rev. Lett.* **2003**, *90*, 167205. [CrossRef] [PubMed]
97. Cuccoli, A.; Roscilde, T.; Tognetti, V.; Vaia, R.; Verrucchi, P. Quantum Monte Carlo Study of S = 1/2 Weakly Anisotropic Antiferromagnets on the Square Lattice. *Phys. Rev. B* **2003**, *67*, 104414. [CrossRef]
98. Pereira, A.R.; Pires, A.S.T. Dynamics of Vortices in a Two-Dimensional Easy-Plane Antiferromagnet. *Phys. Rev. B* **1995**, *51*, 996–1002. [CrossRef]
99. Cuccoli, A.; Gori, G.; Vaia, R.; Verrucchi, P. Phase Diagram of the Two-Dimensional Quantum Antiferromagnet in a Magnetic Field. *J. Appl. Phys.* **2006**, *99*, 08H503. [CrossRef]
100. de Groot, H.J.M.; de Jongh, L.J. Phase Diagrams of Weakly Anisotropic Heisenberg Antiferromagnets, Nonlinear Excitations (Solitons) and Random-Field Effects. *Physica B & C* **1986**, *141*, 1–36. [CrossRef]
101. Landau, D.P.; Binder, K. Phase Diagrams and Critical Behavior of a Two-Dimensional Anisotropic Heisenberg Antiferromagnet. *Phys. Rev. B* **1981**, *24*, 1391–1403. [CrossRef]
102. Holtschneider, M.; Selke, W.; Leidl, R. Two-Dimensional Anisotropic Heisenberg Antiferromagnet in a Magnetic Field. *Phys. Rev. B* **2005**, *72*, 064443. [CrossRef]
103. Zhou, C.; Landau, D.P.; Schulthess, T.C. Hidden Zero-Temperature Bicritical Point in the Two-Dimensional Anisotropic Heisenberg Model: Monte Carlo Simulations and Proper Finite-Size Scaling. *Phys. Rev. B* **2006**, *74*, 064407. [CrossRef]
104. Pelissetto, A.; Vicari, E. Multicritical Behavior of Two-Dimensional Anisotropic Antiferromagnets in a Magnetic Field. *Phys. Rev. B* **2007**, *76*, 024436. [CrossRef]
105. Pires, A.S.T. Kosterlitz-Thouless Transition in a Two-Dimensional Isotropic Antiferromagnet in a Uniform Field. *Phys. Rev. B* **1994**, *50*, 9592–9594. [CrossRef]
106. Cuccoli, A.; Roscilde, T.; Vaia, R.; Verrucchi, P. Field-Induced XY Behavior in the S = 1/2 Antiferromagnet on the Square Lattice. *Phys. Rev. B* **2003**, *68*, 060402. [CrossRef]
107. Cuccoli, A.; Roscilde, T.; Vaia, R.; Verrucchi, P. XY Behaviour of the 2D Antiferromagnet in a Field. *J. Magn. Magn. Mater.* **2004**, *884*, 272–276. [CrossRef]
108. Sengupta, P.; Batista, C.D.; McDonald, R.D.; Cox, S.; Singleton, J.; Huang, L.; Papageorgiou, T.P.; Ignatchik, O.; Herrmannsdörfer, T.; Manson, J.L.; et al. Nonmonotonic Field Dependence of the Néel Temperature in the Quasi-Two-Dimensional Magnet [Cu(HF$_2$)(pyz)$_2$]BF$_4$. *Phys. Rev. B* **2009**, *79*, 060409. [CrossRef]

109. Tsyrulin, N.; Xiao, F.; Schneidewind, A.; Link, P.; Rønnow, H.M.; Gavilano, J.; Landee, C.P.; Turnbull, M.M.; Kenzelmann, M. Two-Dimensional Square-Lattice S = 1/2 Antiferromagnet Cu(pz)$_2$(ClO$_4$)$_2$. *Phys. Rev. B* **2010**, *81*, 134409. [CrossRef]

110. Fortune, N.A.; Hannahs, S.T.; Landee, C.P.; Turnbull, M.M.; Xiao, F. Magnetic-Field-Induced Heisenberg to XY Crossover in a Quasi-2D Quantum Antiferromagnet. *J. Phys. Conf. Ser.* **2014**, *568*, 042004. [CrossRef]

111. Glazkov, V.N.; Krasnikova, Y.; Rodygina, I.; Tarasenko, R.; Orendáčová, A. Low-Temperature Antiferromagnetic Resonance of Quasi-2D Magnet Cu(en)(H$_2$O)$_2$SO$_4$. In Proceedings of the International Symposium Spin Waves 2018, St. Petersburg, Russia, 3–8 June 2018.

112. Kubo, Y.; Kurihara, S. Tunable Rotons in Square-Lattice Antiferromagnets under Strong Magnetic Fields. *Phys. Rev. B* **2014**, *90*, 014421. [CrossRef]

113. Powalski, M.; Uhrig, G.S.; Schmidt, K.P. Roton Minimum as a Fingerprint of Magnon-Higgs Scattering in Ordered Quantum Antiferromagnets. *Phys. Rev. Lett.* **2015**, *115*, 207202. [CrossRef]

114. Orendáčová, A.; Kajňaková, M.; Černák, J.; Park, J.-H.; Čižmár, E.; Orendáč, M.; Vlček, A.; Kravchyna, O.V.; Anders, A.G.; Feher, A.; et al. Hydrogen Bond Mediated Magnetism in [Cu^{II}(en)$_2$(H$_2$O)][Cu^{II}(en)$_2$Ni$_2$Cu$^{I}_2$(CN)$_{10}$]·2H$_2$O. *Chem. Phys.* **2005**, *309*, 115–125. [CrossRef]

115. Brown, S.; Cao, J.; Musfeldt, J.L.; Conner, M.M.; McConnel, A.C.; Southerland, H.I.; Manson, J.L.; Schlueter, J.A.; Phillips, M.D.; Turnbull, M.M.; et al. Hydrogen Bonding and Multiphonon Structure in Copper Pyrazine Coordination Polymers. *Inorg. Chem.* **2007**, *46*, 8577–8583. [CrossRef] [PubMed]

116. Legut, D.; Sýkora, R.; Wdowik, U.D.; Orendáčová, A. Mechanical Properties of the Quasi-One-Dimensional Antiferromagnet Cu(en)(H$_2$O)$_2$SO$_4$. *J. Nanosci. Nanotechnol.* **2018**, *18*, 3016–3018. [CrossRef] [PubMed]

117. Perren, G.; Möller, J.S.; Hüvonen, D.; Podlesnyak, A.A.; Zheludev, A. Spin Dynamics in Pressure-Induced Magnetically Ordered Phases in (C$_4$H$_{12}$N$_2$)Cu$_2$Cl$_6$. *Phys. Rev. B* **2015**, *92*, 054413. [CrossRef]

118. O'Neal, K.R.; Zhou, J.; Cherian, J.G.; Turnbull, M.M.; Landee, C.P.; Jena, P.; Liu, Z.; Musfeldt, J.L. Pressure-Induced Structural Transition in Copper Pyrazine Dinitrate and Implications for Quantum Magnetism. *Phys. Rev. B* **2016**, *93*, 104409. [CrossRef]

119. Cini, A.; Mannini, M.; Totti, F.; Fittipaldi, M.; Spina, G.; Chumakov, A.; Rüffer, R.; Cornia, A.; Sessoli, R. Mössbauer Spectroscopy of a Monolayer of Single Molecule Magnets. *Nat. Commun.* **2018**, *9*, 480. [CrossRef]

120. Campbell, V.E.; Tonelli, M.; Cimatti, I.; Moussy, J.-B.; Tortech, L.; Dappe, Y.J.; Rivie, E.; Guillot, R.; Delprat, S.; Mattana, R.; et al. Engineering the Magnetic Coupling and Anisotropy at the Molecule–Magnetic Surface Interface in Molecular Spintronic Devices. *Nat. Commun.* **2016**, *7*, 13646. [CrossRef]

121. Laskowska, M.; Kityk, I.; Dulski, M.; Jędryka, J.; Wojciechowski, A.; Jelonkiewicz, J.; Wojtyniak, M.; Laskowski, Ł. Functionalized Mesoporous Silica Thin Films as a Tunable Nonlinear Optical Material. *Nanoscale* **2017**, *9*, 12110–12123. [CrossRef]

crystals

MDPI

Article

Thermodynamics and Magnetic Excitations in Quantum Spin Trimers: Applications for the Understanding of Molecular Magnets

Amelia Brumfield and Jason T. Haraldsen *

Department of Physics, University of North Florida, Jacksonville, FL 32224, USA; n00874260@ospreys.unf.edu
* Correspondence: j.t.haraldsen@unf.edu; Tel.: +1-904-620-2235

Received: 21 January 2019; Accepted: 4 February 2019; Published: 12 February 2019

Abstract: Molecular magnets provide a playground of interesting phenomena and interactions that have direct applications for quantum computation and magnetic systems. A general understanding of the underlying geometries for molecular magnets therefore generates a consistent foundation for which further analysis and understanding can be established. Using a Heisenberg spin-spin exchange Hamiltonian, we investigate the evolution of magnetic excitations and thermodynamics of quantum spin isosceles trimers (two sides J and one side αJ) with increasing spin. For the thermodynamics, we produce exact general solutions for the energy eigenstates and spin decomposition, which can be used to determine the heat capacity and magnetic susceptibility quickly. We show how the thermodynamic properties change with α coupling parameters and how the underlying ground state governs the Schottky anomaly. Furthermore, we investigate the microscopic excitations by examining the inelastic neutron scattering excitations and structure factors. Here, we illustrate how the individual dimer subgeometry governs the ability for probing underlying excitations. Overall, we feel these calculations can help with the general analysis and characterization of molecular magnet systems.

Keywords: molecular magnets; spin clusters; Heisenberg exchange Hamiltonian; thermodynamics; inelastic neutron scattering; exact diagonalization

1. Introduction

For over half a century, technology has been governed by the manipulation of the electron and its properties of charge and spin. Whether discussing the first transistor or the promise of quantum computation, the future of technology relies on the ability to take advantage of smaller quantum mechanical interactions for faster computer components, higher density memory, and larger data storage [1–4]. One method for data storage and quantum computation, is the property of spin, where the quest for smaller magnetic systems has lead to significant advances in the area of molecular magnets, which provide small nanomagnetic systems that can be used for quantum computation, spintronics, and magnetic storage [5–7]. However, the study of quantum spin clusters has been expanding and growing due to the possibilities of technological applications, quantum tunneling phenomenon, and interesting anisotropic effects.

Molecular magnets can be described as isolated magnetic clusters that are either part of a two- or three-dimensional lattice or be surrounded by non-magnetic ligands [8,9]. The latter are called single molecule magnets. In either case, molecular magnets produce low-energy, dispersionless excitations that provide avenues for many basic magnetic systems and devices [1]. Molecular magnetic systems are interesting due to their competition between ferromagnetic or antiferromagnetic interactions, which can lead to numerous varying ground states and quantum phase transitions [10,11].

While synthesis and characterization of molecular magnets are not trivial, throughout the last two decades, there have been hundreds of different molecular magnets studied through various techniques [12,13]. A variety of phenomena are probed through these techniques ranging from bulk properties (i.e., heat capacity, magnetic susceptibility, and the like) through magnetometry and electron spin resonance through microscopic properties (anisotropy and spin dynamics) using inelastic neutron scattering (INS) and optical spectroscopy [14–21]. On the theoretical end, there have been a number of experimental and theoretical studies on spin clusters and molecular magnets studying interactions and coupling aspects of these systems [22–32]. Furthermore, there have been advances in computational techniques to help identify and investigate interactions in molecular magnets through density functional theory [33]. Recently, computational databases have started to develop intricate tools to help identify new and distinct molecular magnets [34].

Recently, it has been shown that the larger molecule-based magnets have excitations that are characterized by the smaller subgeometries of the system [35–37]. This observation is important because it provides an opportunity to determine closed-form analytic solutions for systems that are typically solved numerically since most molecular magnetic systems are large clusters composed of high-spin atoms [30,32]. One can gain insight into larger clusters by thoroughly understanding the evolving excitations and transitions in smaller clusters [38].

In this study, we examine the effects of a variable dimer exchange embedded in a quantum spin trimer and verify that, regardless of spin and trimer geometry, the inelastic neutron scattering excitations for any trimer can be determined by the excitations of the individual bases. Furthermore, we examine the evolution of the thermodynamics as a function of spin and dimer exchange and show the quantum phase transition points in the heat capacity. These phase transitions allow for the possible tunability of quantum trimers with external fields (magnetic, strain, etc.)

2. The Heisenberg Trimer Model

A trimer system consists of three interacting spins, where two of the interacting spins create a magnetic dimer. Figure 1 shows various magnetic isosceles trimer geometries with different dimer exchanges. Here, two sides of the trimer have an interaction of J, while the third side is a variational dimer exchange of αJ. When $\alpha = 0$, the trimer is linear. However, as $|\alpha|$ is increased the system will produce a weak dimer, equilateral ($\alpha = 1$), and a strong dimer ($|\alpha| > 1$). It should be noted that the structural geometry is not dependent on the magnetic geometry.

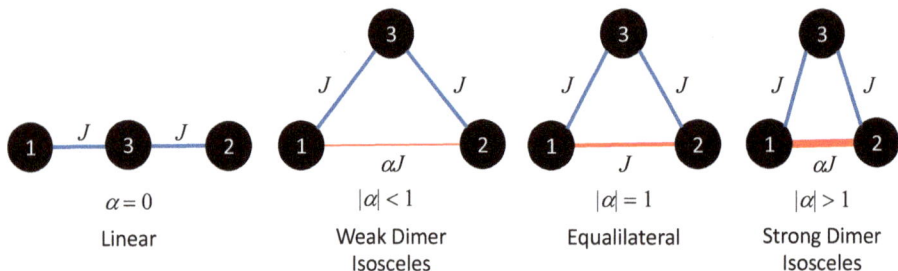

Figure 1. Various magnetic isosceles trimer geometries with different dimer exchanges. Here, two sides of the trimer have an interaction of J, while the third side is a variational dimer exchange of αJ. When $\alpha = 0$, the trimer is linear. However, as $|\alpha|$ is increased the system will produce a weak dimer, equilateral ($\alpha = 1$), and a strong dimer ($|\alpha| > 1$). It should be noted that the structural geometry is not dependent on the magnetic geometry.

Investigating the spin model for this geometry can include many different interactions and variations. However, in this study, the isotropic Heisenberg model is examined for various exchange interactions. Therefore, the Hamiltonian for the isosceles trimer we studied is

$$\mathcal{H} = \alpha J \vec{S}_1 \cdot \vec{S}_2 + J\,(\vec{S}_2 \cdot \vec{S}_3 + \vec{S}_3 \cdot \vec{S}_1) \tag{1}$$

where J is the superexchange constant, α is the dimer exchange parameter, and S_i is the spin. For interactions in antiferromagnets $J > 0$, while ferromagnets have $J < 0$.

With the Hamiltonian, one can use exact diagonalization to find the energy eigenstates as well as the eigenfunctions for the Hamiltonian matrix. The energy states are critical for the thermodynamic properties, while the eigenfunctions or wavefunctions are needed to determine the inelastic neutron scattering cross sections. Due to the symmetry in this system, the energy of each magnetic state for the isosceles trimer can be written exactly using a spin total and spin dimer basis, where

$$E(S_t, S_d) = \frac{J}{2}\Big[S_t(S_t + 1) + S_d(S_d + 1)(\alpha - 1) - \alpha\Big(S_1(S_1 + 1) + S_2(S_2 + 1)\Big) - S_3(S_3 + 1)\Big], \tag{2}$$

Here, S_t is the total spin of the trimer and S_d is the total spin for the dimer state. S_i is the spin for each of the magnetic atoms making the trimer. Therefore, from this, the energy for each state can be determined by knowing the total spin of trimer and dimer states for a given trimer of spin S.

To understand the spin components for each system, one must examine the spin decomposition of the trimer spin tensor product $(S \otimes S \otimes S)$. From the trimer tensor product, one can decompose to an individual spin interacting with a spin dimer $(S \otimes (2S \oplus 2S - 1 \oplus 2S - 2 \oplus ... \oplus 0))$. Therefore, the general trimer spin decomposition can be written as

$$3S \oplus (3S - 1)^2 \oplus (3S - 1)^3 \oplus ... \oplus (3S - 2S)^{2S+1} \oplus (S - 1)^{2S-1} \oplus (S - 2)^{2S-3} \oplus ... \oplus (S - N)^{2(S-N)-1}, \tag{3}$$

where N is $S - 3/2$ for half-integer spins and $S - 1$ for integer spin. The superscript denotes the multiplicity of the state. This summation can be simplified as

$$\sum_0^{2S}(3S - N)^{N+1} \oplus \sum_0^{[[S-1]]}(S - N - 1)^{2(S-N)-1}, \tag{4}$$

where the summations are sums over all direct sums and the second summation is for spins greater than 1/2. From this general direct sum, the total number of states can correctly be deduced from

$$\sum_0^{2S}(2(3S - N) + 1)(N + 1) + \sum_0^{[[S-1]]}(2(S - N - 1) + 1)(2(S - N) - 1). \tag{5}$$

Once the sums are evaluated, the total number of states reduces to $(2S + 1)^3$. From this spin decomposition, we can determine the spin state for any spin system. In this paper, we only examine up to spin 5/2, where the spin decomposition for each can be found in Table 1. However, the general S equations can be used for any spin.

In the S = 1/2 case, you have a S = 1/2 dimer which has spin states of (1 and 0). Adding a third spin to make the trimer produces three spin total states one $S_t = 3/2$ state and two $S_t = 1/2$ states, where the two S = 1/2 states have different dimer bases of $S_d = 1$ and 0. The energy states for the S = 1/2, 1, 3/2, 2, and 5/2 trimers are given in Table 1.

Table 1. Spin decomposition and energy levels for the quantum spin trimers.

System	Dimer/Trimer States	Energy States ($E_{S_t,S_d}/J$)	
$\frac{1}{2} \otimes \frac{1}{2} \otimes \frac{1}{2}$	$1 \oplus 0$ $\frac{3}{2} \oplus \frac{1}{2}^2$	$E_{3/2,1}/J = \frac{1}{4}(2+\alpha)$ $E_{1/2,1}/J = \frac{1}{4}(\alpha-4)$ $E_{1/2,0}/J = -\frac{3}{4}\alpha$	
$1 \otimes 1 \otimes 1$	$2 \oplus 1 \oplus 0$ $3 \oplus 2^2 \oplus 1^3 \oplus 0$	$E_{3,2}/J = 2+\alpha$ $E_{2,2}/J = \alpha-1$ $E_{1,2}/J = \alpha-3$ $E_{2,1}/J = 1-\alpha$ $E_{1,1}/J = -1-\alpha$ $E_{1,0}/J = -2\alpha$ $E_{0,1}/J = -2-\alpha$	
$\frac{3}{2} \otimes \frac{3}{2} \otimes \frac{3}{2}$	$3 \oplus 2 \oplus 1 \oplus 0$ $\frac{9}{2} \oplus \frac{7}{2}^2 \oplus \frac{5}{2}^3 \oplus \frac{3}{2}^4 \oplus \frac{1}{2}^2$	$E_{9/2,3}/J = \frac{9}{4}(2+\alpha)$ $E_{7/2,3}/J = \frac{9}{4}\alpha$ $E_{5/2,3}/J = \frac{9}{4}(\alpha-\frac{14}{9})$ $E_{3/2,3}/J = \frac{9}{4}\alpha-6$ $E_{7/2,2}/J = 3-\frac{3}{4}\alpha$ $E_{5/2,2}/J = -\frac{1}{2}-\frac{3}{4}\alpha$	$E_{3/2,2}/J = -3-\frac{3}{4}\alpha$ $E_{1/2,2}/J = -\frac{9}{2}-\frac{3}{4}\alpha$ $E_{5/2,1}/J = \frac{3}{2}-\frac{11}{4}\alpha$ $E_{3/2,1}/J = -1-\frac{11}{4}\alpha$ $E_{1/2,1}/J = -\frac{5}{2}-\frac{11}{4}\alpha$ $E_{3/2,0}/J = -\frac{15}{4}\alpha$
$2 \otimes 2 \otimes 2$	$4 \oplus 3 \oplus 2 \oplus 1 \oplus 0$ $6 \oplus 5^2 \oplus 4^3 \oplus 3^4 \oplus 2^5 \oplus 1^3 \oplus 0$	$E_{6,4}/J = 4(2+\alpha)$ $E_{5,4}/J = 2(1+2\alpha)$ $E_{4,4}/J = 4\alpha-3$ $E_{3,4}/J = 4\alpha-7$ $E_{2,4}/J = 4\alpha-10$ $E_{5,3}/J = 6$ $E_{4,3}/J = 1$ $E_{3,3}/J = -3$ $E_{2,3}/J = -6$ $E_{1,3}/J = -8$	$E_{4,2}/J = 4-3\alpha$ $E_{3,2}/J = -3\alpha$ $E_{2,2}/J = -3-3\alpha$ $E_{1,2}/J = -5-3\alpha$ $E_{0,2}/J = -6-3\alpha$ $E_{3,1}/J = 2-5\alpha$ $E_{2,1}/J = -1-5\alpha$ $E_{1,1}/J = -3-5\alpha$ $E_{2,0}/J = -6\alpha$
$\frac{5}{2} \otimes \frac{5}{2} \otimes \frac{5}{2}$	$5 \oplus 4 \oplus 3 \oplus 2 \oplus 1 \oplus 0$ $\frac{15}{2} \oplus \frac{13}{2}^2 \oplus \frac{11}{2}^3 \oplus \frac{9}{2}^4 \oplus \frac{7}{2}^5 \oplus \frac{5}{2}^6 \oplus \frac{3}{2}^4 \oplus \frac{1}{2}^2$	$E_{15/2,5}/J = \frac{25}{4}(2+\alpha)$ $E_{13/2,5}/J = 5+\frac{25}{4}\alpha$ $E_{11/2,5}/J = -\frac{3}{2}+\frac{25}{4}\alpha$ $E_{9/2,5}/J = \frac{25}{4}\alpha-7$ $E_{7/2,5}/J = \frac{25}{4}\alpha-\frac{23}{4}$ $E_{5/2,5}/J = \frac{25}{4}\alpha-15$ $E_{13/2,4}/J = 10+\frac{5}{4}\alpha$ $E_{11/2,4}/J = \frac{7}{2}+\frac{5}{4}\alpha$ $E_{9/2,4}/J = \frac{5}{4}\alpha-2$ $E_{7/2,4}/J = \frac{5}{4}\alpha-\frac{13}{2}$ $E_{5/2,4}/J = \frac{5}{4}\alpha-10$ $E_{3/2,4}/J = \frac{5}{4}\alpha-\frac{25}{2}$ $E_{11/2,3}/J = \frac{15}{2}-\frac{11}{4}\alpha$ $E_{9/2,3}/J = 2-\frac{11}{4}\alpha$	$E_{7/2,3}/J = -\frac{5}{2}-\frac{11}{4}\alpha$ $E_{5/2,3}/J = -6-\frac{11}{4}\alpha$ $E_{3/2,3}/J = -\frac{17}{2}-\frac{11}{4}\alpha$ $E_{1/2,3}/J = -10-\frac{11}{4}\alpha$ $E_{9/2,2}/J = 2-\frac{11}{4}\alpha$ $E_{7/2,2}/J = \frac{1}{2}-\frac{23}{4}\alpha$ $E_{5/2,2}/J = -3-\frac{23}{4}\alpha$ $E_{3/2,2}/J = -\frac{11}{2}-\frac{23}{4}\alpha$ $E_{1/2,2}/J = -7-\frac{23}{4}\alpha$ $E_{7/2,1}/J = \frac{5}{2}-\frac{31}{4}\alpha$ $E_{5/2,1}/J = -1-\frac{31}{4}\alpha$ $E_{3/2,1}/J = -\frac{7}{2}-\frac{31}{4}\alpha$ $E_{5/2,0}/J = -\frac{35}{4}\alpha$

3. Evolution of the Thermodynamic Properties

The following sections show the calculations for the spin states of isosceles trimers of $S = 1/2$, $S = 1$, $S = 3/2$, $S = 2$, and $S = 5/2$. As the spin increases, the spin states drastically increase in size and complexity. These calculations are used in the sections following to find the energy levels, partition function, heat capacity, and later the magnetic susceptibility.

Using the energy eigenstates and eigenvalues in Table 1, we can determine the partition function of the system using

$$Z = \sum_{i=1}^{N} e^{-\beta E_i} = \sum_{E_i} (2S_{tot}+1) e^{-\beta E_i}, \qquad (6)$$

where the sum $i = 1 \ldots N$ is over all N independent energy eigenstates (including magnetic substates), the sum $\sum_{E_i}$ is over energy levels only, and $\beta = 1/k_b T$. In practice, we will employ the usual set of $\hat{z}$-polarized magnetic basis states.

While the general definition of the heat capacity can be simply described as the amount of energy required to raise the temperature of a material by a small amount (assuming adiabatic conditions) [39]. The more rigorous definition can be found by through the manipulation of the partition function shown as

$$C = k_B \beta^2 \frac{d^2 \ln(Z)}{d\beta^2} . \tag{7}$$

Although the heat capacity is an important property, the overall size of the functions makes them difficult to report. Most mathematical packages, however, would be able to calculate this property once the partition function is determined from the energy states. The heat capacity has a distinct feature, known as the Schottky anomaly, which has a characteristic λ shape, that denotes a change in entropy of the quantum system, which happens in the transition from an antiferromagnetic or ferromagnetic ground state to a thermally excited state at higher temperatures. This thermal activity eventually leads to a paramagnetic state due to the thermal fluctuations of the spins.

In most magnetic systems, the transition is a second-order continuous transition. Because of this, if one integrates the ratio of the heat capacity and β over all temperatures, then the total entropy of the system can be calculated. From statistical mechanics, the total entropy will be equal to the ratio of the dimensionality of the Hilbert space (N) over the degeneracy of the ground state (N_0) [31,39], which is written as

$$S = \int_0^\infty \frac{C}{\beta} d\beta = k_B \ln\left(\frac{N}{N_0}\right) . \tag{8}$$

This allows one to verify the energy eigenstates through a determination of the entropy.

When a magnetic field is applied to a material, the magnetic moments in a lattice arrangement at a certain temperature T react by either lining up to point in the same direction (ferromagnetic) or lining up in opposite directions (antiferromagnetic). The response of the material to the magnetic field applied is its magnetic susceptibility [39]. The energy eigenvalues can be used to find the magnetic susceptibility, which is given by

$$\chi = \frac{\beta}{Z} \sum_{i=1}^N (M_z^2)_i \, e^{-\beta E_i} = \frac{1}{3} (g\mu_B)^2 \frac{\beta}{Z} \sum_{E_i} (2S_{tot} + 1) \, (S_{tot} + 1) \, S_{tot} \, e^{-\beta E_i} , \tag{9}$$

In these formulas $M_z = mg\mu_B$ where $m = S_{tot}^z/\hbar$ is the integral or half-integral magnetic quantum number, and g is the electron g-factor. The magnetic susceptibility allows one to clearly determine the total spin of the ground state; the larger the total spin the larger the magnetic susceptibility at $T = 0$.

Figures 2 and 3 show the calculated heat capacity, energy levels, and magnetic susceptibility as a function of temperature and dimer exchange (α) for $S = 1/2, 1, 3/2, 2$, and $5/2$ spin trimers with antiferromagnetic and ferromagnetic, respectively.

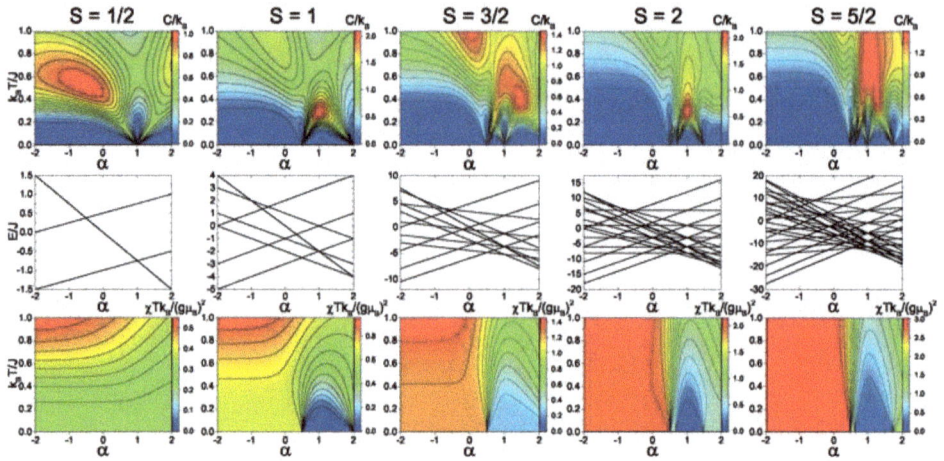

Figure 2. The heat capacity (**top** panels), energy levels (**middle** panels), and magnetic susceptibility (**bottom** panels) for antiferromagnetic (J = 1) isosceles trimers of spin $1/2, 1, 3/2, 2$, and $5/2$ as functions of temperature and dimer exchange α.

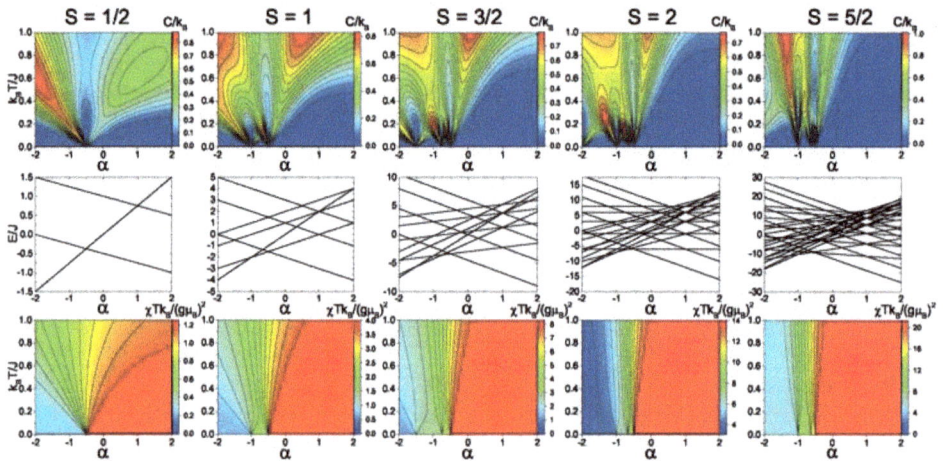

Figure 3. The heat capacity (**top** panels), energy levels (**middle** panels), and magnetic susceptibility (**bottom** panels) for ferromagnetic (J = −1) isosceles trimers of spin $1/2, 1, 3/2, 2$, and $5/2$ as functions of temperature and dimer exchange α.

For the antiferromagnetic spin-$1/2$ model (Figure 2), there is a distinct quantum phase transition at α = 1. At this transition point, the system does not shift its spin total state of $1/2$. However, the spin dimer state does shift from spin 1 to spin 0. The phase transition is clearly shown in the heat capacity due to the shift in the temperature dependence of the Schottky anomaly at the transition point. Since the quantum phase transition is not a shift in total spin, the magnetic susceptibility does not have any major shift. In the ferromagnetic spin-$1/2$ model (Figure 3), the quantum transition shifts to α = −0.5. Unlike the antiferromagnetic spin-$1/2$ model, however, the ferromagnetic case has a distinct shift in total spin from spin $1/2$ to spin $3/2$, which is evident in both the heat capacity and magnetic susceptibility.

In the antiferromagnetic spin-1 model (Figure 2), there are two quantum phase transition points which correspond to the shifts in the total spin S_t = 1 to 0 (α = 0.5) and from S_t = 0 to 1 (α = 2).

This transition is shown in both the heat capacity and magnetic susceptibility, where the magnetic susceptibility of the spin-0 state goes to zero as the temperature approaches zero. In the ferromagnetic spin-1 model (Figure 3), there are also two phase transitions that have a ground states shift from $S_t = 1$ to 2 ($\alpha = -1.0$) and from $S_t = 2$ to 3 ($\alpha = -0.5$), which is evident from the magnetic susceptibility.

In the spin-3/2, spin-2, and spin-5/2 models, there are multiple quantum phase transitions in the energy spectra that correspond to distinct changes in the heat capacity and magnetic susceptibility. From Figures 2 and 3, it is clear that the peak of the Schottky anomaly is directly related to the energy gap between the ground state and first excited state of the given magnetic system. Overall, these transitions allow for possible switch points in systems where α is variable with applied external fields. By switching the ground state with an external field, the change in the magnetization can lead to an increase or decrease in magnetic moment, which will designate the ground state and provide the state, therefore acting as a spin switch.

4. Inelastic Neutron Scattering Structure Factors

While the heat capacity and magnetic susceptibility can provide detailed information about the total spin states and quantum transition in the trimer system, inelastic neutron scattering (INS) allows for the examination of the magnetic states on a microscopic level, where it measures characteristic geometries and excitations for the trimer systems. In addition, therefore, to the calculated bulk properties, we also examined results for inelastic neutron scattering intensities.

As discussed in [31], in "spin-only" magnetic neutron scattering at T = 0, the inelastic magnetic scattering of an incident neutron will have a differential cross section that is proportional to the unpolarized neutron scattering structure factor

$$S_{ba}(\vec{q}, \omega) = \int_{-\infty}^{\infty} \frac{dt}{2\pi} \sum_{\vec{x}_i, \vec{x}_j} e^{i\vec{q} \cdot (\vec{x}_i - \vec{x}_j) + i\omega t} \langle \Psi_j | S_b^\dagger(\vec{x}_j, t) S_a(\vec{x}_i, 0) | \Psi_i \rangle. \tag{10}$$

where the system has an initial state $|\Psi_i\rangle$, with momentum transfer $\hbar\vec{q}$ and energy transfer $\hbar\omega$. Here, the site sums in Equation (10) run over all magnetic ions in one unit cell, where a, b are the spatial indices of the spin operators and $\vec{x}_i$ are the spatial vectors for the spin sites.

Since spin clusters have transitions between discrete energy levels, the time integral in the energy transfer gives a trivial delta function $\delta(E_f - E_i - \hbar\omega)$. We can therefore shift to an "exclusive structure factor" for the individual excitation of states λ within a specific magnetic multiplet, $|\Psi_f(\lambda_f)\rangle$, from the given initial state $|\Psi_i\rangle$. This structure factor can be written as

$$S_{ba}^{(fi)}(\vec{q}) = \sum_{\lambda_f} \langle \Psi_i | V_b^\dagger | \Psi_f(\lambda_f) \rangle \langle \Psi_f(\lambda_f) | V_a | \Psi_i \rangle, \tag{11}$$

where the vector $V_a(\vec{q})$ is a sum of spin operators over all magnetic ions in a unit cell,

$$V_a = \sum_{\vec{x}_i} S_a(\vec{x}_i) e^{i\vec{q} \cdot \vec{x}_i}. \tag{12}$$

Assuming an isotropic magnetic Hamiltonian with a spherical basis for the spin operators S_a, the tensor $S_{ba}^{(fi)}(\vec{q})$ is diagonal and has entries that are proportional to a universal function of $\vec{q}$ times a product of Clebsch–Gordon coefficients [21,31]. The structure factor can then be simplified by examining the unpolarized result $\langle S_{ba}^{(fi)}(\vec{q}) \rangle$, which is obtained by averaging over all polarizations. Since the unpolarized $\langle S_{ba}^{(fi)}(\vec{q}) \rangle$ is proportional to δ_{ab}, the structure factor can be given by the function $S(\vec{q})$;

$$\langle S_{ba}^{(fi)}(\vec{q}) \rangle = \delta_{ab} S(\vec{q}) \propto \sum_{\lambda_i, \lambda_f} \langle \Psi_i(\lambda_i) | V_b^\dagger | \Psi_f(\lambda_f) \rangle \langle \Psi_f(\lambda_f) | V_a | \Psi_i(\lambda_i) \rangle. \tag{13}$$

This function provides us with the overall functional form of the INS structure factors.

Since many molecular magnetic systems are not a single crystal sample, but typically powder, the result given above can be reduced to the powder average by integrating the structure factor over all angles [21,31], which is given by

$$\bar{S}(q) = \int \frac{d\Omega_{\hat{q}}}{4\pi} S(\vec{q}) . \tag{14}$$

While we use this methodology to determine the complete INS structure factor, we are really more interested in the functional form of the structure factor as it provides information on excitations of the system according to spatial parameters and the excited subgeometries for the trimer system.

In Table 2, we show the calculated energy excitations and single-crystal structures factors for various transitions in $S = 1/2, 1$, and $3/2$ isosceles trimers. In bold, we have inferred structures based on the individual subgeometry excitations. From this table, it is clear that when the subgeometry is excited, the structure factor takes on the functional form for that subgeometry. For example, the spin-1/2 trimer has an excitation from $|\Psi_{1/2,0}\rangle \rightarrow |\Psi_{1/2,1}\rangle$. Since the excitation is a transition from $S_d = 0$ to $S_d = 1$, it is a transition of the spin dimer and will have a structure factor that is characteristic of the spin dimer ($\frac{1}{6}(1 - \cos(\vec{q} \cdot \vec{r}_{12}))$). Even when the total spin is changed, if the excitation is due to the dimer transition, then the structure factor will take on that functional form (as shown in the $|\Psi_{0,1}\rangle \rightarrow |\Psi_{1,1}\rangle$ transition.) The third transition of the spin-1/2 trimer, however, does not change the dimer subgeometry, but only the trimer or total spin. This therefore takes on the trimer functional form of $\frac{1}{3}(3 + \cos(\vec{q} \cdot \vec{r}_{12}) - 2\cos(\vec{q} \cdot \vec{r}_{23}) - 2\cos(\vec{q} \cdot \vec{r}_{31}))$.

Since Table 2 shows the single crystal functional forms of the INS structure factors, it should be mentioned that the powder-average intensity is determined through an integration of the single crystal structure factor over the solid angle. This integration simply converts the $\cos(\vec{q} \cdot \vec{r}_{12})$ into a zeroth-order Bessel function $j_0(\vec{q} \cdot \vec{r}_{12})$, where $j_0(x) = sin(x)/x$. The functions in the table can therefore be easily adapted for powder systems.

This analysis reveals a simple and straightforward manner to determine or approximate inelastic neutron scattering structure factors for molecular magnets using an examination of the individual subgeometries. This analysis has been shown to be useful especially for larger molecular magnets. The larger the system, however, the more hidden geometries can play a role, which was shown in the case of $MgCr_2O_4$ [36,37], where a spin heptamer can be exactly determined using a trimer and hexamer basis while some excitations were pentamer excitations.

Table 2. Inelastic neutron scattering (INS) energy excitations and structure factors.

System	Transition $\lvert\Psi_{S_t,S_d}\rangle \to \lvert\Psi_{S_t,S_d}\rangle$	ΔE [1]	Structure Factors [2,3]
Spin 1/2 Trimer	$\lvert\Psi_{1/2,0}\rangle \to \lvert\Psi_{1/2,1}\rangle$	$(\alpha-1)J$	$\frac{1}{6}(1 - \cos(\vec{q}\cdot\vec{r}_{12}))$
	$\lvert\Psi_{1/2,0}\rangle \to \lvert\Psi_{3/2,1}\rangle$	$(\frac{1}{2}+\alpha)J$	$\frac{1}{3}(1 - \cos(\vec{q}\cdot\vec{r}_{12}))$
	$\lvert\Psi_{1/2,1}\rangle \to \lvert\Psi_{3/2,1}\rangle$	$\frac{3}{2}J$	$\frac{1}{3}(3 + \cos(\vec{q}\cdot\vec{r}_{12}) - 2\cos(\vec{q}\cdot\vec{r}_{23}) - 2\cos(\vec{q}\cdot\vec{r}_{31}))$
Spin 1 Isosceles Trimer	$\lvert\Psi_{0,1}\rangle \to \lvert\Psi_{1,2}\rangle$	$(2\alpha-1)J$	$\frac{5}{9}(1 - \cos(\vec{q}\cdot\vec{r}_{12}))$
	$\lvert\Psi_{0,1}\rangle \to \lvert\Psi_{1,1}\rangle$	J	$\frac{1}{3}(3 + \cos(\vec{q}\cdot\vec{r}_{12}) - 2\cos(\vec{q}\cdot\vec{r}_{23}) - 2\cos(\vec{q}\cdot\vec{r}_{31}))$
	$\lvert\Psi_{0,1}\rangle \to \lvert\Psi_{1,0}\rangle$	$(2-\alpha)J$	$\frac{4}{9}(1 - \cos(\vec{q}\cdot\vec{r}_{12}))$
	$\lvert\Psi_{1,2}\rangle \to \lvert\Psi_{1,1}\rangle$	$(2-2\alpha)J$	$\frac{5}{36}(1 - \cos(\vec{q}\cdot\vec{r}_{12}))$
	$\lvert\Psi_{1,2}\rangle \to \lvert\Psi_{1,0}\rangle$	$(3-3\alpha)J$	0
	$\lvert\Psi_{1,2}\rangle \to \lvert\Psi_{2,2}\rangle$	$2J$	$\frac{3}{40}(3 + \cos(\vec{q}\cdot\vec{r}_{12}) - 2\cos(\vec{q}\cdot\vec{r}_{23}) - 2\cos(\vec{q}\cdot\vec{r}_{31}))$
	$\lvert\Psi_{1,2}\rangle \to \lvert\Psi_{2,1}\rangle$	$(4-2\alpha)J$	$\frac{1}{180}(1 - \cos(\vec{q}\cdot\vec{r}_{12}))$
	$\lvert\Psi_{1,1}\rangle \to \lvert\Psi_{1,0}\rangle$	$(1-\alpha)J$	$1 - \cos(\vec{q}\cdot\vec{r}_{12})$
	$\lvert\Psi_{1,1}\rangle \to \lvert\Psi_{2,2}\rangle$	$2\alpha J$	$1 - \cos(\vec{q}\cdot\vec{r}_{12})$
	$\lvert\Psi_{1,1}\rangle \to \lvert\Psi_{2,1}\rangle$	$2J$	$3 + \cos(\vec{q}\cdot\vec{r}_{12}) - 2\cos(\vec{q}\cdot\vec{r}_{23}) - 2\cos(\vec{q}\cdot\vec{r}_{31})$
	$\lvert\Psi_{1,0}\rangle \to \lvert\Psi_{2,2}\rangle$	$(3\alpha-1)J$	0
	$\lvert\Psi_{1,0}\rangle \to \lvert\Psi_{2,1}\rangle$	$(1-3\alpha)J$	$1 - \cos(\vec{q}\cdot\vec{r}_{12})$
Spin 3/2 Isosceles Trimer	$\lvert\Psi_{1/2,2}\rangle \to \lvert\Psi_{1/2,1}\rangle$	$(2-2\alpha)J$	$(1 - \cos(\vec{q}\cdot\vec{r}_{12}))$
	$\lvert\Psi_{1/2,2}\rangle \to \lvert\Psi_{3/2,3}\rangle$	$(3\alpha-\frac{3}{2})J$	$\frac{21}{40}(1 - \cos(\vec{q}\cdot\vec{r}_{12}))$
	$\lvert\Psi_{1/2,2}\rangle \to \lvert\Psi_{3/2,2}\rangle$	$\frac{3}{2}J$	$\frac{3}{8}(3 + \cos(\vec{q}\cdot\vec{r}_{12}) - 2\cos(\vec{q}\cdot\vec{r}_{23}) - 2\cos(\vec{q}\cdot\vec{r}_{31}))$
	$\lvert\Psi_{1/2,2}\rangle \to \lvert\Psi_{3/2,1}\rangle$	$(\frac{7}{2}-2\alpha)J$	$\frac{3}{40}(1 - \cos(\vec{q}\cdot\vec{r}_{12}))$
	$\lvert\Psi_{1/2,2}\rangle \to \lvert\Psi_{3/2,0}\rangle$	$(\frac{9}{2}-3\alpha)J$	0
	$\lvert\Psi_{1/2,1}\rangle \to \lvert\Psi_{3/2,3}\rangle$	$(5\alpha-\frac{7}{2})J$	0
	$\lvert\Psi_{1/2,1}\rangle \to \lvert\Psi_{3/2,2}\rangle$	$(2\alpha-\frac{1}{2})J$	$1 - \cos(\vec{q}\cdot\vec{r}_{12})$
	$\lvert\Psi_{1/2,1}\rangle \to \lvert\Psi_{3/2,1}\rangle$	$\frac{3}{2}J$	$3 + \cos(\vec{q}\cdot\vec{r}_{12}) - 2\cos(\vec{q}\cdot\vec{r}_{23}) - 2\cos(\vec{q}\cdot\vec{r}_{31})$
	$\lvert\Psi_{1/2,1}\rangle \to \lvert\Psi_{3/2,0}\rangle$	$(\frac{5}{2}-\alpha)J$	$1 - \cos(\vec{q}\cdot\vec{r}_{12})$
	$\lvert\Psi_{3/2,3}\rangle \to \lvert\Psi_{3/2,2}\rangle$	$(3-3\alpha)J$	$\frac{21}{100}(1 - \cos(\vec{q}\cdot\vec{r}_{12}))$
	$\lvert\Psi_{3/2,3}\rangle \to \lvert\Psi_{3/2,1}\rangle$	$5-5\alpha)J$	0
	$\lvert\Psi_{3/2,3}\rangle \to \lvert\Psi_{3/2,0}\rangle$	$(6-6\alpha)J$	0
	$\lvert\Psi_{3/2,3}\rangle \to \lvert\Psi_{5/2,3}\rangle$	$\frac{5}{2}J$	$\frac{1}{5}(3 + \cos(\vec{q}\cdot\vec{r}_{12}) - 2\cos(\vec{q}\cdot\vec{r}_{23}) - 2\cos(\vec{q}\cdot\vec{r}_{31}))$
	$\lvert\Psi_{3/2,3}\rangle \to \lvert\Psi_{5/2,2}\rangle$	$(\frac{11}{4}-3\alpha)\alpha J$	$\frac{1}{200}(1 - \cos(\vec{q}\cdot\vec{r}_{12}))$
	$\lvert\Psi_{3/2,3}\rangle \to \lvert\Psi_{5/2,1}\rangle$	$(\frac{15}{2}-5\alpha)J$	0
	$\lvert\Psi_{3/2,2}\rangle \to \lvert\Psi_{3/2,1}\rangle$	$(2-2\alpha)J$	$1 - \cos(\vec{q}\cdot\vec{r}_{12})$
	$\lvert\Psi_{3/2,2}\rangle \to \lvert\Psi_{3/2,0}\rangle$	$(3-3\alpha)J$	0
	$\lvert\Psi_{3/2,2}\rangle \to \lvert\Psi_{5/2,3}\rangle$	$(3\alpha-\frac{1}{2})J$	$1 - \cos(\vec{q}\cdot\vec{r}_{12})$
	$\lvert\Psi_{3/2,2}\rangle \to \lvert\Psi_{5/2,2}\rangle$	$\frac{5}{2}J$	$3 + \cos(\vec{q}\cdot\vec{r}_{12}) - 2\cos(\vec{q}\cdot\vec{r}_{23}) - 2\cos(\vec{q}\cdot\vec{r}_{31})$
	$\lvert\Psi_{3/2,2}\rangle \to \lvert\Psi_{5/2,1}\rangle$	$(\frac{9}{2}-2\alpha)J$	$1 - \cos(\vec{q}\cdot\vec{r}_{12})$

[1] Due to the INS structure factor being zero, some transitions cannot be seen in INS; [2] Structure factors in bold have not been calculated and are inferred from the excitations of the geometries; [3] The calculation of the powder intensity replaces cosine function with a zeroth-order Bessel function.

Figure 4 $S_t = 1$ isosceles trimer single-crystal structure factors as a function of E/J vs. qa/π with $S_t = 1$ ground state at $\alpha = 0.2$ (a) and with $S_t = 0$ ground state at $\alpha = 0.8$(b). Here, $r_{12} = r_{13}/2 = r_{23}/2 = a$, where a is the spatial distance of the dimer. We also include a generic magnetic form factor to illustrate the damping effect that will be expressed at larger q [40]. Overall, the different structure factor functions allow for experimentalists to easily distinguish between dimer and trimer excitations, as well as extract geometric parameters through a fitting of the intensity profile.

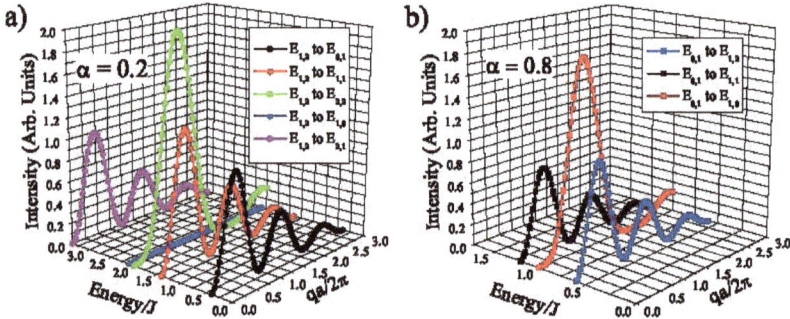

Figure 4. (a) S = 1 isosceles trimer single-crystal structure factors as a function of E/J vs. q with $S_t = 1$ ground state at $\alpha = 0.2$ (**a**) and with $S_t = 0$ ground state at $\alpha = 0.8$ (**b**). Here, $r_{12} = r_{13}/2 = r_{23}/2 = a$, where a is the spatial distance of the dimer. We also include a generic magnetic form factor to illustrate the damping effect that will be expressed at larger q [40]. Note: Different colors were used to differentiate the different transitions for that particular α.

Another point of interest is the change in ground state at $\alpha = 0.5$, which brings a discontinuity to the neutron scattering intensities and helps to clarify the spin states of the ions. The sudden change from a $S_t = 0$ to a $S_t = 1$ ground state in the S = 1 trimer immediately allows two more transitions that are described in Figure 4, where at $\alpha = 0.8$ there are only three possible transitions and at $\alpha = 0.2$ there are five possible transitions. As α decreases below 0.5 for the S = 1 case, the trimer is allowed to order itself into a less frustrated state with an overall spin of 1, but a dimer spin state of 2 (both ions are in spin up states. Whereas with $\alpha > 0.5$, the trimer experiences a greater frustration that pushes the spins into a $S_d = 1$ dimer state (one spin and one neutral), which means the overall spin of the trimer is zero. This transition is illustrated in Figure 5. The S = 3/2 trimer has an analogous state flip at $\alpha = 0.5$. The use of the spin dimer basis states therefore allows for the spin configurations to be known and a better understanding of the frustration of the trimer becomes clear. The nature of the change in ground state can l

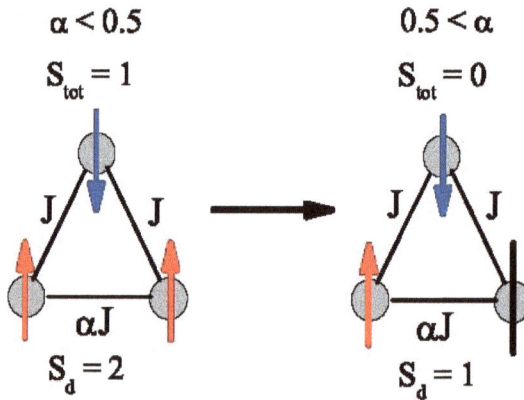

Figure 5. The transition as α is increased through the spin ground state transition. Red arrows indicate spin up states, blue arrows are spin down states, black double headed arrow is the spin neutral state.

Furthermore, an analysis of the inelastic neutron scattering structure factors reveals a hidden selection rule for magnetic clusters and molecular magnets. From Table 2, some of the transitions have no structure factor or inelastic scattering intensity, even though they have the standard neutron scattering transition of $\Delta S_t = \pm 1$ or 0. This zero intensity is due to the transition of the dimer being greater than ± 1. This therefore clarifies the selection rule for magnetic clusters by extending it to the individual subgeometries.

5. Conclusions

We have presented calculations of the thermodynamics, inelastic neutron scattering excitations, and structure factors for general isosceles trimers. Using a Heisenberg spin-spin exchange model, we determined general relationships for the quantum energy levels, spin states, and spin decomposition of isosceles trimers for any spin. Through a calculation of the partition function, we determined the heat capacity and magnetic susceptibility as functions of temperature and dimer exchange coupling (α). An analysis of the thermodynamics revealed multiple quantum phase transitions, where the temperature dependence of the Schottky anomaly indicated the proximity to this transition in α. These transitions provided the potential for active spin switching in molecular clusters through a dynamic transition from external fields.

Furthermore, we examined the calculated neutron scattering structure factors (scattering intensities) for various excitations and showed that the spin total transitions did not just govern molecular magnet excitations, but the spin transition of the individual subgeometries. INS has a typical selection rule for magnetic states of $\Delta S = \pm 1$ or 0, which means that neutron scattering will only examine transitions that are of a small spin variation. We presented here that the selection rules of neutron scattering of finite clusters should include another selection rule that is $\Delta S_d = \pm 1$ or 0, where ΔS_d is that of that spin dimer basis. Because each spin state has a specific dimer basis associated with it, the different states are also bound by that basis. For example, the spin-1 equilateral trimer has three $S = 1$ excited states and two $S = 2$ excited states. Under standard INS selection rules, the transition from any of the $S = 1$ to the two $S = 2$ excited states is allowed. When the structure factors for these transitions are calculated, however, it becomes evident that not all the transitions are allowed. When examined closer, the $S = 1$ excited states have dimer bases of $S_d = 0, 1, 2$, and the $S = 2$ excited states have dimer bases of $S_d = 1, 2$.

Overall, the examination of spin trimers helps to provide insights into the foundational characteristics of molecular magnets. Through an understanding of the individual excitations of smaller clusters, one can determine the properties of larger, more complex clusters, which can help in the determination and realization of new and exciting spintronic materials and device applications.

In this model, we examined the isotropic Heisenberg Hamiltonian and the standard interactions to show the variation as spin evolves. Further studies could include other interactions and parameters (e.g., anisotropy, external magnetic field, biquadratic term, etc.) to examine the effect on the quantum phase transitions and the excitations that occur. Overall, these other terms should not have any impact on the functional form of the INS structure factors. Although, parameters like the external magnetic field will produce a Zeeman splitting of the excitations and anisotropy may shift the overall ground states. Further comprehensive studies can examine these effects.

Author Contributions: Conceptualization, J.T.H.; methodology, J.T.H. and A.B.; validation, J.T.H. and A.B.; formal analysis, J.T.H. and A.B.; investigation, A.B. and J.T.H.; resources, J.T.H.; data curation, J.T.H.; writing—original draft preparation, A.B. and J.T.H.; writing—review and editing, J.T.H. and A.B.; visualization, J.T.H.; supervision, J.T.H.; project administration, J.T.H.; funding acquisition, J.T.H.

Funding: This research was partially funded by the Institute for Materials Science at Los Alamos National Laboratory (J.T.H.).

Conflicts of Interest: The authors declare no conflicts of interest.

References

1. Kahn, O. *Molecular Magnetism*; VCH Publishers: New York, NY, USA, 1993.
2. Bogani, L.; Wernsdorfer, W. Molecular spintronics using single-molecule magnets. *Nat. Mater.* **2008**, *7*, 179–186. [CrossRef] [PubMed]
3. Bartolomé, J.; Luis, F.; Fernández, J. (Eds.) *Molecular Magnets. NanoScience and Technology*; Springer: Berlin/Heidelberg, Germany, 2014.
4. Holynska, M. (Ed.) *Single-Molecule Magnets: Molecular Architectures and Building Blocks for Spintronics*; Wiley: Hoboken, NJ, USA, 2018.
5. DiVincenzo, D.P.; Loss, D. Quantum computers and quantum coherence. *J. Magn. Magn. Mater.* **1999**, *200*, 202. [CrossRef]
6. Nielsen, M.A.; Chuang, I.L. *Quantum Computation and Quantum Information*; Cambridge University Press: Cambridge, UK, 2000.
7. Leuenberger, M.N.; Loss, D. Quantum computing in molecular magnets. *Nature* **2001**, *410*, 789–793. [CrossRef] [PubMed]
8. Dagotto, E.; Rice, T.M. Surprises on the Way from One- to Two-Dimensional Quantum Magnets: The Ladder Materials. *Science* **1996**, *271*, 618–623. [CrossRef]
9. Rojas-Dotti, C.; Moliner, N.; Lloret, F.; Martínez-Lillo, G. Ferromagnetic Oxime-Based Manganese(III) Single-Molecule Magnets with Dimethylformamide and Pyridine as Terminal Ligands. *Crystals* **2019**, *9*, 23. [CrossRef]
10. Whangbo, M.H.; Koo, H.J.; Dai, D. Spin exchange interactions and magnetic structures of extended magnetic solids with localized spins: Theoretical descriptions on formal, quantitative and qualitative levels. *J. Solid State Chem.* **2003**, *176*, 417–481. [CrossRef]
11. Waldmann, O. Q dependence of the inelastic neutron scattering cross section for molecular spin clusters with high molecular symmetry. *Phys. Rev. B* **2003**, *68*, 174406. [CrossRef]
12. Goswami, S.; Mondala, A.K.; Konar, S. Nanoscopic molecular magnets. *Inorg. Chem. Front.* **2015**, *2*, 687–712. [CrossRef]
13. Gatteschi, D.; Sessoli, R. Molecular nanomagnets: The first 10 years. *J. Magn. Magn. Mater.* **2004**, *272–276*, 1030–1036. [CrossRef]
14. Barra, A.L.; Caneschi, A.; Cornia, A.; Frabrizi de Biani, F.; Gatteschi, D.; Sangregorio, C.; Sessoli, R.; Sorace, L. Single-Molecule Magnet Behavior of a Tetranuclear Iron(III) Complex. The Origin of Slow Magnetic Relaxation in Iron(III) Clusters. *J. Am. Chem. Soc.* **1999**, *121*, 5302. [CrossRef]
15. Furukawa, Y.; Luban, M.; Borsa, F.; Johnston, D.C.; Mahajan, A.V.; Miller, L.L.; Mentrup, D.; Schnack, J.; Bino, A. Magnetism and spin dynamics in the cluster compound $Cr_4S(O_2CCH_3)_8(H_2O)_4](NO_3)_2H_2O$. *Phys. Rev. B* **2000**, *61*, 8635. [CrossRef]
16. Luban, M.; Borsa, F.; Bud'ko, S.; Canfield, P.; Jun, S.; Kap Jung, J.; Kögerler, P.; Menrup, D.; Müller, A.; Modler, R.; et al. Heisenberg spin triangles in V6-type magnetic molecules: Experiment and theory. *Phys. Rev. B* **2002**, *66*, 054407. [CrossRef]
17. Bouwen, A.; Caneschi, A.; Gatteschi, D.; Goovaerts, E.; Schoemaker, D.; Sorace, L.; Stefan, M. Single-Crystal High-Frequency Electron Paramagnetic Resonance Investigation of a Tetranuclear Iron(III) Single-Molecule Magnet. *J. Phys. Chem.* **2001**, B105, 2658–2663. [CrossRef]
18. Cornia, A.; Sessoli, R.; Sorace, L.; Gatteschi, D.; Barra, A.L.; Daiguebonne, C. Origin of Second-Order Transverse Magnetic Anisotropy in Mn12-Acetate. *Phys. Rev. Lett.* **2002**, *89*, 257201. [CrossRef] [PubMed]
19. Fitta, M.; Pelka, R.; Konieczny, P.; Balanda, M. Multifunctional Molecular Magnets: Magnetocaloric Effect in Octacyanometallates. *Crystals* **2019**, *9*, 9. [CrossRef]
20. Hennion, M.; Pardi, L.; Mirebeau, I.; Suard, E.; Sessoli, R.; Caneschi, A. Neutron study of mesoscopic magnetic clusters: $Mn_{12}O_{12}$. *Phys. Rev. B* **1997**, *56*, 8819. [CrossRef]
21. Squires, G.L. *Introduction to the Theory of Thermal Neutron Scattering*; Dover: New York, NY, USA, 1996.
22. Mentrup, D.; Schnack, J.; Luban, M. Spin dynamics of quantum and classical Heisenberg dimers. *Phys. A* **1999**, *272*, 153–161. [CrossRef]
23. Mentrup, D.; Schmidt, H.-J.; Schnack, J.; Luban, M. Transition from quantum to classical Heisenberg trimers: Thermodynamics and time correlation functions. *Phys. A* **2000**, *278*, 214–221. [CrossRef]

24. Cifta, O. The irregular tetrahedron of classical and quantum spins subjected to a magnetic field. *J. Phys. A* **2001**, *34*, 1611. [CrossRef]

25. Ameduri, M.; Klemm, R.A. Time correlation functions of three classical Heisenberg spins on an isosceles triangle and on a chain. *Phys. Rev. B* **2002**, *66*, 224404. [CrossRef]

26. Orendáčová, A.; Tarasenko, R.; Tkáč, V.; Cizmár, E.; Orendáč, M.; Feher, A. Interplay of Spin and Spatial Anisotropy in Low-Dimensional Quantum Magnets with Spin 1/2. *Crystals* **2019**, *9*, 6.

27. Kirchner, N.; van Slageren, J.; Tsukerblat, B.; Waldmann, O.; Dressel, M. Antisymmetric exchange interactions in Ni4 clusters. *Phys. Rev. B* **2008**, *78*, 094426. [CrossRef]

28. Waldmann, O.; Ako, A.M.; Gudel, H.U.; Powell, A.K. Assessment of the anisotropy in the molecule Mn19 with a high-spin ground state S = 83/2 by 35 GHz electron paramagnetic resonance. *Inorg. Chem.* **2008**, *47*, 3486–3488. [CrossRef] [PubMed]

29. Waldmann, O. A criterion for the anisotropy barrier in single molecule magnets. *Inorg. Chem.* **2007**, *46*, 10035. [CrossRef] [PubMed]

30. Haraldsen, J.T.; Barnes, T.; Sinclair, J.W.; Thompson, J.R.; Sacci, R.L.; Turner, J.F.C. Magnetic properties of a Heisenberg coupled-trimer molecular magnet: General results and application to spin-1/2 vanadium clusters. *Phys. Rev. B* **2009**, *80*, 064406. [CrossRef]

31. Haraldsen, J.T.; Barnes, T.; Musfeldt, J.L. Neutron scattering and magnetic observables for S=1/2 spin clusters and molecular magnets. *Phys. Rev. B* **2005**, *71*, 064403. [CrossRef]

32. Haraldsen, J.T. Evolution of thermodynamic properties and inelastic neutron scattering intensities for spin-1/2 antiferromagnetic quantum ring. *Phys. Rev. B* **2016**, *94*, 054436. [CrossRef]

33. Postnikov, A.V.; Kortus, J.; Pederson, M.R. Density functional studies of molecular magnets. *Phys. Status Solidi B* **2006**, *243*, 2533–2572. [CrossRef]

34. Borysov, S.S.; Geilhufe, R.M.; Balatsky, A.V. Organic materials database: An open-access online database for data mining. *PLoS ONE* **2017**, *12*, e0171501. [CrossRef]

35. Haraldsen, J.T. Heisenberg Pentamer: Insights into Inelastic Neutron Scattering on Magnetic Clusters. *Phys. Rev. Lett.* **2011**, *107*, 037205. [CrossRef]

36. Gao, S.; Guratinder, K.; Stuhr, U.; White, J.S.; Mansson, M.; Roessli, B.; Fennell, T.; Tsurkan, V.; Loidl, A.; Ciomaga Hatnean, M.; et al. Manifolds of magnetic ordered states and excitations in the almost Heisenberg pyrochlore antiferromagnet MgCr$_2$O$_4$. *Phys. Rev. B* **2018**, *97*, 134430. [CrossRef]

37. Roxburgh, A.; Haraldsen, J.T. Thermodynamics and spin mapping of quantum excitations in a Heisenberg spin heptamer. *Phys. Rev. B* **2018**, *98*, 214434. [CrossRef]

38. Houchins, G.; Haraldsen, J.T. Generalization of polarized spin excitations for asymmetric dimeric systems. *Phys. Rev. B* **2015**, *91*, 014422. [CrossRef]

39. Blundell, S.J.; Blundell, K.M. *Concepts in Thermal Physics*; Oxford University Press Inc.: New York, NY, USA, 2010.

40. Dianoux, A.J.; Lander, G. *Neutron Data Booklet*; OCP Science: Philadelphia, PA, USA, 2003.

crystals

MDPI

Review

The Effect of Pressure on Magnetic Properties of Prussian Blue Analogues

Maria Zentkova * and Marian Mihalik

Technological Laboratory, Department of Magnetism, Institute of Experimental Physics SAS, Watsonova 47, 04001 Kosice, Slovakia; mihalik@saske.sk
* Correspondence: zentkova@saske.sk

Received: 8 January 2019; Accepted: 16 February 2019; Published: 20 February 2019

Abstract: We present the review of pressure effect on the crystal structure and magnetic properties of $Cr(CN)_6$-based Prussian blue analogues (PBs). The lattice volume of the *fcc* crystal structure space group $Fm\bar{3}m$ in the Mn-Cr-CN-PBs linearly decreases for $p \leq 1.7$ GPa, the change of lattice size levels off at 3.2 GPa, and above 4.2 GPa an amorphous-like structure appears. The crystal structure recovers after removal of pressure as high as 4.5 GPa. The effect of pressure on magnetic properties follows the non-monotonous pressure dependence of the crystal lattice. The amorphous like structure is accompanied with reduction of the Curie temperature (T_C) to zero and a corresponding collapse of the ferrimagnetic moment at 10 GPa. The cell volume of Ni-Cr-CN-PBs decreases linearly and is isotropic in the range of 0–3.1 GPa. The Raman spectra can indicate a weak linkage isomerisation induced by pressure. The Curie temperature in Mn^{2+}-Cr^{III}-PBs and Cr^{2+}-Cr^{III}-PBs with dominant antiferromagnetic super-exchange interaction increases with pressure in comparison with decrease of T_C in Ni^{2+}-Cr^{III}-PBs and Co^{2+}-Cr^{III}-PBs ferromagnets. T_C increases with increasing pressure for ferrimagnetic systems due to the strengthening of magnetic interaction because pressure, which enlarges the monoelectronic overlap integral S and energy gap Δ between the mixed molecular orbitals. The reduction of bonding angles between magnetic ions connected by the CN group leads to a small decrease of magnetic coupling. Such a reduction can be expected on both compounds with ferromagnetic and ferrimagnetic ordering. In the second case this effect is masked by the increase of coupling caused by the enlarged overlap between magnetic orbitals. In the case of mixed ferro–ferromagnetic systems, pressure affects $\mu(T)$ by a different method in Mn^{2+}–N≡C–Cr^{III} subsystem and Cr^{III}–C≡N–Ni^{2+} subsystem, and as a consequence T_{comp} decreases when the pressure is applied. The pressure changes magnetization processes in both systems, but we expect that spontaneous magnetization is not affected in Mn^{2+}-Cr^{III}-PBs, Ni^{2+}-Cr^{III}-PBs, and Co^{2+}-Cr^{III}-PBs. Pressure-induced magnetic hardening is attributed to a change in magneto-crystalline anisotropy induced by pressure. The applied pressure reduces saturated magnetization of Cr^{2+}-Cr^{III}-PBs. The applied pressure $p = 0.84$ GPa induces high spin–low spin transition of cca 4.5% of high spin Cr^{2+}. The pressure effect on magnetic properties of PBs nano powders and core–shell heterostructures follows tendencies known from bulk parent PBs.

Keywords: Prussian blue analogues; effect of high pressure; crystal structure; magnetic properties; superexchange interaction

---

1. Introduction

The main interest for reinvestigation of PBs was dominantly driven by the vision of new molecule-based magnetic materials with a temperature of magnetic ordering higher than 300 K and simultaneously very sensitive to external parameters like high pressure or light. Huge versatility in tuning its physical properties using external perturbations-stimuli, confirmed in many papers [1], opened the possibility of applications in various research branches. We will mention several of the most popular topics active at the present time.

A very sound application possibility is connected with the use of Prussian blue, its analogues, and derivatives as electrodes for alkali ion batteries. This application possibility is based on the fact that Prussian blue analogues with large three dimensional networks have large spaces to host large alkali ions, such as Li^+, Na+, and K+. When applied to rechargeable batteries, its large channels and interstices in the open framework make PB a prominent candidate for cathode materials with long cycle life and fast charge transfer kinetics. Additionally, the easy preparation process and composition variety of PB serves as a good basis for the complex metal oxides employed as catalysts for metal-air batteries [2,3].

As it is generally known that oxygen oxidation eases 142 kJ/g, that means much less energy than the amount obtained from gasoline (47.5 kJ/g). It is thus not surprising that the hydrogen is considered as an alternative to fossil fuel derivatives. The advantage of hydrogen is that it can be produced from water cleavage and its oxidation by-product is water. On the other hand, the classical fossil fuel derivatives release carbon dioxide, which is undoubtedly responsible for global warming and corresponding climate changes. One of the hydrogen storage means is H_2 adsorption in nanoporous solids. The transition metal centres, located at the surface of cavities, are supposed to act as prototype materials for H_2 storage, and PBs belong to this group of materials as well [4–6].

A very successful story is related to application of Prussian blue and its derivatives as biosensors, defined as "self-contained integrated devices, which are capable of providing specific quantitative or semi-quantitative analytical information using a biological recognition element (biochemical receptor), which is retained in direct spatial contact with an electrochemical transduction element" [7]. During the last decade, PBs-modified biosensors have been successfully applied to blood, serum, and urine samples [8].

Prussian blue analogues are sensitive to different external physical impulses, such as light and pressure. There are several mechanisms to explain the effect of the pressure on magnetic properties of selected PBs. The first mechanism is connected with the intermetallic charge transfer reported on few PBs transition metal systems, e.g., $Co^{II}_x[Fe^{III}(CN)_6]_y \cdot zH_2O$ and its alkali-doped analogues were reported to undergo photo-caused magnetization and thermally-caused demagnetization [9] and pressure-caused structural phase transitions [10]—all of which are the result of an externally driven electron transfer between the Co and Fe sites. PBs are materials, in which one can observe the switching between electronic states within the use of an external stimulus, may be referred to as switchable or tuneable. The switching phenomena reported on selected PBs develop as a result of molecular instability [11]. Magnetic measurements obtained by Mösbauer spectroscopy on the series of Co-Fe PBs have revealed how the presence of an alkali ion in the lattice affects pressure driven electron transfer between Co and Fe [12]. The study of the charge transfer effects, as well as spin and magnetic order induced by high hydrostatic pressure, which was applied to the two bimetallic multifunctional PBs, $K_{0.1}Co_4[Fe(CN)_6]_{2.7} \cdot 18H_2O$ and $K_{0.5}Mn_3[Fe(CN)_6]_{2.14} \cdot 6H_2O$, was reported in [13]. The goal of the investigation was comparison of two opposite tendencies of variation in magnetic properties under pressure: (i) the magnetization reduction and magnetic order disappearing for the $K_{0.1}Co_4[Fe(CN)_6]_{2.7} \cdot 18H_2O$ magnets and (ii) an enhancement of the magnetization and variation of the sign of exchange coupling for the $K_{0.5}Mn_3[Fe(CN)_6]_{2.14} \cdot 6H_2O$ material. Magnetization measurements under pressure pointed out that the external hydrostatic pressure considerably enlarges the ferrimagnetic Curie temperature, T_C, for $A_2Mn[Mn(CN)_6]$ (A = K, Rb, Cs). For this monoclinic system with A = K and Rb, dT_C/dp amounts are 21.2 and 14.6 K GPa^{-1}, respectively. The cubic A = Cs compound also exhibits a monotonous enlargement with an initial rate of 4.22 K GPa^{-1} and about 11.4 K GPa^{-1} above 0.6 GPa. The enlargement in T_C we associate with deformation of the structure, such that the $Mn^{II}-N≡C$ angle reduces with increasing pressure. The smaller the alkali cation, the greater the reduction in the $Mn^{II}-N≡C$ angle caused by pressure and the larger the enlargement of T_C/dp. The large rise in T_C for the A = K compound is the highest class among several cyano-bridged metal complexes. The tuning of the transition temperature by pressure may result in additional applications, such as switching devices [14].

The pressure as a very efficient tool for tuning of magnetic properties in Prussian blue analogues was confirmed to be responsible for linkage isomerization [15], for magnetic pole inversion [16], and influenced spin crossover in [17,18]. In our paper, we will concentrate on the general mechanism present for the pressure effect on superexchange interaction in TM-Cr(CN)$_6$−PBs and K-M-Cr(CN)$_6$−PBs, where TM = Mn, Ni, Co, Cr, and M = Mn, Ni.

2. Magnetic Prussian Blue Analogues

The history of Prussian blue itself goes back to 18th century, when draper Diesbach synthesized Prussian blue (ferric hexacyanoferrate) as the first known synthetic blue pigment, which was more easily available and cheaper than other blue pigments available in that time. The first record in literature about that fact was published in [19]. Prussian blue was not only used for paintings but also for coloring of fabrics, and it is still used for this purpose when sold under the commercial name iron blue [20]. Since the discovery, its chemical stoichiometric composition became an intricate puzzle for chemists. The first ideas concerning its crystal structure were published by Keggin and Miles in [21]. Buser et al. in 1970 published a detailed crystal structure and confirmed the composition as Fe$_4$[Fe(CN)$_6$]$_3$.zH$_2$O (z = 14–16) in [22]. They concluded that PB formed a perovskite-like disordered structure with cubic unit cell dimensions of 10.2 Å. The cubic symmetry and the existence of a disordered structure lead to the fact that the lattice can accommodate wide variety of ions. When both positions of iron in the lattice are replaced by different ions, the resultant compound belongs to the family of Prussian blue analogues (PBs). Prussian blue and its analogues form a large family of compounds due to the fact that it is easy to modify the synthetic conditions and consequently to obtain compound of desired stoichiometric composition, shape, and size. Versatility in tuning its physical properties using external perturbations-stimuli opened the possibility of applications in various research branches.

2.1. Crystal Structure of Prussian Blue Analogues

The formulas of Prussian Blue and its analogues are often written as M$_x$A[B(CN)$_6$]$_z$·nH$_2$O, where M is an alkali metal [1]. For z = 1, the face-centred cubic (*fcc*) structure with unit cells comprising eight octants, with the interiors referred to as interstitial, tetrahedral, or cuboctahedral sites, are adopted by the Prussian Blue analogues (PBs). Two different types of octahedral metal sites are present in the lattice: C6 strong ligand-field sites and N6 weak ligand-field sites. For z < 1, it was determined that the B sites are fractionally occupied, and that the A centres surrounding the vacant sites are randomly filled with one or more water molecules in their coordination spheres. As the water in the lattice is present also at interstitial sites (crystal water), the structure of PBs is supposed to be stabilized by the hydrogen bond network as well [1].

The Prussian Blue Fe^{II}$_4$[Fe^{III}(CN)$_6$]$_3$·14H$_2$O crystallizes in a cubic structure where the space group is *Fm$\bar{3}$m*; when taking into account also the weak reflections, the space group is changed to *Pm$\bar{3}$m*. The crystallization conditions strongly influence distribution of the [Fe(CN)$_6$] vacancies in the lattice. They can be either disordered in the crystal, giving an apparent high-symmetry structure with a fractional occupancy (3/4) of the [Fe(CN)$_6$] sites, or ordered, thereby lowering the symmetry. Due to the stoichiometric reasons, the presence of vacancies in the lattice is intrinsic to PBs whenever z < 1. The structure of PBs resembles a typical perovskite structure ABO$_3$ that means the structure, as such, is CaTiO$_3$. There are, however, at least two important differences between PBs and classical perovskite materials. The first evident difference is the different chemical composition of links between metal ions; instead of oxygen atoms, the octahedral metal centres are connected by cyanide bridges to form the cubic framework. The second difference is the fact that the [B(CN)$_6$]p− units in the solid are the same as the hexacyanometalate reactant used in the synthesis [B(CN)$_6$]p− units, which are stable as well in solution. The fact that Prussian blue and its analogues can be synthesized directly from prebuilt molecular precursors results in the fact that they can be considered as molecule-based materials. The ideal structural model of Prussian blue contains a linear A–N≡C-B unit and the geometries of metal atoms in the A and B sites are perfectly octahedral. In reality, it is however a rare situation. The X-ray diffraction Rietveld analyses show that the [B(CN)$_6$] units are often tilted (Figure 1). The tilting results in A–N≡C angles that are less than 180°. There are many reasons for tilting and distortions, starting with conditions of synthesis as well as effect of external perturbations. Observed distortions were, not surprisingly, found to have significant impact on the overlap of local wave functions, and thus on the efficiency of the exchange interaction pathway.

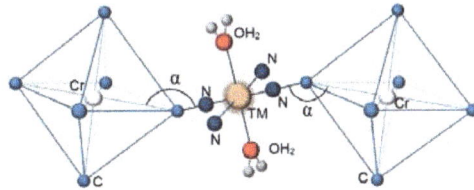

Figure 1. Schematic tilting of the octants in Prussian Blue analogues.

2.2. Magnetic Properties

Despite the fact that the first paper dealing with magnetic properties of Prussian blue showed very low ordering temperature T_C = 5.6 K, nearly 10 years later, T_C reached room temperature for some of its analogues, also with the help of external stimuli [23]. Early work of Bozorth et al. clearly demonstrated the direct relationship between magnetic ordering temperatures and type of paramagnetic TM centers, as demonstrated by the rather low magnetic ordering temperature of Prussian Blue, in which only half the TM ions are paramagnetic [24]. The pronounced increase in the Curie temperature of PBs is referred to in the 1980s, when Babel et al. discovered CsMn[Cr(CN)$_6$], a ferrimagnet with T_N = 90 K, and proposed a superexchange mechanism to account for the magnetic properties of PBs [25]. Superexchange refers to the magnetic coupling of next-to-nearest paramagnetic neighbors through a non-magnetic anion. Superexchange interactions are usually explained in the terms of the Goodenough-Kanamori-Anderson rules and are typically indirect interactions between 3d metal ions [26–28]. On the basis of these rules, a magnetic ion-ligand-magnetic ion, including an angle of 90°, where the singly occupied d-orbitals of the paramagnetic ions are orthogonal to one another and there is zero orbital overlap, is anticipated to give rise to weak ferromagnetic exchange interactions, which are assigned as a potential exchange mechanism. For a magnetic ion-ligand-magnetic ion, including an angle of 180°, the singly occupied d-orbitals of the paramagnetic ions are non-orthogonal to one another and there is direct orbital overlap. Consequently, the unpaired electrons of the magnetic ions align antiparallel to one another, giving rise to strong antiferromagnetic exchange interactions in a superexchange mechanism, known as kinetic exchange. In the case of PBs, the metal d orbitals split into t_{2g} and e_g sets by the CN ligands. The magnetic interactions in these materials are given by the super-exchange interaction between metal ions A^{2+} and B^{III} interposed through a three-dimensional network of C≡N bridges, as resultant 3D magnetic ordering with Curie temperatures T_C up to 376 K depending on the nature of the metal ions is created [27]. Magnetic features of Prussian Blue analogues are usually interpreted taking into account two assumptions: **(1)** the super-exchange coupling only between next A^{2+} metal ion and B^{III} (A^{2+}–N≡C-B^{III}) ion; **(2)** in the case when the magnetic orbital symmetries of the metal ions are identical, the dominant super-exchange coupling is antiferromagnetic (J_{AF}) and on the other hand the super-exchange coupling is ferromagnetic (J_F) for magnetic orbitals with different symmetries. The B^{III} ion, surrounded by the carbon atoms of six cyanide ligands, experiences a large ligand field. Consequently, all familiar [B^{III}(CN)$_6$] units are invariably low spin and have electrons only in the t_{2g} orbitals. A^{2+} ions in Prussian Blue are usually high spin and can have both e_g and t_{2g} magnetic orbitals. When A has only e_g magnetic orbitals, all the exchange coupling with the t_{2g} magnetic orbitals of the [B(CN)$_6$] will be ferromagnetic. When only t_{2g} magnetic orbitals are present on A, all the exchange interactions with the t_{2g} magnetic orbitals present on $[B(CN)_6]$ will be antiferromagnetic. When both t_{2g} and e_g magnetic orbitals are simultaneously present on A, ferromagnetic and antiferromagnetic interactions with the t_{2g} magnetic orbitals on $[B(CN)_6]$ coexist and compete. The conclusions are straightforward: the $t_{2g}(B)$–$e_g(A)$ pathways lead to ferromagnetic (F) interactions; the $t_{2g}(B)$–$t_{2g}(A)$ pathways lead to antiferromagnetic (AF) interactions [1].

It cannot be overlooked that the huge variety of interesting physical phenomena in the family of PBs are closely related to the fact that the cubic symmetry of the lattice can accommodate a lot of transition metal ions as hexacyanometalate anions. Therefore, the Prussian Blue analogues can be prepared with a number of metals in the B sites [1]. As examples of paramagnetic ions, according to [1] we can mention:

- $[B(CN)_6]^{4-}$ for B = V^{II}, Mn^{II}
- $[B(CN)_6]^{3-}$ for B = Ti^{III}, Cr^{III}, Mn^{III}, Fe^{III}
- $[B(CN)_6]^{2-}$ for B = Mn^{IV}

All above-mentioned paramagnetic anions are low spin due to the large ligand field splitting induced by the cyanide ligand, with anions spins varying from $S = 1/2$ for the Ti^{III}, Mn^{II}, and Fe^{III} derivatives, to $S = 1$ for the Mn^{III} derivative, up to $S = 3/2$ for the V^{II}, Cr^{III}, and Mn^{IV} derivatives. The paramagnetic ions placed in the A site are usually high spin due to the fact that N-coordinated cyanide and water molecules are weak field ligands. As examples of the ions placed on the A positions in the lattice, V^{II}, Cr^{II}, Mn^{II}, Fe^{II}, Co^{II}, Ni^{II}, Cu^{II}, and Fe^{III} ions can be mentioned. Corresponding spin values range from $S = 1/2$ for Cu^{II} to $S = 5/2$ for Mn^{II} and Fe^{III} [1].

3. Probing of Magnetocrystalline Correlations Using External Pressure

This simple model mentioned above has already been tested on the $TM^{2+}{}_3[Cr^{III}(CN)_6]_2 \cdot zH_2O$ and $KM^{2+}[Cr(CN)_6]$ systems, where $TM^{2+} = Cr^{2+}$, Mn^{2+}, Ni^{2+}, Co^{2+}, and $M^{2+} = Mn^{2+}$, Ni^{2+} [29–40]. Our paper is focused on the pressure effect on crystal structure and magnetic properties.

3.1. Crystal Structure of TM^{2+}-Cr^{III}-PBs

The crystal structure of selected PBs was investigated by X-ray and neutron powder diffraction techniques. We studied $(Ni_xMn_{1-x})_3[Cr(CN)_6]_2 \cdot zD_2O$ ($x = 0$, 0.38 and 1) materials by powder neutron diffraction techniques in temperature ranges higher or lower than the Curie temperature T_C [30]. Our study enabled us to determine the entire crystal structure, including D-sites. We determined the crystal lattice of $Mn_3[Cr(CN)_6]_2 \cdot zH_2O$ and $Ni_3[Cr(CN)_6]_2 \cdot zH_2O$ molecular magnets with the help of the Rietveld technique by application software program FullProf, and the crystal structure was confirmed as a cubic space group $Fm\bar{3}m$ (No 225) with lattice parameters $a = 10.754\,04(6)$ Å and $a = 10.4341(3)$ Å [30]. However the lattice parameter of $Co_3[Cr(CN)_6]_2 \cdot 10H_2O$ are as follows: $a = 1.04905(3)$ nm and $a = 1.03805(9)$ nm for $Cr_3[Cr(CN)_6]_2 \cdot 2H_2O$ [33]. Doping with Ni for Mn leads to the reduction in the volume of elementary cells, which is approximately equal to 3.8 percent compared to $Mn_3[Cr(CN)_6]_2$ material [34], and $(Ni_xMn_{1-x})_3[Cr(CN)_6]_2$ compound follows Vegard's law and the enlargement of x results in a progressive reduction of the lattice parameter [34]. Lattice parameters decrease nearly linearly with substitution of Cu for Mn in the case of $(Cu_xMn_{1-x})_3[Cr(CN)_6]_2 \cdot zH_2O$ metalo-complexes: $a = 10.51909$ Å, 10.49812 Å, 10.50418 Å, 10.49833 Å, 10.4887 Å, 10.3851 Å for $x = 0.2$, 0.25, 0.3, 0.35, 0.4, and 1.0, respectively [36]. Both $KMn[Cr(CN)_6]$ and $KNi[Cr(CN)_6]$ crystallizes in the *fcc* system, with lattice constants $a = 10.786793$ Å and $a = 10.48679(5)$ Å, respectively [39,40]. The examples of x-ray diffraction and neutron diffraction patterns are displayed in Figure 2.

Neutron powder diffraction (ND) patterns were obtained on two different samples of $(Ni_xMn_{1-x})_3[Cr(CN)_6]_2 \cdot zD_2O$ mixed ferri-ferromagnetic material. The H atoms were substituted by D in the process of samples synthesis. The z number varies between 12 and 15, but in most events is equal to 12. The higher amount of background present in ND measurements points to the fact that D_2O molecules are replaced by H_2O molecules in the process of the ageing of the samples and incoherent scattering from H supplies to background (Figure 2b). The incoherent scattering caused pronounced enlargement of the time of measurement, but finally we obtained good statistics and the crystal structure, including D-sites, was completely described by the Rietveld refinement procedure. Prior to the neutron experiment, we left the experimental material in wet conditions to elude the aging, e.g., substitution of D_2O by H_2O. Our procedure resulted in a lower background but a second phase (ice made of D_2O) was indicated at low temperature [30]. Determination of the positions of building elements (CrC_6 octants, D_2O molecules, and other atoms were considered as isolated) in the cell was calculated by the direct-space method by application of reverse Monte-Carlo approach (Figure 3). The incoherent scattering led to extension of the experimental time, but in the end a good statistic was reached and the crystal structure, including D-sites, was determined by the Rietveld refinement.

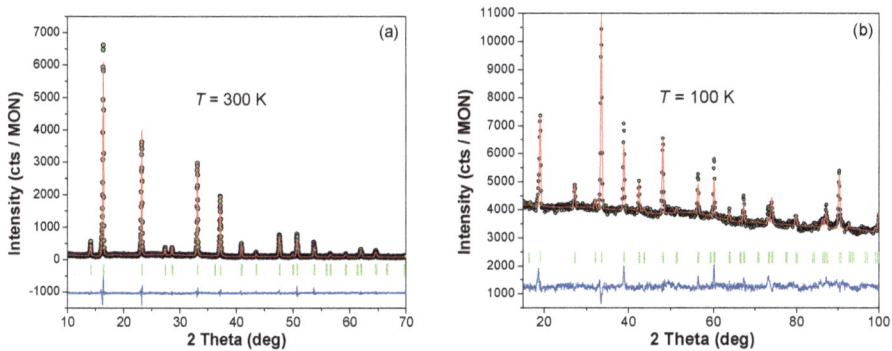

Figure 2. X-ray and neutron diffraction patterns for $Mn_3[Cr(CN)_6]_2 \cdot zD_2O$ powders at different temperatures. Circles represent the measured data and line shows the fitted model. The positions of all possible Bragg reflections are marked by the vertical marks in the middle and the lower curve shows the difference between the observed and calculated intensities. (**a**) X-ray powder diffraction pattern: a = 10. 788 Å, reliability factors of the refinement procedure R_p = 8.36, R_{wp} = 10.7, R_{exp} = 8.12, χ_2 = 1.73; (**b**) neutron diffraction pattern: a = 10.730 Å, R_{wp} = 29.6.

3.2. *Effect of Pressure on the Crystal Structure of* TM^{2+}-Cr^{III}-*PBs*

Effects of pressure on the structure and magnetic properties of 3-D cyanide bridged bimetallic coordination polymer magnets, $Mn^{2+}Cr^{III}$ ferrimagnet $[Mn(en)]_3[Cr(CN)_6]_2 \cdot 4H_2O$ (*a* en = ethylenediamine), $Ni^{2+}Cr^{III}$ ferromagnet $[Ni(dipn)]_3[Cr(CN)_6]_2 \cdot 3H_2O$ (*b* dipn = N, N-di(3-aminopropyl)amine were methodically studied in hydrostatic pressure up to 4.7 GPa by application of a piston-cylinder-type pressure cell and a diamond anvil cell [41,42]. The lattice of the ferrimagnet (*a*) was reduced in an isotropic manner for $p \leq 1.7$ GPa, and the lattice volume linearly decreased. At higher pressure, the *bc* frame and total volume decreased only slightly, but the stacking along the *a* axis tended to tilt with the expansion of the *a* axis anisotropically. The reduction of lattice size saturated with a total volume reduction of about 9% at 3.2 GPa. An amorphous structure was observed at above 4.2 GPa. The pressure induced reduction of the lattice size is completely reversible. The original crystal structure recovered after releasing 4.5 GPa, indicating sufficient elasticity of Cr-CN-Mn. The compound (*b*) showed a different response to pressure. The cell volume decreased linearly and isotropically in the range 0–3.1 GPa with about 15% compression at 4.7 GPa, extending of the vertexes, and no amorphization, which demonstrated the structural strength. The shrinkage ratio of (*b*) is twice as large as that of (*a*) [42].

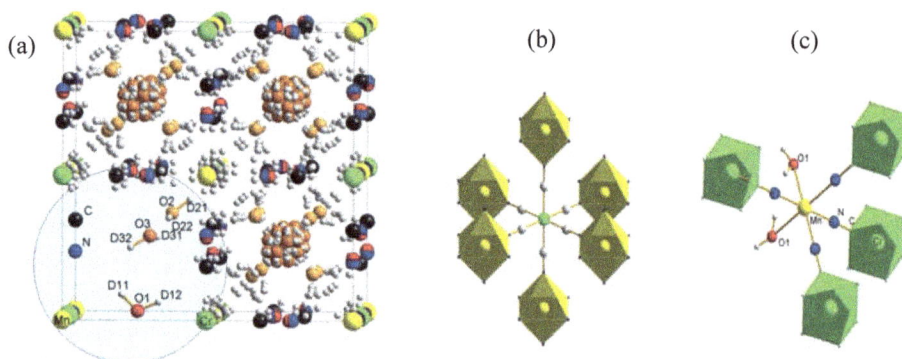

Figure 3. (a) The crystal structure is characterized by significant compositional disorder. The marked region describes possible water molecules distribution; (b) each Cr cation is linked via C≡N groups to six Mn cations and (c) each Mn cation is surrounded on average by four [Cr(CN)$_6$] complexes and by two D$_2$O(1) molecules.

The pressure-caused change of crystal structure can be identified indirectly from the variation of pressure effect on T_C [43]. The first dT_C/dp = 25 K/GPa of Mn$_3$[Cr(CN)$_6$]$_2$·zH$_2$O PBs was alike to the amounts published in the study at pressures below 1 GPa [29]. At higher pressures, an essential but nonlinear increase in T_C from 63 K at ambient pressure to a maximum T_C = 93 K at 3.2 GPa pressure was found. This reflects an enlargement of the superexchange coupling between the transition metal cations as the bonds to cyanide are reduced by hydrostatic pressure. The ferrimagnetic moment falls down as T_C is enlarged, which may point to the higher degree of tilting of the Mn(CN)$_6$ and Cr(NC)$_6$ octants. At pressures above 3.2 GPa, the magnetic transition is broader, and T_C is reduced to zero near 10 GPa, with a correspondent break down of the ferrimagnetic moment. The finding coincides with a pressure-caused amorphisation of the material, as was reported on similar PBs [42].

Reference [44] resumes 0 to 0.6 GPa neutron diffraction data obtained on a nickel hexacyanochromate PBs K$_{0.25}$Ni[Cr(CN)$_6$]$_{0.75}$(D$_2$O)$_{0.25}$·2.1D$_2$O having *fcc* Prussian blue structure. Small thermal contraction is a characteristic feature of this material. On the other hand, pressure causes pronounced consequences on the top positions, enabling determination of the bulk modulus, $K = -VdV/dp$. The 110 K phase exhibits a variation from 10.477 Å at ambient pressure to 10.410 Å at high pressure, such that K = 31.43 GPa. In like manner, at 5 K the usage of pressure causes a reduction from 10.468 to 10.413 Å that leads to almost the same amount of K = 31.94 GPa. The assumption of linear volume reduction with pressure is evidenced by the alike K amounts for 0.5 and 0.6 GPa. Second, the widths of the peaks provide information about particle size and strain, and these results exhibit pressure-caused anisotropic widening. Third, the strength of the tops provides information about the fractional coordinates of atoms within the unit cell and their positional distributions. A real-space visualization of how pressure varies the ND scattering length density (SLD) identifies an antisymmetric change rising near the Cr sites while, simultaneously reducing by a similar value at the Ni site. The isomerization of cyanide linker from C bonding to Cr to C bonding to Ni makes these data clear. However, a shift of the coordinated water and the cyanide linker will also multiply an antisymmetric change in SLD between the metal ions [44].

The effect of pressure on Raman spectra of molecule-based magnets KNiCr(CN)$_6$ and KMnCr(CN)$_6$ was studied in [40]. The M–N≡C–Cr linkages M = Ni, Mn form the *fcc* lattice that may contain vacancies. It is generally accepted that the cyanide bridge can be extra susceptible to its surroundings, including the oxidation state and the spin state of the magnetic ions in the lattice. We reported on the effect of pressure on the v [C≡N] vibration band, which is placed into the 2100–2200 cm^{-1} spectral region [40]. The location of the band passes under pressure nonlinearly beginning at 2162.7 cm^{-1} for 6.50 GPa for KMnCr(CN)$_6$ (Figure 4a). The CN frequency of KNiCr(CN)$_6$

falls down at first from 2195.7 cm^{-1} for zero applied pressure to 2189.1 cm^{-1} for 0.23 GPa, and then rises to 2211.9 cm^{-1} for 6.2 GPa (Figure 4b). A thin arm (an arrow in Figure 4b) happens to be visible in the pressure region from 0.23 to 2.3 GPa, moving its location from 2188.4 cm^{-1} to 2193.9 cm^{-1}, with rising pressure pointing to the existence of two different CN bridge neighbourhoods. The hydrostatic pressure-susceptible arm can point to the linkage isomerisation, suggesting that Cr^{III}–N≡C–Ni^{II} fragments grow at the expense of the original Cr^{III}–C≡N–Ni^{II} units [44,45]. However, the effect can also be superimposed by other contributions, such as the presence of surface non-bridging cyanides or coupling between alkali cations.

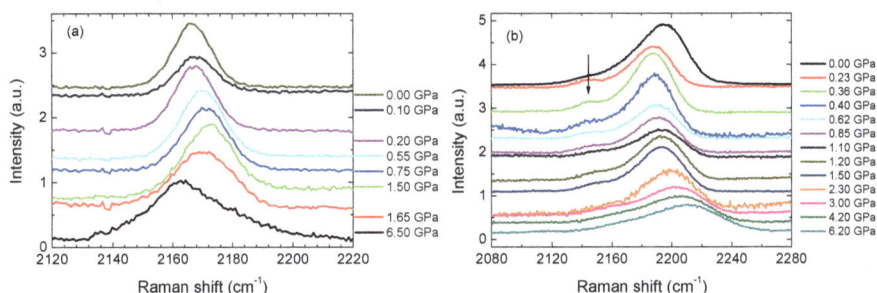

Figure 4. (**a**) Raman spectra of KMnCr(CN)$_6$ for various pressures. (**b**) Raman spectra of KNiCr(CN)$_6$ for various pressures. The arrow points to only very small evidence of linkage isomerism.

3.3. Magnetic Structure of TM^{2+}-Cr^{III}-PBs

Magnetic structure was determined by powder ND experiment, which was performed on two samples of (Ni$_x$Mn$_{1-x}$)$_3$[Cr(CN)$_6$]$_2$·zD$_2$O on E9 (λ = 1.79734 Å, HZ-Berlin) for z = 0, 1 at T = 2 K, 100 K and on G4.1 (λ = 2.42500 Å, LLB Saclay) for z = 0, 0.38, and 1 at different temperatures below T = 70 K. The pronounced magnetic contribution we found only in the case of the Mn$_3$[Cr(CN)$_6$]$_2$ magnet, and in this study we focused merely on the mentioned material [30]. A very poor magnetic signal was detected on Ni$_3$[Cr(CN)$_6$]$_2$ contribution at 1.6 K, perhaps because of a distribution of magnetically ordered domains [30,44]. Magnetization measurements of (Ni$_{0.38}$Mn$_{0.62}$)$_3$[Cr(CN)$_6$]$_2$ mixed ferro-ferri magnet indicated a maximum at about 60 K and low value of M at T = 2 K [34]. ND measurements undertaken on this material at 60 K pointed out an almost weak indication of magnetic signal. A strong magnetic signal was detected on Mn$_3$[Cr(CN)$_6$]$_2$ under 60 K and emerged only on allowed nuclear peaks—no additional signal was detected on forbidden nuclear Bragg summits. Diffraction apexes (222), (400), (422), and (440), with magnetic signal compared to experimental error, were excluded from refinement of magnetic contribution I_{2K}-I_{70K} (Figure 5a). Magnetization measurements and elastic ND of Mn$_3$[Cr(CN)$_6$]$_2$ clearly point out a magnetic structure consisting of Mn and Cr sublattices, with antiparallel magnetic moments μ_{Mn} = 3.790 μ_B and μ_{Cr} = −1.375 μ_B leading to overall ferromagnetic ordering below the Curie temperature T_C = 63 K. The simplified model of possible ferrimagnetic ordering of Mn$_3$[Cr(CN)$_6$]$_2$·zD$_2$O is plotted in Figure 5b. On the other hand, the magnetic structure of the Dy[Fe(CN)$_6$]·4D$_2$O molecule-based magnet with the orthorhombic crystal structure (*Cmcm* space group) was refined using Rietveld technique [46,47]. ND indicated that the magnetic structure is formed by Fe and Dy sublattices, which are merged antiferromagnetically, leading to overall ferrimagnetic arrangement with the magnetic phase transition at T_C = 3.7 K. While in the case of Fe-atoms the y-component of magnetic moment is large and the z-component is nominal, in the case of Dy-atoms, the x-and the y-magnetic moment components are large and the ordering of magnetic moments on Dy-sublattice is non-collinear [47].

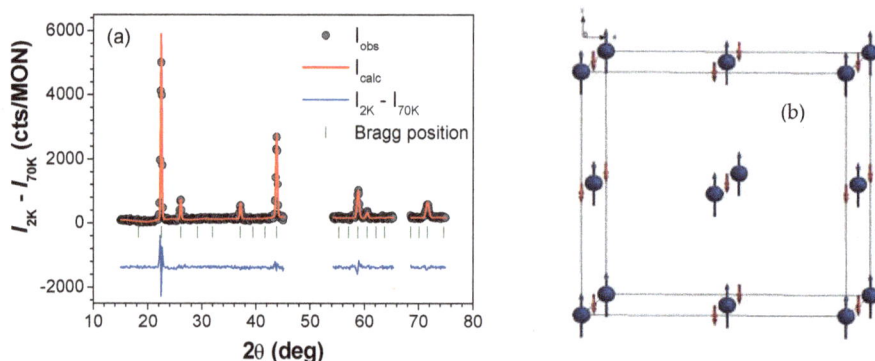

Figure 5. (a) Results of Rietveld refinement of data taken on $Mn_3[Cr(CN)_6] \cdot zD_2O$. (b) Simple model of ferrimagnetic structure of $Mn_3[Cr(CN)_6] \cdot zD_2O$, Mn (4a) (0,0,0): blue; Cr (4b) ($\frac{1}{2}, \frac{1}{2}, \frac{1}{2}$): red; μ_{Mn} = 3.790 μ_B a μ_{Cr} = −1.375 μ_B.

3.4. Pressure Effect on Magnetic Properties of Polycrystaline Samples

The above-mentioned magnetic model was tested on the $TM^{+2}{}_3[Cr^{III}(CN)_6]_2 \cdot zH_2O$ system [29–38] and $KTM^{2+}Cr(CN)_6$ [39,40], where TM^{2+} is a 3d ion and the following papers [29,32,34,38–40] are focused on the pressure effect on magnetic properties of these two types of PBs. The Cr^{III} in the low spin anion $[Cr^{III}(CN)_6]^{3-}$ has $(t_{2g})^3$ orbital resulting in six ferromagnetic (FM) and nine antiferromagnetic (AFM) pathways, with $(t_{2g})^3(e_g)^2$ orbitals of Mn^{2+} leading to J_{AF} interaction. Compared to the $(t_{2g})^3$ orbital of Cr^{III}, we have six F pathways with $(e_g)^2$ orbitals of Ni^{2+} leading to overall J_F interaction (Figure 6).

$$Mn^{2+}{}_3[Cr^{III}(CN)_6]_2 \cdot 15H_2O \qquad\qquad Ni^{2+}{}_3[Cr^{III}(CN)_6]_2 \cdot 15H_2O$$

$$Mn^{2+}(S = 5/2) \qquad\qquad [Cr^{III}(CN)_6]^{3-} (S=3/2) \qquad\qquad Ni^{2+} (S = 1)$$

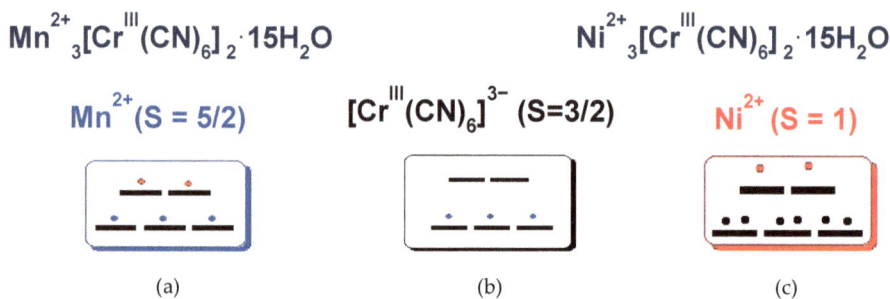

(a)   (b)   (c)

Figure 6. (a) There are six F and nine AF pathways with $(t_{2g})^3(e_g)^2$ orbitals of Mn^{2+} leading to overall J_{AF} in interaction; (b) Cr^{III} in anion $[Cr^{III}(CN)_6]^{3-}$ is low spin and has only $(t_{2g})^3$ orbitals; (c) there are six F pathways with $(e_g)^2$ orbitals of Ni^{2+} leading to overall J_F interaction.

Measurements of magnetic susceptibility (Figure 7) indicate a ferrimagnetic ordering below $T_C \sim 65$ K for the Mn-sample with μ_{eff} = 10.48 μ_B and θ = −39.5 K. The magnetic moment μ_s saturates to an amount of 8 μ_B/f.u. at T = 2 K and μ_0H = 5 T. The Ni-sample orders ferromagnetically below $T_C \sim 56$ K, μ_{eff} = 8.6 μ_B, θ = 72 K. The magnetic moment saturates to a higher value of μ_s = 10.3 μ_B/f.u at T = 2K and μ_0H = 5 T [29,30].

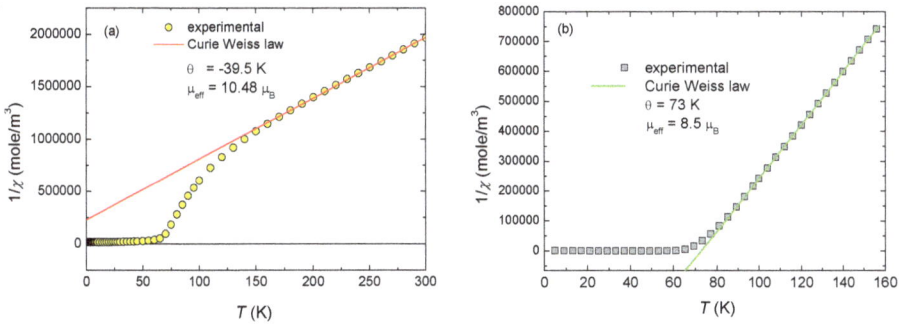

Figure 7. The Curie Weiss law proposes the type of magnetic interactions; (**a**) AFM coupling for $Mn_3[Cr(CN)_6]_2 \cdot 15H_2O$ and (**b**) FM coupling for $Ni_3[Cr(CN)_6]_2 \cdot 15H_2O$.

Figure 8a shows magnetic isotherms for Mn^{2+}-Cr^{III}-PBs obtained by measurements at various pressures [29]. The magnetic transition at T_C was defined as the inflection point of $M(T)$ curve in this region. Magnetization as a function of hydrostatic pressure saturates in a very steep way, already reaching independence on the applied magnetic field for the values higher than $\mu_0 H = 200$ mT (Figure 8a). There is no magnetic hysteresis, with very small remnant magnetization μ_r and coercive field H_c. The effect of applied pressure on magnetization curves is pronounced only in the region of small magnetic fields. It was observed that the shape of magnetic isotherms changes and magnetization saturates at higher magnetic fields, as it is supposed that effective barriers for domain wall motion induced under pressure negatively influence its free motion in the irreversible part of magnetic hysteresis [29,44]. The second possible explanation for the effect of applied pressure on magnetic isotherms can be ascribed to the change in magneto crystalline anisotropy [44], as its shape resembles magnetization curves typical for the hard magnetization axis. Both magnetic ions are placed in octahedral positions, and therefore magnetic moments are aligned along the axes of octahedron. Even without applied external pressure, the octahedrons are tilted (Figure 1). Hydrostatic pressure enlarges the extent of tilting and octants are deformed. Both effects influence the efficiency of the magnetic exchange pathway and lead to magnetic anisotropy and reduction of the magnetic moment. The magnetic isotherms of Ni^{2+}-Cr^{III}-PBs measured for various applied pressures [29] are plotted in Figure 8b. The magnetization saturates at a higher magnetic field, even at ambient pressure. In the region of small magnetic fields below $\mu_0 H = 5$ mT, initial magnetization increases linearly, followed by a steeper S-shape increase and consequently showing almost linear dependence again above $\mu_0 H = 1$T. The remnant magnetization and coercive field are small but more pronounced. It is expected that the hydrostatic pressure leaves spontaneous magnetization unchanged, and the magnetization processes are strongly changed only at low temperatures below 8 K. The magnetization saturates in a higher magnetic field, the interval of linearity for initial magnetization extends to $\mu_0 H = 10$ mT and the remnant magnetization M_r is a little bit higher. Pressurization of the crystal lattice results in deformation, including small rotation of octants leading to reorientation of magnetic moments of 3d magnetic ions followed by decrease of magnetization in the saturated state.

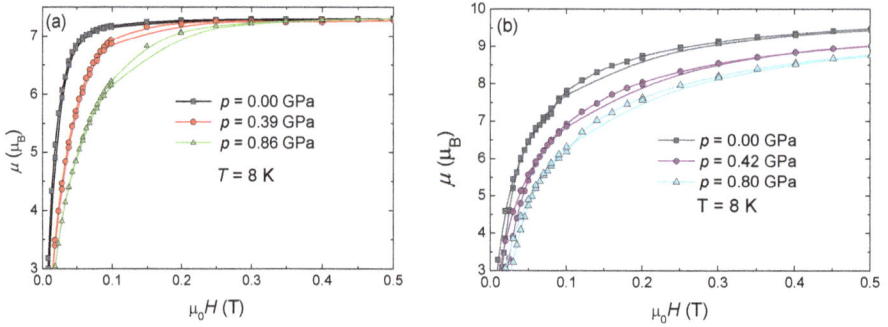

Figure 8. Effect of pressure on magnetic isotherms for (**a**) $Mn_3[Cr(CN)_6]_2 \cdot 15H_2O$ and (**b**) for $Ni_3[Cr(CN)_6]_2 \cdot 15H_2O$.

The effect of applied pressure on the temperature of the transition to magnetically ordered state—the Curie temperature T_C is shown in the Figure 9. The Curie temperature T_C, defined as the inflection point of the $M(T)$ curve, increases almost linearly in the whole range of applied pressures. The pressure = coefficient $\Delta T_C/\Delta p$ = 25.5 K/GPa is one of the highest positive changes of T_C with pressure so far published for any PBs [29]. The positive pressure coefficients $\Delta T_C/\Delta p$ ≈ 13 K/GPa and $\Delta T_C/\Delta p$ ≈ 11 K/GPa were reported for [Mn(en)]$_3$[Cr(CN)$_6$]$_2 \cdot$4H$_2$O [41,42] and for Mn$_3^{2+}$[Mn^{III}(CN)$_6$]$_2 \cdot$12H$_2$O·1.7(CH$_3$OH) [48] Mn-based PB ferromagnetic materials. The typical property of majority PBs, hysteretic behaviour between ZFC and FC magnetization for small magnetic fields, is shown for Mn^{2+}-Cr^{III}-PBs in Figure 9. ZFC and FC magnetization curves merge very well in the range above the magnetic phase transition and show hysteretic behaviour below the bifurcation temperature T_b. The hysteretic behaviour between ZFC and FC regimes defines an interval where irreversible magnetization processes determine shape of magnetization curve. A small maximum below T_C in the region where ZFC and FC regimes have large hysteresis is very often accompanied with freezing temperature T_f of the cluster-glass system, which is often present in PBs. The hydrostatic pressure enlarges this hysteretic behaviour, which is observed for ZFC and FC regimes. Smaller magnetization $M(T)$ observed on the ZFC curve points to the variation in magneto-crystalline anisotropy caused by pressure. It points out to the fact that the cluster-glass-like feature is enhanced under applied pressure.

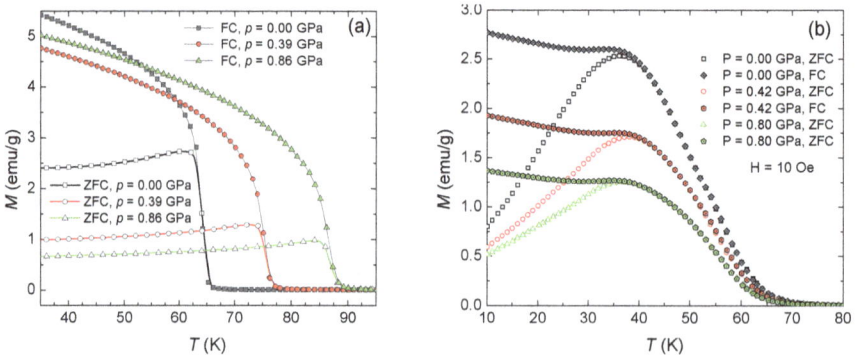

Figure 9. Effect of pressure on magnetization for (**a**) Mn^{2+}-Cr^{III}-PBs and (**b**) Ni^{2+}-Cr^{III}-PBs.

The real part of AC susceptibility $\chi'(T)$ for the Mn-sample, which we obtained for non-DC magnetic field, modulation with the amplitude of 3 Oe and frequency f = 0.4 Hz confirmed very similar behaviour to the one described earlier on the $M(T)$ measurements under pressure in ZFC regime

(Figure 10a). The pronounced maximum in the imaginary part of susceptibility $\chi''(T)$ at about T_C indicates an energy-dissipative process, which is connected with forming of a magnetically-ordered state (Figure 10b). Hydrostatic pressure gradually shifts the peak at T_C to higher temperatures, which indicates the enlargement of T_C. The gradient of the T_C rises with pressure, estimated from a steep rise of $\chi'(T)$ or as a peak of $\chi''(T)$, is similar to $\Delta T_C/\Delta p = 25.5$ K/GPa, determined from $M(T)$ measurements. Mn^{2+} has five unpaired d-electrons $(t_{2g})^3(e_g)^2$ with S = 5/2 and Cr^{III} has three d-electrons $(t_{2g})^3$ with S = 3/2 in the Mn^{2+}-Cr^{III}-PBs. In the ligand-field model, each t_{2g} orbital (*a*) of Mn^{2+} can find t_{2g} orbital (*b*) in Cr^{III}, which it can strongly interact with in a three-dimensional network [1]. The super-exchange interaction of this pair will be antiferromagnetic. From the extended Hückel calculations follows that the antiferromagnetic contribution to the magnetic coupling J is determined approximately by the expression 2 S $(\Delta 2 - \delta 2)1/2$, where δ is the energy gap between the (unmixed) a and b orbitals, Δ is the energy gap between the molecular orbitals formed by them, and S is the mono-electronic overlap integral between *a* and *b*. The antiferromagnetic term can be rewritten as $(\Delta^2 - \delta^2) = (\Delta - \delta)(\Delta + \delta)$; the strength of the interaction is gauged by the term $(\Delta - \delta)$ and the stabilization of charge-transfer states, in which an electron being transferred from one magnetic orbital to the other is gauged by the term $\Delta + \delta$. The most general effect of applied pressure, namely shortening of the distance between the magnetic ions in the lattice, leads to the increase of the amount of the overlap integral S. The enlargement of T_C in Mn^{2+}-Mn^{III}-PBs can, therefore, be ascribed to an enlargement of the overlap integrals between d_π (Mn^{III}) and $\pi^*(CN^-)$ and d_π (Mn^{2+}) and $\pi^*(CN^-)$. The hydrostatic pressure enlarges Δ, resulting in an enlargement of both expressions $(\Delta - \delta)$, $(\Delta + \delta)$ and gives rise to the enhancement of the antiferromagnetic interactions. The term $(\Delta^2 - \delta^2)$ can be determined for number of TM^{2+} and TM^{III}, with the precondition that the spacing between magnetic ions is identical. In the case that the coupling between a single pair of orbitals is known, one can sum all the combinations presented by a three-dimensional network to obtain the consequential contribution. In Mn^{2+}-Cr^{III}-PBs with nine AF and six F pathways, the antiferromagnetic coupling will be preferred. The rise of T_C caused by hydrostatic pressure was associated with a rise of super-exchange coupling J, determined by the enlargement of S and Δ in Mn^{2+}-Cr^{III}-PBs.

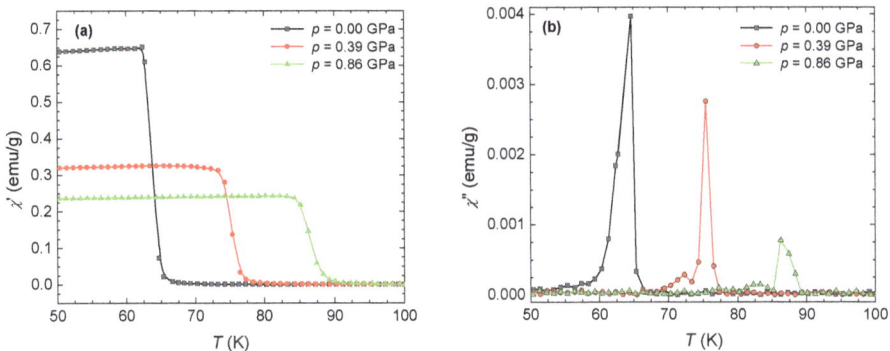

Figure 10. AC susceptibility for Mn^{2+}-Cr^{III}-PBs under pressures: (**a**) in phase and (**b**) out of phase.

However, the situation is different for the case of Ni^{2+}-Cr^{III}-PBs, where magnetization and AC susceptibility measurements indicate a slight decrease in T_C with pressure $\Delta T_C/\Delta p = -3.0$ K/GPa, and later more detailed experiments have shown that the T_C is not modified prominently under hydrostatic pressure [29]. Ni^{2+} has two unpaired d-electrons $(e_g)^2$ with S = 1 and six paired d-electrons $(t_{2g})^6$, which do not contribute to magnetic coupling. Magnetic interactions between the (e_g) orbital of Ni and (t_{2g}) of Cr are therefore dominantly ferromagnetic, the overlap of t_{2g} and e_g orbitals is practically equal to 0 ($\Delta = 0$), the pressure does not change J, and we do not expect any change in magnetic coupling. The hydrostatic pressure subtly affects bonding angles between magnetic ions interposed by the CN group. A slight deviation from perfect 180° of the bonding angle reduces the power of magnetic interactions, and consequently, T_C decreases.

Our study [29] revealed principally separate effects of pressure on the magnetic phase transition in Mn^{2+}-Cr^{III}-PBs with prevailing antiferromagnetic super-exchange coupling and a Ni^{2+}-Cr^{III}-PBs compound with prevailing ferromagnetic coupling. The applied pressure powered the super-exchange antiferromagnetic coupling J_{AF}. The Curie temperature T_C increased with the hydrostatic pressure, $\Delta T_C / \Delta p$ = 25.5 K/GPa, for the ferrimagnetic $Mn_3[Cr(CN)_6]_2$ compound as a result of powered magnetic interaction determined by enlarged amount of the monoelectronic overlap integral S and energy gap Δ between the mixed molecular orbitals. However, the hydrostatic pressure does not change J_F exchange magnetic coupling in the $Ni_3[Cr(CN)_6]_2$ ferromagnetic material. The bonding angle between magnetic ions differs from the perfect amount of 180° for any Prussian Blue analogue. The hydrostatic pressure affects bonding angles between magnetic ions interposed by the cyano-bridge and reduces the strength of magnetic interaction. Reduction of the magnetic coupling due to variations of bonding angle is general feature for both types of molecule magnets.

However, the enhanced exchange interaction of $Mn_3[Cr(CN)_6]_2$ caused by enhanced overlapping between magnetic orbitals masks this effect. The magnetization processes are modified by the pressure in both magnets, but we expect that the hydrostatic pressure does not affect the spontaneous magnetization. Pressure-induced magnetic hardening can be interpreted as a variation of magneto-crystalline anisotropy due to pressure. The mentioned effect was observed mainly for $Ni_3[Cr(CN)_6]_2$, where saturated magnetization μ_s decreased in the entire temperature range, while μ_s remains unchanged for $Mn_3[Cr(CN)_6]_2$ at low temperatures. The enlargement of μ_s for Mn^{2+}-Cr^{III}-PBs near to T_C was associated with the increase of magnetic interactions [29].

The applied pressure linearly shifts T_C of Co^{2+}-Cr^{III}-PBs down in the used interval of hydrostatic pressures (Figure 11a) with the determined negative coefficient $\Delta T_C / \Delta p$ = −1.8 K/GPa. The overlap of t_{2g} and e_g orbitals is principally zero and J_F is not changed by pressure. The hydrostatic pressure slightly affects bonding angles between magnetic ions mediated by the cyano-bridge, and this is the reason for reduction of T_C. Magnetization of Co^{2+}-Cr^{III}-PBs is reduced only weakly (Figure 11b). We expect that the spontaneous magnetization is unaffected hydrostatically by hydrostatic pressure, and only the magnetization processes are strongly affected even at low temperatures. The hydrostatic pressure increases the degree of deformation in the crystal structure, which leads to miss-orientation of magnetic moments placed on magnetic ions and to reduction of saturated magnetization [32].

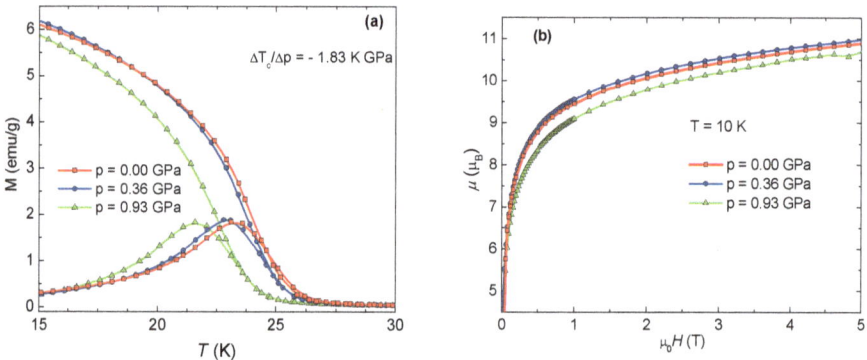

Figure 11. Effect of pressure on magnetization Co^{2+}-Cr^{III}-PBs: (**a**) temperature dependence of magnetization M(T) measured under different pressures in ZFC and FC regimes in magnetic field μ_0H = 2 mT; (**b**) magnetic isotherms $\mu(\mu_0H)$.

In the Prussian blue analogue $Cr_3[Cr(CN)_6]_2 \cdot 2H_2O$ the anion $[Cr^{III}(CN)_6]^{3-}$ has only $(t_{2g})^3$ orbitals and there are 3 ferromagnetic (F) and 9 antiferromagnetic (AF) pathways with $(t_{2g})^3(e_g)^2$ orbitals of

Cr^{2+} in high spin state leading to overall J_{AF} interaction. On the other hand, the $(t_{2g})^3$ orbitals of Cr^{III} have 6 AF pathways with $(t_{2g})^2$ orbitals of Cr^{2+} in low spin state leading to overall J_{AF} interaction (Figure 12). The transition to a magnetically ordered state is accompanied by a steep increase in $M(T)$ (Figure 13). The hydrostatic pressure shifts T_C of Cr^{2+}-Cr^{III}-PBs to higher temperatures almost linearly in this range of applied pressures. The estimated positive coefficient $\Delta T_C/\Delta p = 29$ K/GPa is the highest positive change of T_C with pressure which has so far been published for any PBs [32]. The applied pressure reduces the length of exchange path Cr^{2+}–N≡C–Cr^{III} and increases overlap integral S. The hydrostatic pressure increases Δ resulting in the enlargement of both terms $(\Delta - \delta)$ and $(\Delta + \delta)$ powers the antiferromagnetic coupling [1]. The variance between $M(T)$ determined for ZFC and FC measurements identifies the temperature range where irreversible magnetization processes take place (Figure 13a). A flat peak near to T_C in the range where large differences between ZFC and FC magnetic curves take place is very often attributed to freezing temperature T_f of the cluster glass system, which is a very common feature of the majority of PBs. The pressure enlarges the gap between magnetization obtained in ZFC and FC regimes for Cr^{2+}-Cr^{III}-PBs. Reduced value of magnetization $M(T)$ in ZFC regime indicates a change in ligand field leading to change of magnetocrystalline anisotropy.

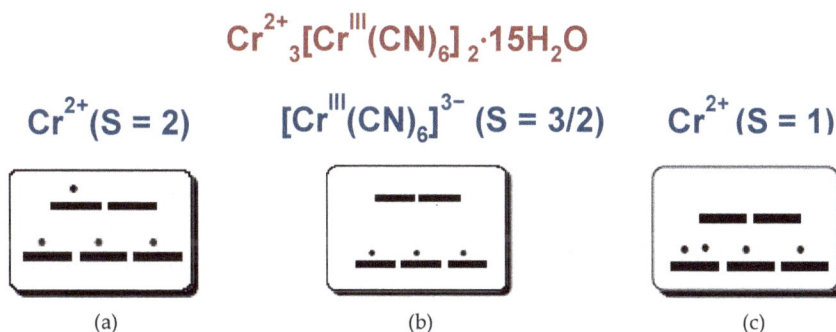

Figure 12. (**a**) There are three F and nine AF pathways with $(t_{2g})^3(e_g)^1$ orbitals of Cr^{2+} leading to overall J_{AF} interaction; (**b**) Cr^{III} in anion $[Cr^{III}(CN)_6]^{3-}$ is low spin and has only $(t_{2g})^3$ orbitals; (**c**) there are six AF pathways with $(e_g)^2$ orbitals of Cr^{2+} in low spin stay leading to overall J_{AF} interaction.

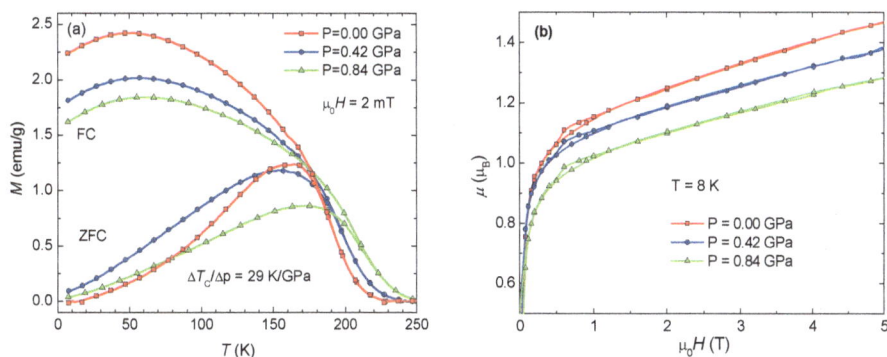

Figure 13. Effect of pressure on magnetization for Cr^{2+}-Cr^{III}-PBs: (**a**) ZFC and FC curves; (**b**) magnetic isotherms $\mu(\mu_0 H)$.

The magnetic isotherms of Cr^{2+}-Cr^{III}-PBs determined at various pressures are present in Figure 13b. Magnetization goes up at first very steeply, at very low field starts to saturate, and above $\mu_0 H = 1$ T the magnetization grows almost by a linear way. The applied pressure reduces saturated magnetization of Cr^{2+}-Cr^{III}-PBs. The Cr^{III} in anion $[Cr^{III}(CN)_6]^{3-}$ is low spin and exhibits only $(t_{2g})^3$ orbital and the spin S = 3/2. However, we suppose for cation Cr^{2+} two possibilities: at first Cr^{2+} is high spin, having S = 2 and magnetic orbitals are $(t_{2g})^3(e_g)^1$, implying three F and nine AF pathways, or another opportunity when Cr^{2+} is low spin S = 1 and magnetic orbitals are $(t_{2g})^2$, implying six AF pathways [1]. In the case that all Cr^{2+} are high spin, the expected theoretical amount of spontaneous magnetization μ_s = g[3S(Cr^{2+}) − 2S(Cr^{III})] = 6 μ_B; g = 2 is the Lande factor. The experimentally estimated amount μ_s(exp) = 1.73 μ_B [32] is widely less than theoretical one, pointing out that a portion of Cr^{2+} is low spin (S = 1), yielding total compensation of spins μ_s = 0 μ_B. It seems that only approximately 30% of Cr^{2+} is high spin. The applied pressure p = 0.84 GPa causes high spin–low spin transition of cca 4.5% of high spin Cr^{2+}.

Magnetic properties of mixed ferro-ferrimagnet $(Ni_xMn_{1-x})_3[Cr(CN)_6]_2·zH_2O$ and the compensation temperature T_{comp} for different amounts of x were first mentioned and interpreted by molecular field theory in [49]. This happened because the negative magnetization due to the Mn^{2+}–N≡C–Cr^{III} subsystem with dominant J_{AF} and the positive magnetizations due to Cr^{III}–C≡N–Ni^{2+} subsystem with dominant J_F have different temperature dependences of magnetization $\mu(T)$, and at T_{comp}, which is below T_C, the sign of the magnetization is reversed [50]. The possibility that the spontaneous magnetization might change sign at particular T_{comp} was envisaged by Néel in the classical theory of ferrimagnets [51,52].

The effect of pressure on magnetization features of $(Ni_{38}Mn_{62})_3[Cr(CN)_6]_2·zH_2O$ was studied in pressures up to 0.8 GPa [34]. Both ferrimagnetic J_{AF} and ferromagnetic coupling J_F are present in this type of magnet and magnetization inversion is observed at the compensation temperature T_{comp}. Our study revealed that J_{AF} is prevailing in this material. The Curie temperature T_C of the magnet rises with hydrostatic pressure, dT_C/dp = 10.6 KGPa^{-1}, due to enhancement of J_{AF}. The rise of J_{AF} is associated with an enlarged amount of the single electron overlapping integral S and an energy gap Δ between the mixed molecular orbitals t_{2g} (Mn^{2+}) and t_{2g} (Cr^{III}) caused by pressure. Magnetization process is also influenced by pressure—magnetization saturates at a higher magnetic field and saturated magnetization decreases with pressure. The compensation temperature T_{comp} falls down under hydrostatic pressure. The $\mu(T)$ curves plotted in Figure 14a were obtained on the material with the very low amount of μ_s compared to parent compounds [34]. The same sample were used for high pressure measurements. The smooth shape of ZFC and FC magnetization curves together with nearly the same compensation temperature T_{comp} ~ 12 K for both ZFC and FC regimes confirm the high quality of the sample. Magnetic susceptibility follows the Curie–Weiss law above 100 K, with the effective magnetic moment μ_{eff} = 9.32 μ_B and the paramagnetic Curie temperature θ = −18.8 K (Figure 14b). The negative value of θ indicates that J_{AF} is dominant but the shape of $1/\chi(T)$ below 100 K points out to ferrimagnetic ordering.

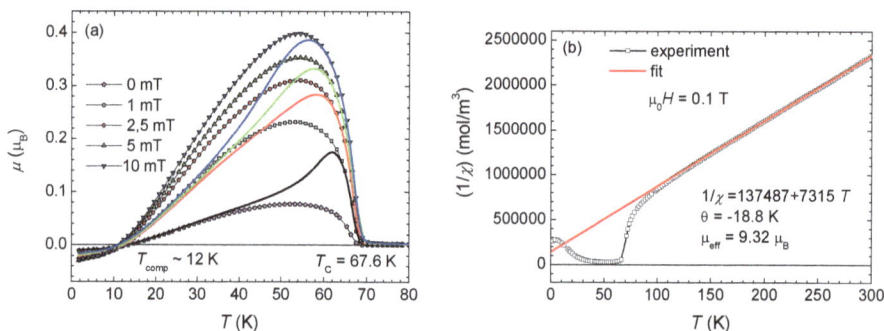

Figure 14. Temperature dependence of magnetization and inverse susceptibility for $(Ni_{0.38}Mn_{0.62})_3$ $[Cr(CN)_6]_2·zH_2O$; (a) ZFC (full lines) and FC (symbols) curve (b) susceptibility follows the Curie-Weiss law above T = 100 K.

The magnetic phase transition at 67.6 K of this mixed ferri-ferro magnet takes place at a higher temperature than for both Ni$_3$[Cr(CN)$_6$]$_2$·zH$_2$O (T_C = 56 K) and Mn$_3$[Cr(CN)$_6$]$_2$·zH$_2$O (T_C = 65 K). Simultaneously, with increase of Ni for Mn substitution, the decrease of lattice parameters was observed. The increase of T_C correlates with the chemical pressure induced by this substitution. An unusually high value of magnetic Grüneisen parameter ε in the range 9.03–9.97, which was reported for PBs [46], points to a very strong magneto-structural correlation in PBs. The material-dependent parameter ε can be, in accordance to the definition in a phenomenological model, ascribed to the intersite distances, and consequently to the interaction strength J.

The general tendencies of the hydrostatic pressure effect on T_C remain the same. The magnetic phase transition is shifted to higher temperature (Figure 15a) but dT_C/dp = 10.6 K/GPa (Figure 15b) is significantly decreased compared to Mn$_3$[Cr(CN)$_6$]$_2$·zH$_2$O, where dT_C/dp = 25.5 K/GPa [29]. Despite weakening of the effect, the interpretation is again related to changes of electronic structure, as in Mn^{2+}–N≡C–Cr^{III}. The single electron overlapping integral S, as well the energy gap Δ of the mixed molecular orbitals t$_{2g}$ (Mn^{2+}) and t$_{2g}$ (Cr^{III}) become enhanced, and J_{AF} interaction is more effective [1,29]. On the contrary, the ferromagnetic Cr^{III}–C≡N–Ni^{2+} subsystem is weakly affected due to no overlap of magnetic orbitals, and therefore J is given Mn^{2+}–N≡C–Cr^{III} -part. The number of exchange paths sensitive to pressure is reduced in this ferri-ferro-mixed magnet, resulting in the lower value of dT_C/dp in (Ni$_{0.38}$Mn$_{0.62}$)$_3$[Cr(CN)$_6$]$_2$.

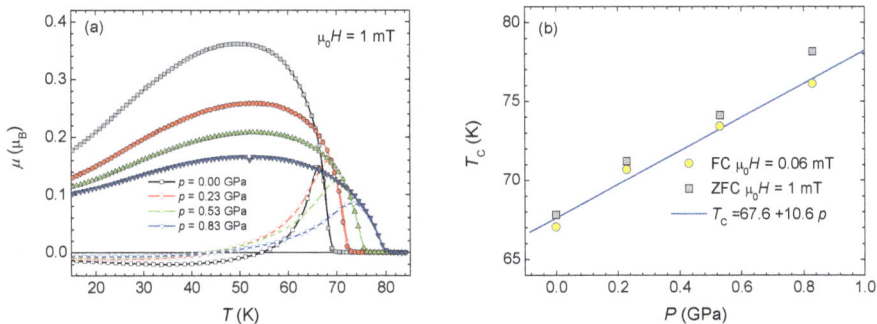

Figure 15. Pressure effect on (**a**) ZFC and FC curves and (**b**) T_C of (Ni$_{0.38}$Mn$_{0.62}$)$_3$[Cr(CN)$_6$]$_2$·zH$_2$O.

The characteristic feature of ZFC and FC curves, the compensation temperature T_{comp}, appears because the algebraic sum of $\mu(T)$ of Mn^{2+}–N≡C–Cr^{III} part and Cr^{III}–C≡N–Ni^{2+} part is 0 at T_{comp} [51]. It seems that T_{comp} is not an intrinsic property of the system and depends on treatment. Heat treatment removes T_{comp} from FC curves but not from ZFC curves [34]. The hydrostatic pressure p_1= 0.23 GPa reduces T_{comp}. The application of higher pressure does not have any effect on T_{com} (Figure 15a).

Both types of magnetic ions (Mn^{2+}, Ni^{2+}) and (Cr^{III}) are located in the octants, which are basic motives of the crystal structure. A very stable octant around Cr^{III} is formed from six CN groups and coordinated to Cr^{III} via C atom. The coordination of Mn^{2+}, Ni^2 ions via N and H atoms is realized by four CN and two H$_2$O groups forming the octahedrons that can be deformed easier. Both types of octants can rotate and can be tilted, as it was shown for vanadium-based PBs [1,33]. As a consequence, the bonding angle between magnetic ions deviates from 180 degrees, and that is why J_{AF} will be reduced. Simultaneously, this reduction of J_{AF} competes with an increase of J_{AF} resulting from the enhanced value of S and Δ induced by pressure, leading to the smaller value of dT_C/dp. In these two cases the octants are tilted, the alignment of moments on magnetic ions is not strictly parallel or anti-parallel, and as a consequence μ saturates at a higher magnetic field (Figure 16a). The number of domain barriers, like [Cr^{III}(CN)$_6$] vacancies, which are not filled by H$_2$O, can be reduced by pressure. and the lower coercive force is expected for a system with a smaller number of domains (Figure 16b).

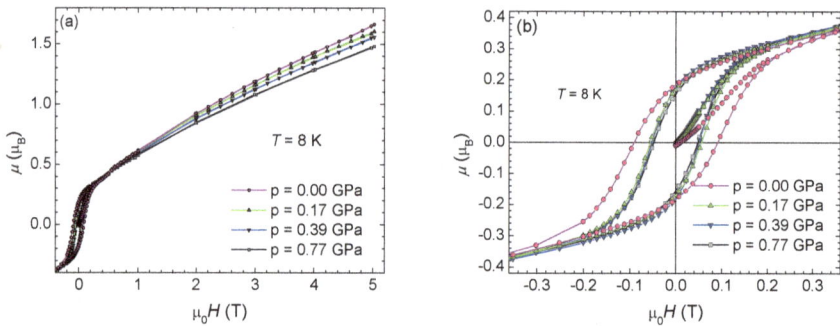

Figure 16. Magnetic isotherms of $(Ni_{0.38}Mn_{0.62})_3[Cr(CN)6]_2 \cdot zH_2O$; (**a**) magnetization saturates at higher magnetic field under pressure, (**b**) effect on remnant magnetization and coercive field.

Both magnetic isotherms and thermo-magnetic curves of KMnCr(CN)$_6$ are affected by hydrostatic pressure, as it was demonstrated in our paper [39]. Our results confirmed that the applied hydrostatic pressure raises the enhancement of magnetic coupling in this material, because a linear increase of T_C was observed under pressure (Figure 17a). A step on $\mu(T)$ in the vicinity of magnetic phase transition was generated by the applied pressure (Figure 17a) and $d\mu(T)/dT$ curve shows two minima associated with two magnetic transitions. Corresponding pressure coefficients of the two magnetic phases were estimated as $\Delta T_{C1}/\Delta p$ =18.18 KGPa^{-1} and $\Delta T_{C2}/\Delta p$ = 26.62 KGPa^{-1}. The basic magnetic characteristics are very similar to those earlier mentioned for Mn$_3$[Cr(CN)$_6$]$_2$ compound. Magnetization reached saturation at low temperatures for low magnetic fields μ_0H = 200 mT. Remnant magnetization μ_r and coercive field H_c are close to zero. This situation is changed under the pressure. The magnetic isotherms alter their curve with the hydrostatic pressure (Figure 17b) and magnetization saturates at a higher magnetic field. The hydrostatic pressure does not change the saturated magnetization μ_s, remnant magnetization μ_r, and coercive force H_c at low temperatures in a pronounced way.

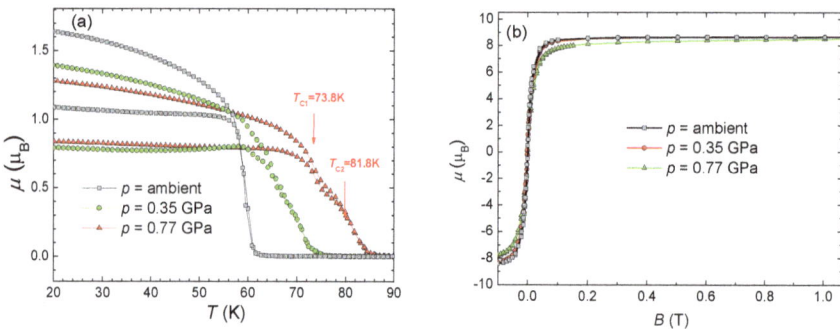

Figure 17. Magnetization curves for KMnCr(CN)$_6$ under pressure: (**a**) μ (T) s in ZFC and FC regimes; (**b**) $\mu(B)$ taken at temperature T = 8 K.

Our study of KMnCr(CN)$_6$ and KNiCr(CN)$_6$ molecule-based magnets presented in the paper [39] confirmed the main differences in pressure effect on T_C between Mn-PBs with prevailing antiferromagnetic magnetic interaction and Ni-PBs showing ferromagnetic properties. In the case of of the KMnCr(CN)$_6$ magnet, the pressure enhances the super-exchange antiferromagnetic coupling J_{AF} and that is why T_C simultaneously increases, due to enlarged monoelectronic overlap integral S and energy gap Δ between the mixed molecular orbitals. The hydrostatic pressure does not affect J_F

exchange interaction in KNiCr(CN)$_6$ and the magnetic phase transition does not change. The bonding angle between magnetic ions differs a little from the ideal value of 180° in real systems.

The bonding angle between magnetic ions represented by the CN group softly decreases with pressure, which leads to reduction of the magnetic coupling. This effect is present on both types of magnets, but it is masked on ferromagnetic materials because pressure strengthens exchange interaction resulting from enlarged overlapping of magnetic orbitals. The magnetization processes are influenced by the pressure in both systems, but the spontaneous magnetization does not change. Pressure-induced magnetic hardening was ascribed to variations in magneto-crystalline anisotropy induced by pressure. The rise of μ_s in Mn^{2+}-Cr^{III}-PBs near T_C was associated with the enhancement of magnetic interactions.

3.5. Magnetic Properties of Magnetic Nanoparticles

In the last few years, there has been considerable interest in preparation and investigation of magnetic nanoparticles (NAP) because of their potential applications in high density recording media, but also for the reasons of macroscopic tunnelling [53] and quantum computing [54]. After discovery of single–molecule magnet behaviour in Mn$_{12}$-acetate with highest spin ground state $S = 51/2$ [55], one of the important issues of magnetism is the study of objects with magnetic moments intermediately between this value and the value of metallic NAP with $S \geq 1000$ [56]. NAP based on Prussian blue analogues (PBs) prepared by reverse micelle technique with $S < 1000$ are promising candidates. The first report on application of this technique for preparation of NAP based on PBs has been made by Vaucher in [57]. Later in works [58,59], authors referred to preparation of cyanide bridged Cr^{III}-Ni^{2+} nanoparticles. The reverse micelle technique described in [56] was used [37,38]. The content of organic surfactants in the samples was estimated to about 22 wt.% Ni-NAP and 28 wt.% Mn-NAP. The average size of about 4.5 nm determined from X-ray measurements corresponds with TEM results (Figure 18).

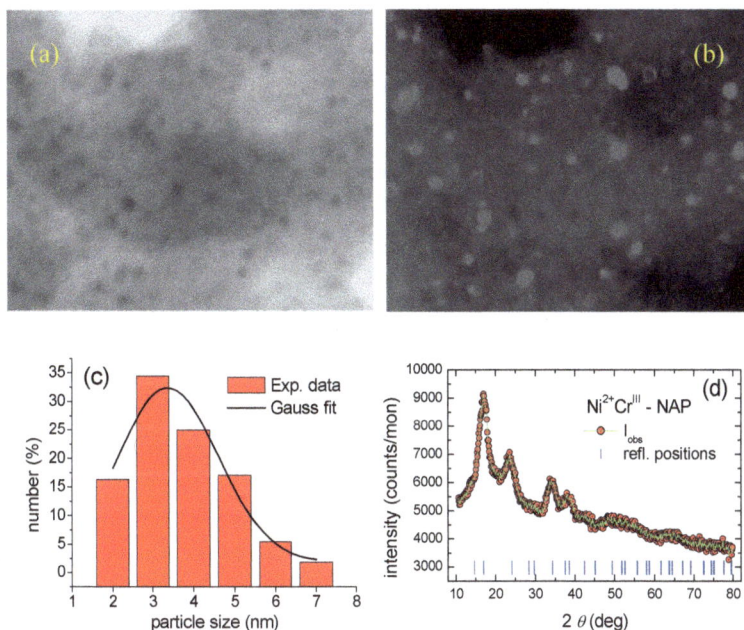

Figure 18. The Ni^{2+}-Cr^{III}-NAP embedded in an organic matrix; (**a**) dark-field TEM image; (**b**) bright-field TEM image; (**c**) distribution of NAP determined from TEM by an analysis of the picture; (**d**) XRD patterns.

We expect that both systems Ni-NAP and Mn-NAP behave as systems of strongly interacting magnetic particles. The super-exchange interaction is dominant in intra-NAP magnetic interaction. The dipole-dipole interaction is dominant in inter-NAP magnetic interaction. The dispersion of NAP into an organic matrix leads to a dilution of the mother PBs, and consequently the Curie temperature is reduced from T_C = 56 K for Ni-PBs to T_C = 21 K for Ni-NAP system and from T_C = 65 K for Mn-PBs to T_C = 38 K for Mn-NAP system. In the case of Mn-NAP, we found two satellite minima in dM/dT. Each of these minima indicates T_C of Mn-PBs sub-systems and represents two extra regions with different concentration of magnetic NAP [37]. Pressure induced changes of the saturated magnetization μ_s and the Curie temperature T_C are reversible. We expect that pressure affects inter-NAP distance (high compressibility of an organic matrix) and changes magnetic properties of NAP. The applied pressure strengthens the magnetic super-exchange interaction of Mn-NAP with the increased overlapping of magnetic orbitals [38]. As it is shown in Figure 19b, pressure decreases T_C (dT_C = dp = −3 K/GPa) for the system of Ni-NAP with dominant JF, which is mainly attributed to less effective magnetic coupling due to the reduction in the bonding angle Ni^{2+}–N≡C–Cr^{III} from an ideal value of 180°. Figure 20a demonstrates a dilution effect. The shape of the magnetization curve for Ni-NAP is the same as for Ni-PBs and the reduction of the saturated magnetization μ_s is mainly due to reduced ratio of mother PBs (only 65 wt.%) in the system. The applied pressure increases the magnetic moment μ_s (Figure 20b). This behavior is opposite to the behavior of the mother Ni-PBs and is attributed to the reduction of inter-NAP distances by compression of the organic matrix [38].

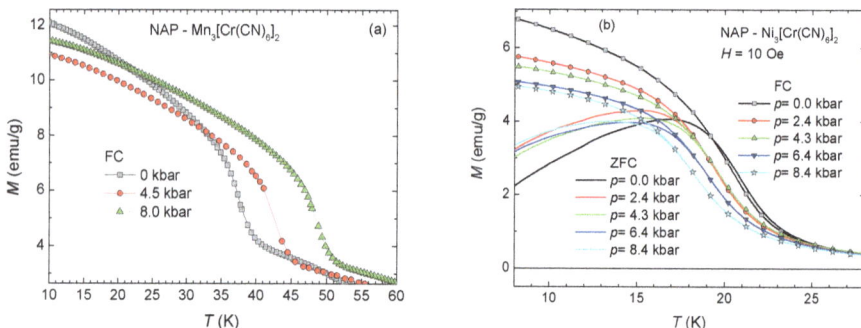

Figure 19. Magnetization measurements M(T) under pressure for: (**a**) Mn3[Cr(CN)6]2-NAP; (**b**) Ni3[Cr(CN)6]2-NAP.

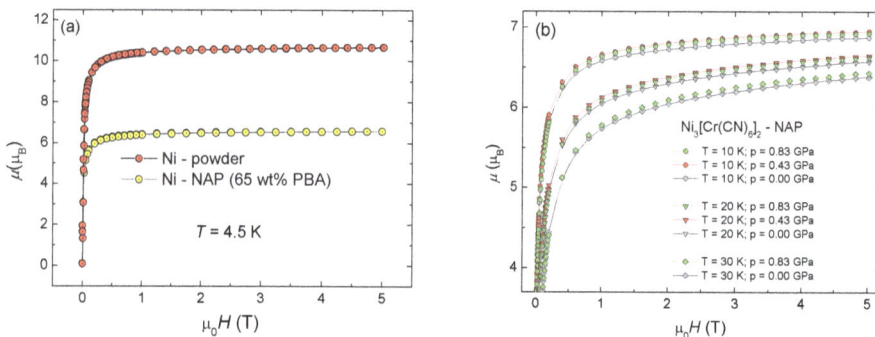

Figure 20. The field dependence of magnetization: (**a**) comparison of mother Ni-PBs and Ni-NAP; (**b**) the pressure effect on magnetization of Ni-NAP system at different temperatures.

3.6. Cr^{III}-PBs Core–Shell Heterostructures under Pressure

Magnetic studies on the PBs with chemical formula $Li_xCu_y[Fe(CN)_6]_z \cdot zH_2O$ (LiCuFe–PBs) and $Li_jNi_k[Cr(CN)_6]_l \cdot zH_2O$ (LiNiCr–PBs), as well as LiCuFe@LiNiCr–PBs core–shell heterostructures, have been conducted under pressures ranging from ambient to ≈ 1.4 GPa and at temperatures of 2–90 K. The results for the single component CuFe–PBs indicate robust magnetic properties under the range of pressures studied, where a $T_C = 20$ K was observed at all pressures [60]. Pressure studies of single component NiCr–PBs are consistent with previously published results by other workers below 1.0 GPa. However, at pressures above 1.0 GPa, the decrease in magnetization is accompanied by a decrease in the T_C, an indication of changes in the superexchange value. The results obtained with the single component samples can be mapped onto the observations of the heterostructures [60].

Cubic core@shell particles of Prussian blue analogues, composed of a shell (≈ 80 nm thick) of ferromagnetic $K_{0.3}Ni[Cr(CN)_6]_{0.8} \cdot 1.3H_2O$ **(A)**, $T_C \approx 70$ K, surrounding a bulk (≈ 350 nm) core of photoactive ferrimagnetic $Rb_{0.4}Co[Fe(CN)_6]_{0.8} \cdot 1.2H_2O$ **(B)**, $T_C \approx 20$ K, have been studied [61]. Below nominally 70 K, these CoFe@NiCr samples exhibit a persistent photo induced decrease in low-field magnetization, and these results resemble data from other BA core@shell particles and analogous ABA heterostructured films [44]. This net decrease suggests that the photo induced lattice expansion in the B layer generates a strain-induced decrease in the magnetization of the NiCr layer, similar to a pressure-induced decrease observed in a similar, pure NiCr material and in CoFe@NiCr cubes [10]. Upon further examination, the data also reveal a significant portion of the NiCr shell whose magnetic superexchange, J, is perturbed by the photo induced strain from the CoFe constituent [61].

The photo and thermal responses of the magnetism of CoFe@CrCr-PBs core@shell nanoparticles were investigated down to 5 K and up to 0.5 GPa in 100 G in a home-made, anvil pressure-cell, which was adapted to a probe suitable for use with a commercial magnetometer. The effect of pressure on the Curie temperatures of the CoFe-PBs core (≈ 25 K) and the CrCr-PBs shell (≈ 200 K) together with a change of the relaxation temperature of the photo-CTIST (charge-transfer-induced spin-transition) of the CoFe-PBs core (≈ 125 K) were alike to the operation already presented for the single-phase materials. Specifically, although the magnetic ordering temperature of the CrCr-PBs shell shifts to higher temperatures, the relaxation temperature of the photo-CTIST of the CoFe-PBs core goes down to lower temperatures when the pressure was enlarged, thereby lowering the temperature range in which the CrCr-PBs component is photo switchable [62].

4. Conclusions

The isotropic contract of the Mn-Cr-CN-PBs lattice with the linear reduction of the volume was observed for $p \leq 1.7$ GPa. The additional enlargement of the pressure leads to almost no decrease of the volume. The reduction of the crystal lattice size saturates at about 3.2 GPa reaching a volume decrease of approximately 9%. The evidence of an amorphous structure was obtained at above 4.2 GPa. The pressure does not induce any linkage isomerisation. The pressure induced reduction of the lattice size is completely reversible and the original crystal structure recovered if the pressure 4.5 GPa was released, which indicates that the framework extended by Cr-CN-Mn linkages had sufficient elasticity. The effect of pressure on magnetic properties follows the non-monotonous pressure dependence of crystal lattice. At pressures above 3.2 GPa the magnetic transition becomes much broader and T_C is suppressed to zero near 10 GPa, with a corresponding collapse of the ferrimagnetic moment. This is consistent with a pressure-induced amorphous structure of the sample, as was observed in a similar PBs.

On the other hand, Ni-Cr-CN Prussian Blue analogue has a quite different response to pressure. The volume of elementary cell narrows by linear way and isotropic in the range of 0–3.1 GPa, reaching compression of about 15% at 4.7 GPa, which is accompanied by broadening of the peaks, but the crystal character is unaffected. The applied pressure-sensitive shoulder in the Raman spectra can point to the linkage isomerisation, indicating that $Cr^{III}–N{\equiv}C–Ni^{II}$ fragments grow at the expense of the original

Cr^{III}–N≡C–Ni^{II} units. However, the effect can be superimposed also by other contributions, such as the presence of surface non-bridging cyanides or shift of the coordinated water.

The comprehensive study of magnetic properties under hydrostatic pressure on PBs revealed principally different behaviour in Mn^{2+}-Cr^{III}-PBs and Cr^{2+}-Cr^{III}-PBs, with prevailing antiferromagnetic super-exchange interaction J_{AF} compared to Ni^{2+}-Cr^{III}-PBs and Co^{2+}-Cr^{III}-PBs with prevailing ferromagnetic interaction J_F. The strengthened magnetic coupling determined by enhanced values of the monoelectronic overlap integral S and energy gap Δ between the mixed molecular orbitals under hydrostatic pressure leads to the enlargement of J_{AF} and the Curie temperature T_C rises. On the contrary, J_F does not change with pressure. In this case, no overlap of magnetic orbitals is present, and that is why the elastic contraction of crystal lattice does not change the overlap. The hydrostatic pressure slightly changes tilting of $Cr^{III}C_6$ and $TM^{2-}$$N4(OH)_2$ octants, which leads to a small change of the bonding angle and reorientation of magnetic moments. Both effects reduce T_C and saturated magnetic moment. This small reduction of basic magnetic characteristics was observed in samples with prevailing J_F, but is present even in samples with dominant J_{AF}; in this case, the effect is masked by a large increase of J_{AF} due to the enhanced overlapping of magnetic orbitals.

The pressure affects magnetization processes in systems with prevailing J_{AF} coupling as well as with prevailing J_F interaction, but we expect that pressure does not change the spontaneous magnetization in Mn^{2+}-Cr^{III}-PBs, Ni^{2+}-Cr^{III}-PBs, and Co^{2+}-Cr^{III}-PBs. Pressure-induced magnetic hardening we attribute to variations of magneto-crystalline anisotropy. The enlargement of the saturated magnetization μ_s of Mn^{2+}-Cr^{III}-PBs near T_C was associated with the enhancement of magnetic interactions. The compensation temperature T_{comp} appears because the algebraic sum of $\mu(T)$ for Mn^{2+}–N≡C–Cr^{III} subsystem and Cr^{III}–C≡N–Ni^{2+} subsystem is zero at T_{comp}. Pressure affects $\mu(T)$ by different ways in these two subsystems, and as a consequence T_{comp} decreases when the pressure is applied. The applied pressure reduces saturated magnetization of Cr^{2+}-Cr^{III}-PBs. We can assume for cation Cr^{2+} two possibilities: at first Cr^{2+} is high spin, or the second possibility when Cr^{2+} is low spin. It seems that only approximately 30% of Cr^{2+} is high spin. The applied pressure $p = 0.84$ GPa induces high spin–low spin transition of cca 4.5% of high spin Cr^{2+}.

The pressure effect on magnetic properties of PB nano-powders and core–shell heterostructures follows tendencies known from bulk parent PBs. We expect that both systems Ni-NAP and Mn-NAP behave as systems of strongly interacting magnetic particles. The super-exchange interaction is dominant intra-NAP magnetic interaction. The dipole-dipole interaction is dominant inter-NAP magnetic interaction. The dispersion of NAP into an organic matrix leads to a dilution of the mother PBs and consequently the Curie temperature is reduced. The applied pressure strengthens magnetic super-exchange interaction of Mn-NAP with dominant J_{AF} and increases T_C and pressure decreases T_C for system of Ni-NAP with dominant J_F, which is mainly attributed to less effective magnetic coupling due to the reduction in the bonding angle. The applied pressure increases magnetic moment μ_s. This behaviour is opposite to the behaviour of the mother PBs and is attributed to the reduction of inter-NAP distances by compression of the organic matrix.

A persistent photo-induced reduction in low-field magnetization is a characteristic feature of CoFe@NiCr artificial structures resembling already-known results from PBs core@shell particles and analogical heterostructured coatings. The plain reduction in low-field magnetization for CoFe@NiCr heterostructures indicates that the photo induced lattice expansion resembles a pressure-induced reduction observed on related net NiCr material. The pressure effect on the Curie temperature of the CoFe-PBs core and the CrCr-PBs shell together with a change of the relaxation temperature of the photo-CTIST of the CoFe-PBs core resembles the behaviors known from the single-phase materials. Despite the Curie temperature of the CrCr-PBs shell shifting to higher temperatures, the relaxation temperature of the photo-CTIST of the CoFe-PBs core decreases to lower temperatures when the pressure rises, thereby lowering the temperature region in which the CrCr-PBs part is photo switchable.

Funding: This research received no external funding.

Conflicts of Interest: The authors declare no conflict of interest.

References

1. Verdaguer, M.; Girolami, G. Magnetic Prussian Blue Analogs. In *Magnetism: Molecules to Materials V*; Miller, J.S., Drilon, M., Eds.; Wiley–VCH & KGaA: Weinheim, Germany, 2004; pp. 283–386. ISBN 3-527-30665-X.
2. Ma, F.; Li, Q.; Wang, T.; Zhang, H.; Wu, G. Energy storage materials derived from Prussian blue analogues. *Sci. Bull.* **2017**, *62*, 358–368. [CrossRef]
3. Wang, B.; Han, Y.; Wang, X.; Bahlawane, N.; Pan, H.; Yan, M.; Jiang, Y. Prussian Blue Analogs for Rechargeable Batteries. *Science* **2018**, *3*, 110–133. [CrossRef] [PubMed]
4. Krap, C.P.; Balmaseda, J.; Zamora, B.; Reguera, E. Hydrogen storage in the iron series of porous Prussian blue analogues. *Int. J. Hydrog. Energy* **2010**, *35*, 10381–10386. [CrossRef]
5. Kaye, S.S.; Long, J.R. The role of vacancies in the hydrogen storage properties of Prussian blue analogues. *Catal. Today* **2007**, *120*, 311–316. [CrossRef]
6. Kumar, A.; Kanagare, A.B.; Banerjee, S.; Kumar, P.; Kumar, M.; Jagannath; Sudarsan, V. Synthesis of cobalt hexacyanoferrate nanoparticles and its hydrogen storage properties. *Int. J. Hydrog. Energy* **2018**, *43*, 7998–8006. [CrossRef]
7. Thevenot, D.R.; Toth, K.; Durst, R.A.; Wilson, G.S. Electrochemical biosensors: Recommended definitions and classifications. *Biosens. Bioelectron.* **2001**, *16*, 121–131. [CrossRef] [PubMed]
8. Salazar, P.; Martin, M.; O'Neill, R.D.; Gonzales-Mora, J.L. In vivo biosensor based on Prussian blue for brain chemistry monitoring > Methodological Review and biological applications. In *In Vivo Neuropharmacology and Neurophysiology, Neuromethods*; Springer Science + Business Media: New York, NY, USA, 2017; Volume 121, pp. 155–179. [CrossRef]
9. Sato, O.; Iyoda, T.; Fujishima, A.; Hashimoto, K. Photoinduced Magnetization of a Cobalt-Iron Cyanide. *Science* **1996**, *272*, 704–705. [CrossRef]
10. Bleuzen, A.; Cafun, J.D.; Bachschmidt, A.; Verdaguer, M.; Munsch, P.; Baudelet, F.; Itie, J.P. CoFe Prussian blue analogues under variable pressure. Evidence of departure from cubic symmetry: X-ray diffraction and absorbtion study. *J. Phys. Chem. C* **2008**, *112*, 17709–17715. [CrossRef]
11. Kahn, O. Molecular Bistability and Information Storage. In Proceedings of the Twelfth Annual International Conference of the IEEE in Engineering in Medicine and Biology Society, Philadelphia, PA, USA, 1–4 November 1990; pp. 1683–1685. [CrossRef]
12. Ksenofontov, V.; Levchenko, G.; Reiman, S.; Gutlich, P.; Bleuzen, A.; Escax, V.; Verdaguer, M. Pressure induced electron transfer in ferromagnetic Prussian blue analogues. *Phys. Rev. B* **2003**, *68*, 024415. [CrossRef]
13. Levchenko, G.G.; Berezhnaya, L.V.; Filimonov, G.G.; Han, W. Charge Transfer, Change the Spin Value and Driving of Magnetic Order by Pressure in Bimetallic Molecular Complexes. *J. Phys. Chem. B* **2018**, *122*, 6846–6853. [CrossRef]
14. Sugimoto, M.; Yamashita, S.; Akutsu, H.; Nakazawa, Y.; DaSilva, J.G.; Kareis, C.M.; Miller, J.S. Increase in the Magnetic Ordering Temperature (T_C) as a Function of the Applied Pressure for $A_2Mn[Mn(CN)_6]$ (A = K, Rb, Cs) Prussian Blue Analogues. *Inorg. Chem.* **2017**, *56*, 10452–10457. [CrossRef] [PubMed]
15. Coronado, E.; Gimenez-Lopez, M.C.; Levchenko, G.; Romero, M.F.; Garcia-Baonza, V.; Milner, A.; Paz-Pasternak, M. Pressure tuning of magnetism and linkage isomerism in iron(II) hexacyanochromate. *J. Am. Chem. Soc.* **2005**, *127*, 4580–4581. [CrossRef] [PubMed]
16. Egan, L.; Kamenev, K.; Papanikolau, D.; Takabayashi, Y.; Margadonna, S. Pressure-induced sequential magnetic pole inversion and antiferromagnetic/ferromagnetic crossover in a trimetallic Prussian blue analogue. *J. Am. Chem. Soc.* **2006**, *128*, 6034–6035. [CrossRef] [PubMed]
17. Papnikolau, D.; Kosaka, W.; Margadonna, S.; Kagi, H.; Ohkoshi, S.-I.; Prassides, K. Piezomagnetic behaviour of the spin crossover Prussian blue analogue $CsFe[Cr(CN)_6]$. *J. Phys. Chem. C* **2007**, *111*, 8086–8091. [CrossRef]
18. Levchenko, G.G.; Khristov, A.V.; Varyukhin, V.N. Spin crossover in iron(II)-containing complex compounds under a pressure. *Low Temp. Phys.* **2014**, *40*, 571–585. [CrossRef]
19. Frisch, J.L. Notitia Caerulei Berolinensis Nuper Inventi. *Miscellanea Berolinensia ad incrementum Scientiorum* **1710**, *1*, 377–378.

20. Kraft, A. On the discovery and history of Prussian blue. *Bull. Hist. Chem.* **2008**, *33*, 61–67.

21. Keggin, J.F.; Miles, F.D. Structures and formulae of the Prussian blues and related compounds. *Nature* **1936**, *137*, 577–578. [CrossRef]

22. Buser, H.J.; Schwarzenbach, D.; Petter, W.; Ludi, A. The crystal structure of Prussian blue: $Fe_4[Fe(CN)_6]_3 \cdot xH_2O$. *Inorg. Chem.* **1977**, *16*, 2704–2710. [CrossRef]

23. Herren, F.; Fischer, P.; Ludi, A.; Hälg, W. Neutron diffraction study of Prussian blue $Fe_4[Fe(CN)_6]_3 \cdot xH_2O$. Location of water molecules and long/range magnetic order. *Inorg. Chem.* **1980**, *19*, 956–959. [CrossRef]

24. Bozorth, R.M.; Williams, H.J.; Walsh, D.E. Magnetic Properties of Some Orthoferrites and Cyanides at Low Temperatures. *Phys. Rev.* **1956**, *103*, 572–578. [CrossRef]

25. Babel, D. Magnetism and Structure: Model Studies on Transition Metal Fluorides and Cyanides. *Comments Inorg. Chem.* **1986**, *5*, 285–320. [CrossRef]

26. Goodenough, J.B. Theory of the Role of Covalence in the Perovskite-Type Manganites [La, M(II)]MnO_3. *Phys. Rev.* **1955**, *100*, 564–573. [CrossRef]

27. Kanamori, J. Superexchange interaction and symmetry properties of electron orbitals. *J. Phys. Chem. Solids* **1959**, *10*, 87. [CrossRef]

28. Anderson, P.W. New approach to the theory of superexchange interactions. *Phys. Rev.* **1959**, *115*. [CrossRef]

29. Zentková, M.; Arnold, Z.; Kamarád, J.; Kavečanský, V.; Lukáčová, M.; Maťaš, S.; Mihalik, M.; Mitróoová, Z.; Zentko, A. Effect of pressure on magnetic properties of $TM_3[Cr(CN)_6]_2 \cdot 12H_2O$. *J. Phys.: Condens. Matter* **2007**, *19*, 266217. [CrossRef] [PubMed]

30. Mihalik, M.; Kavečanský, V.; Maťaš, S.; Zentková, M.; Prokhenko, O.; André, G. Magnetic properties and neutron diffraction study of $(Ni_xMn_{1-x})_3[Cr(CN)_6]_2$ molecule-based magnets. *J. Phys.: Conf. Ser.* **2010**, *200*, 022035. [CrossRef]

31. Zentkova, M.; Mihalik, M.; Kovac, J.; Zentko, A.; Mitroova, Z.; Lukacova, M.; Kavecansky, V.; Kiss, L.F. Magnetic properties of $TM_3[Cr(CN)_6]_2 \cdot nH_2O$. *Phys. Stat. Sol. (b)* **2006**, *243*, 272–276. [CrossRef]

32. Mitróová, Z.; Maťaš, S.; Mihalik, M.; Zentková, M.; Arnold, Z.; Kamarád, J. Effect of Pressure on Magnetic Properties of Hexacyanochromates. *Acta Phys. Polonica A* **2008**, *113*, 469–472. [CrossRef]

33. Lukacova, M.; Kiss, L.F.; Marysko, M.; Mihalik, M.; Mitroova, Z.; Stopka, P.; Zentko, A.; Zentkova, M. New magnetic phenomena in vanadium hexacyanochromates. *Phys. Stat. Sol. (a) Appl. Res.* **2003**, *196*, 240–243. [CrossRef]

34. Zentková, M.; Mihalik, M.; Arnold, Z.; Kamarád, J. Effect of pressure on magnetic properties of mixed ferro—Ferrimagnet $(Ni_{0.38}Mn_{0.62})_3[Cr(CN)_6]_2 \cdot zH_2O$. *J. Phys. Conf. Ser.* **2010**, *200*, 022074. [CrossRef]

35. Bokor, M.; Tompa, K.; Kiss, L.F.; Zentková, M.; Zentko, A.; Mihalik, M.; Maťaš, S.; Mitróová, Z. ^1H NMR on $(Ni_x Mn_{1-x})_3 [Cr(CN)_6]_2 \cdot nH_2 O$. *Acta Phys. Polonica A* **2008**, *113*, 485–488. [CrossRef]

36. Vavra, M.; Antoňák, M.; Jagličic, Z.; Mihalik, M.; Mihalik, M., Jr.; Csach, K.; Zentková, M. Magnetic Properties of $(Cu_xMn_{1-x})_3[Cr(CN)_6]_2 \cdot zH_2O$ Complexes. *Acta Phys. Polonica A* **2010**, *118*, 998–999. [CrossRef]

37. Zentko, A.; Kavečanský, V.; Mihalik, M.; Maťaš, S.; Mitróová, Z.; Zentková, M.; Maryško, M.; Jagličić, Z. Magnetic Relaxation and Memory Effect in Nickel-Chromium Cyanide Nanoparticles. *Acta Phys. Polonica A* **2008**, *113*, 511–514. [CrossRef]

38. Zentko, A.; Zentková, M.; Kavečanský, V.; Mihalik, M.; Mitróová, Z.; Arnold, Z.; Kamarád, J.; Cieslar, M.; Zeleňák, V. Effect of Pressure on Magnetic Properties of $TM_3[Cr(CN)_6]_2 \cdot nH_2O$ Nanoparticles. *Acta Phys. Polonica A* **2008**, *113*, 489–492. [CrossRef]

39. Vavra, M.; Hrabčák, P.; Zentková, M.; Mihalik, M.; Mihalik, M., Jr.; Csach, K. The Effect of Pressure on Magnetic Properties of $KMnCr(CN)_6$. *EPJ Web Conf.* **2013**, *40*, 14001. [CrossRef]

40. Zentková, M.; Vavra, M.; Mihalik, M.; Mihalik, M., Jr.; Lazurová, J.; Arnold, Z.; Kamarád, J.; Kamenev, K.; Míšek, M. Raman spectroscopy and magnetic properties of $KMCr(CN)_6$ under pressure. *High Press. Res.* **2015**, *35*, 22–27. [CrossRef]

41. Maeda, T.; Mito, M.; Deguchi, H.; Takagi, S.; Kaneko, W.; Ohba, M.; Okawa, H. Pressure effects on a dimetallic ferrimagnet [Mn(en)]$_3[Cr(CN)_6]_2 \cdot 4H_2O$. *Polyhedron* **2005**, *24*, 2497–2500. [CrossRef]

42. Ohba, M.; Kaneko, W.; Kitagawa, S.; Maeda, T.; Mito, M. Pressure Response of Three-Dimensional Cyanide-Bridged Bimetallic Magnets. *J. Am. Chem. Soc.* **2008**, *130*, 4475–4484. [CrossRef]

43. Giriat, G.; Wang, W.; Attfield, J.P.; Huxley, A.D.; Kamenev, K.V. Turnbuckle diamond anvil cell for high-pressure measurements in a superconducting quantum interference device magnetometer. *Rev. Sci. Instrum.* **2010**, *81*, 073905–073910. [CrossRef]

44. Pajerowski, D.M.; Conklin, S.E.; Leao, J.; Harriger, L.W.; Phelan, D. High-pressure neutron scattering of the magnetoelastic Ni-Cr Prussian blue analog. *Phys. Rev. B* **2015**, *91*, 094104–094110. [CrossRef]

45. Coronado, E.; Gimenez-Lopez, M.; Korzeniak, T.; Levchenko, G.; Romero, F.; Segura, A.; Garcia-Baonza, V.; Cezar, J.; de Groot, F.M.F.; Milner, A.; et al. Pressure-Induced Magnetic Switching and Linkage Isomerism in $K_{0.4}Fe_4[Cr(CN)_6]_{2.8} \cdot 16H_2O$: X-ray Absorption and Magnetic Circular Dichroism Studies. *J. Am. Chem. Soc.* **2007**, *130*, 15519–15532. [CrossRef] [PubMed]

46. Awaga, K.; Sekine, T.; Okawa, M.; Fujita, W.; Holmes, S.M.; Girolami, G.S. High-pressure effects on a manganese hexacyanomanganate ferrimagnet with $T_N = 29K$. *Chem. Phys. Lett.* **1998**, *293*, 352–356. [CrossRef]

47. Kavečanský, V.; Mihalik, M.; Lukáčová, M.; Mitróová, Z.; Maťaš, S. Neutron Diffraction Study of Crystal and Magnetic Structure of $Dy[Fe(CN)_6] \cdot 4D_2O$. *J. Phys. Suppl. D* **2004**, *54*, D571–D574. [CrossRef]

48. Matas, S.; Kavecansky, V.; Lukacova, M.; Mihalik, M.; Mitroova, Z.; Zentkova, M. The symmetry analysis and magnetic model of $Dy[Fe(CN)_6] \cdot 4D_2O$. *J. Alloy. Compd.* **2008**, *459*, 526–530. [CrossRef]

49. Ohkoshi, S.; Iyoda, T.; Fujishima, A.; Hashimoto, K. Magnetic properties of mixed ferro-ferrimagnets composed of Prussian blue analogs. *Phys. Rev. B* **1997**, *56*, 11642–11652. [CrossRef]

50. Kahn, O. The magnetic turnabout. *Nature* **1999**, *399*, 21–23. [CrossRef]

51. Néel, M.L. Propriétés magnétiques des ferrites; ferrimagnétisme et antiferromagnétisme. *Ann. Phys.* **1948**, *12*, 137–198. [CrossRef]

52. Middlemiss, D.S.; Lawton, L.M.; Wilson, C.C. A solid-state hybrid density functional theory study of Prussian blue analogues and related chlorides at pressure. *J. Phys. Condens. Matter* **2008**, *20*, 335231. [CrossRef]

53. Thomas, L.; Lionti, F.; Ballou, R.; Gatteschi, D.; Sessoli, R.; Barbara, B. Macroscopic quantum tunnelling of magnetization in a single crystal of nanomagnets. *Nature* **1996**, *383*, 145–147. [CrossRef]

54. Lerunberger, M.N.; Loss, D. Quantum computing in molecular magnets. *Nature* **2001**, *410*, 789–793. [CrossRef] [PubMed]

55. Larionova, J.; Gross, M.; Pilkington, M.; Andres, H.; Stoeckli-Evans, H.; Güdel, H.U.; Decurtins, S. High-Spin Molecules: A Novel Cyano-Bridged MnMo Molecular Cluster with a S = 51/2 Ground State and Ferromagnetic Intercluster Ordering at Low Temperatures. *Angew. Chem. Int. Ed* **2000**, *39*, 1605–1609. [CrossRef]

56. Pileni, M.P. Nanocrystal Self-Assemblies: Fabrication and Collective Properties. *J. Phys. Chem. B* **2001**, *105*, 3358–3371. [CrossRef]

57. Vaucher, S.; Li, M.; Mann, S. Synthesis of Prussian Blue Nanoparticles and Nanocrystal Superlattices in Reverse Microemulsions. *Angew. Chem. Int. Ed.* **2000**, *39*, 1793–1796. [CrossRef]

58. Catala, L.; Gacoin, T.; Boilot, J.-P.; Riviere, E.; Paulsen, C.; Lhotel, E.; Mallah, T. Cyanide-Bridged CrIII–NiII Superparamagnetic Nanoparticles. *Adv. Mater.* **2003**, *15*, 826–829. [CrossRef]

59. Catala, L.; Gloter, A.; Stephan, O.; Rogez, G.; Mallah, T. Superparamagnetic bimetallic cyanide-bridged coordination nanoparticles with TB = 9 K. *Chem. Commun.* **2006**, 1018–1020. [CrossRef]

60. Peprah, M.K.; Li, C.H.; Talham, D.R.; Meisel, M.W. Effect of pressure on the magnetic properties of LiCuFe and LiCuFe@LiNiCr Prussian blue analogues. *Polyhedron* **2013**, *66*, 264–267. [CrossRef]

61. Knowles, E.S.; Li, C.H.; Dumont, M.F.; Peprah, M.K.; Andrus, M.J.; Talham, D.R.; Meisel, M.W. Photoinduced perturbations of the magnetic superexchange in core@shell Prussian blue analogues. *Polyhedron* **2013**, *66*, 153–156. [CrossRef]

62. Peprah, M.K.; VanGennep, D.; Quintero, P.A.; Risset, O.N.; Brinzari, T.V.; Li, C.H.; Dumont, M.F.; Xia, J.-S.; Hamlin, J.J.; Talham, D.R.; et al. Pressure-tuning of the photomagnetic response of heterostructured CoFe@CrCr-PBA core@shell nanoparticles. *Polyhedron* **2017**, *123*, 323–327. [CrossRef]

crystals

MDPI

Review

Multifunctional Molecular Magnets: Magnetocaloric Effect in Octacyanometallates

Magdalena Fitta *[ID], Robert Pełka [ID], Piotr Konieczny [ID] and Maria Bałanda [ID]

Institute of Nuclear Physics Polish Academy of Sciences, Radzikowskiego 152, 31-342 Krakow, Poland;
robert.pelka@ifj.edu.pl (R.P.); piotr.konieczny@ifj.edu.pl (P.K.); maria.balanda@ifj.edu.pl (M.B.)
* Correspondence magdalena.fitta@ifj.edu.pl; Tel.: +48-12-662-8374

Received: 23 November 2018; Accepted: 18 December 2018; Published: 22 December 2018

Abstract: Octacyanometallate-based compounds displaying a rich pallet of interesting physical and chemical properties, are key materials in the field of molecular magnetism. The $[M(CN)_8]^{n-}$ complexes, ($M = W^V$, Mo^V, Nb^{IV}), are universal building blocks as they lead to various spatial structures, depending on the surrounding ligands and the choice of the metal ion. One of the functionalities of the octacyanometallate-based coordination polymers or clusters is the magnetocaloric effect (MCE), consisting in a change of the material temperature upon the application of a magnetic field. In this review, we focus on different approaches to MCE investigation. We present examples of magnetic entropy change ΔS_m and adiabatic temperature change ΔT_{ad}, determined using calorimetric measurements supplemented with the algebraic extrapolation of the data down to 0 K. At the field change of 5T, the compound built of high spin clusters $Ni_9[W(CN)_8]_6$ showed a maximum value of $-\Delta S_m$ equal to 18.38 J·K^{-1} mol^{-1} at 4.3 K, while the corresponding maximum $\Delta T_{ad} = 4.6$ K was attained at 2.2 K. These values revealed that this molecular material may be treated as a possible candidate for cryogenic magnetic cooling. Values obtained for ferrimagnetic polymers at temperatures close to their magnetic ordering temperatures, T_c, were lower, i.e., $-\Delta S_m = 6.83$ J·K^{-1} mol^{-1} ($\Delta T_{ad} = 1.42$ K) and $-\Delta S_m = 4.9$ J·K^{-1} mol^{-1} ($\Delta T_{ad} = 2$ K) for $\{[Mn^{II}(pyrazole)_4]_2[Nb^{IV}(CN)_8]\cdot 4H_2O\}_n$ and $\{[Fe^{II}(pyrazole)_4]_2[Nb^{IV}(CN)_8]\cdot 4H_2O\}_n$, respectively. MCE results have been obtained also for other -$[Nb(CN)_8]$-based manganese polymers, showing significant T_c dependence on pressure or the remarkable magnetic sponge behaviour. Using the data obtained for compounds with different T_c, due to dissimilar ligands or other phase of the material, the $\Delta S_m \sim T_c^{-2/3}$ relation stemming from the molecular field theory was confirmed. The characteristic index n in the $\Delta S_m \sim \Delta H^n$ dependence, and the critical exponents, related to n, were determined, pointing to the 3D Heisenberg model as the most adequate for the description of these particular compounds. At last, results of the rotating magnetocaloric effect (RMCE), which is a new technique efficient in the case of layered magnetic systems, are presented. Data have been obtained and discussed for single crystals of two 2D molecular magnets: ferrimagnetic $\{[Mn^{II}(R-mpm)_2]_2[Nb^{IV}(CN)_8]\}\cdot 4H_2O$ (mpm = α-methyl-2-pyridinemethanol) and a strongly anisotropic (tetren)Cu$_4$[W(CN)$_8$]$_4$ bilayered magnet showing the topological Berezinskii-Kosterlitz-Thouless transition.

Keywords: molecular magnets; magnetocaloric effect; octacyanometallates; critical behaviour; coordination polymers

---

1. Introduction

In the quest for novel materials which could be used in modern technologies, molecular substances play an important role, as they may show properties not available in conventional materials. Molecule-based compounds attract much attention as they combine interesting magnetic, electronic, and optical properties together with low weight, transparency, and chemical sensitivity. Molecular magnets represent a vast and still growing family of compounds based on molecular buildings blocks, where organic groups mediate magnetic interactions between metal ions or may also carry their own magnetic moment. Research into molecule-based materials is motivated by their potential functionality due to the properties of the molecular building blocks or the specific character of the resulting molecular network. Thanks to the rational design and advanced chemical syntheses, it is possible to obtain systems with different magnetic dimensionalities: molecular ferro-, ferri-, or antiferromagnets with the substantial ordering temperature T_c, magnetic molecular layers, magnetic molecular chains, or magnetic molecular clusters, the latter two regarded as molecular nanomagnets. In many cases, properties of the systems, like T_c, coercive field, magnetic moment, and colour, may be changed and even controlled with different factors, such as temperature, irradiation with light, external pressure, or sorption of guest molecules. Molecular magnets which show spin crossover transition, photomagnetism, magnetic sponge behaviour, or optical activity are potential candidates for efficient sensors and switches. On the other hand, superparamagnetic character, slow relaxation, and quantum tunneling in anisotropic high-spin clusters, may be used in high-density magnetic storage, spintronics, or quantum computing. Results of extensive studies on molecular magnetism have been reviewed in books and monographies [1–4].

The relatively later explored functionality of molecular magnets has been the magnetocaloric effect (MCE), consisting in change of the material temperature when a magnetic field is applied or removed. MCE is an intrinsic thermodynamic property of all magnetic solids: it occurs in paramagnets, ferro-, and ferrimagnets [5] but also in antiferromagnets at the metamagnetic transition [6]. Investigations of MCE are significant also for the basic reason that its dependence on the magnetic field is related to the critical behaviour of the material. The microscopic description of MCE has been presented in [7–10]. Of particular interest are "magnetic coolers", i.e., substances for which an adiabatic demagnetization provokes the substantial temperature decrease. The value of the effect depends on the temperature derivative of magnetization, thus for paramagnets it is strongest at the lowest temperatures, while for ordered magnets it achieves maximum at the magnetic ordering temperature T_c. The aim is to replace standard, non-ecological refrigeration techniques, studies on magnetocaloric effect in conventional magnetic solids are concentrated on cooling in the room temperature range.

The most suitable candidates for magnetic refrigeration are here systems like $Gd_5Si_2Ge_2$, $Tb_5Si_2Ge_2$, MnAs, or Ni-Mn-In Heusler alloys [11]. Materials with first order magnetic transitions at T_c may be considered as most suitable for large MCE but thermal and magnetic losses appearing in transitions of this type may impede the practical applications. Consequently, materials showing the second order magnetic phase transition are also taken into account. Interest in the magnetocaloric effect in molecular materials and "chilling with magnetic molecules" [12] started about a decade ago. At first, it has been investigated for slowly relaxing molecular nanomagnets with large spin, Single Molecule Magnets (SMM) [2]. Because of the large ground-state spin S of SMM, the entropy associated with the magnetic degrees of freedom, $S_m = R\ln(2S+1)$ (R is the gas constant) should be sizeable. However, molecular anisotropy, which is essential in SMMs as it decides on long relaxation time and blocking temperature, delays magnetization and demagnetization of the system, thus brings about weak MCE. As stated by Evangelisti and Brechin, the ideal molecular refrigerant should have large spin, negligible magnetic anisotropy, prevailing ferromagnetic coupling, and large magnetic density [13]. These conditions are fulfilled in metal–organic frameworks with densely packed magnetic ions, like [Gd(HCOO)$_3$] [14]. Large values of MCE observed in frameworks are not related to transition at T_c as these systems remain paramagnetic or show the long-range order only in the sub-Kelvin range. MCE originates here from the Schottky anomaly consequent on splitting of the energy levels in the field. Very large MCE values have been also observed in the ferromagnetic acetate tetrahydrate Gd^{3+} dimer [15], in the high nuclearity $Gd_{42}Co_{10}$ cluster [16], and in a 24-Gd capsule-like cluster at temperature 2.5 K [17]. An interesting increase in the magnetocaloric effect in Mn^{II} glycolate on transition between the three-dimensional coordination polymer and discrete mononuclear phase induced by water molecules was reported in [18]. It follows that MCE in molecular clusters will play an important role in cooling in the sub-Kelvin temperature range. Advances in the design of magnetic molecules for use as cryogenic magnetic coolants were reviewed in [19].

This paper presents investigations on the magnetocaloric effect performed by us for molecular magnets based on octacyanidometallates. As it has been known from long-lasting studies of Prussian blue analogs, $M_A^{II}[M_B^{III}(CN)_6]_{2/3}\cdot zH_2O$ or $A^I M_A^{II}[M_B^{III}(CN)_6]$ (M_A^{II} and M_B^{III} are 3d metal ions and A^I is the alkali ion), as well as of other hexacyanidometallates, the cyanobridge is able to mediate strong antiferromagnetic or ferromagnetic interactions between the metal moments [20]. Besides high temperatures of magnetic order reaching room temperature, Prussian blue analogs (PBAs) may show many fascinating properties, like light- and pressure switchable magnetism, magneto-optical effects, or chemically controlled growth of nanosized systems [21]. Unlike the hexacyano- $[M(CN)_6]$ blocks of octahedral symmetry, which set up the cubic structure of PBAs, bimetallic octacyanometallates are based on more flexible building blocks offering eight CN-bridges to link the M and M′ metal ions. Most often used $[M(CN)_8]$ blocks are 5d $[W^V(CN)_8]$ or 4d $[Nb^{IV}(CN)_8]$ and $[Mo^V(CN)_8]$ complexes, all of spin $S = 1/2$. Eight possible coordination sites, the proper choice of other metal M′ and of additional organic ligands resulted in a variety of obtained structures and magnetic properties [22–24]. Among octacyanometallates one can find slowly relaxing systems [25–27], guest-molecule absorptive porous networks [28,29], photomagnets [30,31], and magneto-optically active compounds [32], as well as molecular sponges, which change, in a reversible way, the ordering temperature T_c and the coercive field H_c upon hydration/dehydration process [33] and other functional materials [34].

Below, we discuss the magneto-thermal properties of the octacyanometallates showing the different types of crystal architecture. The most numerous group is the family of 3D octacyanonobiate-based networks with different nonmagnetic organic ligands which significantly affect the structure and overall behaviour of the material. Another subject under study is the high-spin dodecanuclear cluster compound $Ni_9[W(CN)_8]_6$, a possible candidate for cryogenic magnetic cooling. Two experimental methods for measuring the MCE data, i.e., calorimetry and/or magnetometry were used. Moreover, the new approach, consisting in measuring MCE for two perpendicular sample orientations (so called Rotating Magnetocaloric Effect RMCE), is reported for a low anisotropy 2D $\{Mn^{II}(R\text{-mpm})_2]_2[Nb^{IV}(CN)_8]\}\cdot 4H_2O$ ferrimagnet, as well as for an anisotropic bilayered 2D $Cu_4[W(CN)_8]_4$ molecular crystal showing the topological phase transition. Conclusions regarding the scaling and critical behaviour in some systems under study are also included.

2. Deriving Magnetocaloric Effect from Calorimetric Data

2.1. Thermodynamic Setup

The magnetocaloric effect (MCE) is quantified either by the isothermal entropy change ΔS_m or the adiabatic temperature change ΔT_{ad} due to the external field change $H_i \rightarrow H_f$. While the former quantity may be derived from magnetometric data by using the integral version of the Maxwell relation $\partial S(T,H)/\partial H = \partial M(T,H)/\partial T$, the latter one can be obtained solely from calorimetric data. In this way, the heat capacity measurements performed without an external magnetic field as well as in a nonzero field represent a more complete set of characteristics allowing one to arrive at both MCE quantities. The step of pivotal importance in such a derivation is the determination of the temperature and field dependence of the entropy thermodynamic function $S(T,H)$. Then, the isothermal entropy change ΔS_m is obtained by a simple subtraction:

$$\Delta S_m(T, \Delta H = H_f - H_i) = S(T, H_f) - S(T, H_i) \tag{1}$$

while the calculation of the adiabatic temperature change ΔT_{ad} is based on the formula:

$$\Delta T_{ad}(T, \Delta H = H_f - H_i) = [T(S, H_f) - T(S, H_i)]_S \tag{2}$$

which requires the inversion of the $S(T,H)$ dependence with respect to variable T. However, the above procedure is not as straightforward as it might seem. The problem lies in the experimental limitations where we can measure the heat capacity $C_p(T,H)$ down to a possibly small but finite temperature T_L never reaching the limit of 0 K, while for the correct determination of the $S(T,H)$ function one needs to know the $C_p(T,H)$ function from zero absolute temperature: $S(T,H) = \int_0^T C_p(T',H)/T' dT'$. A natural way to solve the problem is using a plausible extrapolation scheme. In what follows we will present two such schemes employed in the investigations of molecular magnets. The first scheme is more complete and involves the determination of the so called baseline incorporating the lattice contribution to the heat capacity. Having the baseline, one can extract the magnetic excess heat capacity ΔC_p and extrapolate it down to 0 K by assuming a two parameter algebraic function $\Delta C_p(T,H) = A(H)T^{B(H)}$ which is believed (there are no theoretical accounts of that) to work well for the nonzero-field case as it is well-known to do in the zero-field case (the Bloch law for ferromagnets). The extrapolation scheme consists in two independent steps. Firstly, the algebraic function is fitted to the magnetic excess heat capacity data ΔC_p in the narrow temperature range $[T_L, T_{max}]$ for the different applied field values. Next, the resultant fits are complemented with the extrapolated baseline to yield the final form of the sought-for low temperature extrapolation of $C_p(T,H)$. The other scheme is a simplified one as it obviates the need to determine the normal heat capacity (baseline). However, it does require the extrapolation of the C_p data down to 0 K. The extrapolation is performed assuming again a two parameter function $C_{p,LT}(T,H) = A(H)T^3 + B(H)T^{3/2}$, where the first term corresponds to the lattice contribution, while the second term represents the Bloch law for the magnonic contribution.

The above schemes will be exemplified by three compounds belonging to the class of molecular magnets. The first compound (**1**) with the chemical formula $\{Ni[Ni(4,4'dtbpy)(H_2O)]_8[W(CN)_8]_6\}\cdot 17H_2O$ (4,4'dtby $-$ 4-4-di-tert-butyl-2,2'-bipyridine, $C_{20}H_{24}N_2O_4$) ($Ni_9[W(CN)_8]_6$ in short) consists of unique clusters. The main component of the compound is a fifteen-center cyanido-bridged $Ni(II)_9W(V)_6$ molecule forming a cube-like framework to which the spacious tert-butyl substituted bipyridine ligands are connected. The spin carried by the W(V) ion is $S_W = 1/2$, while that of the Ni(II) ion is $S_{Ni} = 1$. This allows one to calculate the maximal molar magnetic entropy of the cluster as equal to $S_{max,1} = R\ln(2S_{Ni}+1)^9(2S_W+1)^6 \approx 116.79$ J·K^{-1}·mol^{-1}, which makes one anticipate a considerable magnetocaloric response. However, the spins constituting the cluster are ferromagnetically coupled through the cyanide linkages giving rise to a relatively high spin $S = 12$ [35], which leads to the entropy content amounting to $R\ln(2S+1) \approx 26.76$ J·K^{-1}·mol$^{-1} \approx S_{max,1}/5$. It is this reduced value that sets the order of magnitude of MCE at low temperatures, where the intracluster interactions are at play. The detailed study of MCE for this compound was reported in [36]. The second compound (**2**) is a bimetallic molecular magnet $\{[Mn^{II}(pyrazole)_4]_2[Nb^{IV}(CN)_8]\cdot 4H_2O\}_n$ (pyrazole is a five membered ring ligand $C_3H_4N_2$) [37].The Mn^{II} ions carry the spin of $S_{Mn} = 5/2$, while the Nb^{IV} ions possess the spin of $S_{Nb} = \frac{1}{2}$, which implies that the maximal molar magnetic entropy of the system is $S_{max,2} = R\ln(2S_{Mn}+1)^2(2S_{Nb}+1) \approx 35.56$ J·K^{-1} mol^{-1}. This, together with the fact that the compound exhibits the second-order phase transition, makes one anticipate a considerable magnetocaloric response. Yet, an antiferromagnetic coupling between the Mn^{II} and Nb^{IV} ions was suggested by the analysis of the magnetometric data [37], which implies that a more representative spin value per formula unit (at least in the magnetic fields below the decoupling threshold) is $S = 2S_{Mn} - S_{Nb} = 9/2$. The corresponding entropy content is reduced down to $R\ln(2S+1) \approx 19.14$ J·K^{-1} mol^{-1} but still substantial. The detailed analysis of MCE for **2** was reported in [38]. The third compound (**3**) is a bimetallic coordination polymer $\{[Fe^{II}(pyrazole)_4]_2[Nb^{IV}(CN)_8]\cdot 4H_2O\}_n$, isostructural with **2** [37]. The spin of the Fe^{II} ion is $S_{Fe} = 2$, while the spin carried by the Nb^{IV} ion is $S_{Nb} = \frac{1}{2}$, which sets the maximal molar magnetic entropy of the system as equal to $S_{max,3} = R\ln(2S_{Fe}+1)^2(2S_{Nb}+1) \approx 32.53$ J·K^{-1} mol^{-1}. The compound is known to display the second-order phase transition and thus a considerable magnetocaloric effect may be anticipated. On the other hand, the preliminary analysis of the magnetometric data implied an antiferromagnetic coupling between the constituent ions [37], which suggests that the total spin of $S = 2S_{Fe} - S_{Nb} = 7/2$ is a more representative spin value per formula unit, at least in the low temperature and low field regime. The entropy content associated with this spin value amounts to $R \ln(2S + 1) \approx 17.29$ J·K^{-1} mol^{-1}, which is reduced by half but still substantial. The detailed report of MCE for this compound is given in [39].

2.2. Cluster Compound Ni$_9$[W(CN)$_8$]$_6$

The compound crystallizes in a triclinic system, space group $P\bar{1}$. The unit cell comprises one centrosymmetric cluster, the structure of which is shown in Figure 1. Due to the relatively large size of the tert-butyl substituted bpy (bpy = 2,2'-bipyridine = C$_{10}$H$_8$N$_2$) the distances between the cluster centers exceed 20 Å and the π–π interactions in the system are practically absent. Hence the inter-cluster superexchange interactions may be completely neglected in the studied temperature range of 2–300 K.

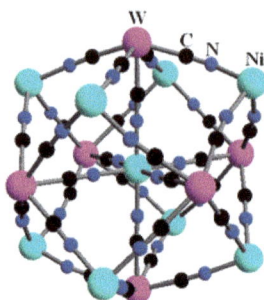

Figure 1. Core of the Ni$_9$W$_6$ clusters: W—magenta; Ni—cyan; N—blue; C—black. The relative sizes of the balls correspond to the atomic radii of the respective atoms.

Heat capacity of the compound was detected on cooling with the PPMS instrument without a field and for an array of applied field values in the temperature intervals 0.4–20 K with the ^{3}He probe and 1.8–100 K with the standard probe cooled with liquid ^{4}He. Due to the different sensitivities of the probes in different temperature regimes, we used the data provided by the former system below 15 K and those provided by the latter system above 15 K. In the zero applied field a broad anomaly of width about 2 K with a very flat maximum around 1.9 K was revealed, see Figure 2. An increasing applied field is apparent to gradually suppress the anomaly leaving no trace of it for the maximal field value of 9T.

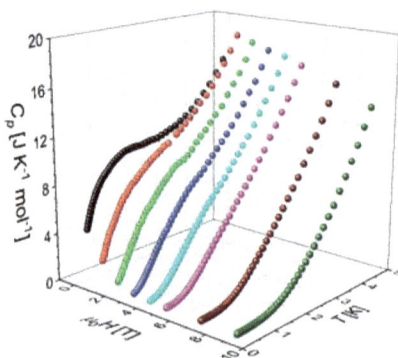

Figure 2. Heat capacity of the compound in the low temperature range. There is a broad anomaly with a flat maximum at about 1.9 K gradually suppressed by the increasing magnetic field.

The zero-field heat capacity of **1** is depicted in Figure 3 in the range of 0.4–100 K. To construct the baseline an approximate approach was employed. The observed C_p values within the specific range 30–63 K above the anomaly, which we shall elucidate in what follows, are assumed to comprise two separate contributions. The first contribution involves the lattice degrees of freedom C_p(lattice) and the second contribution $C_{mag}(H,T)$ is due to the Schottky-like anomaly anticipated for spin clusters coupled via the exchange interactions. While a polynomial approximation involving cubic and higher-order terms is employed to describe the former, the latter is introduced with the term proportional to T^{-2}. Thus the experimental heat capacity values are assumed to be represented within the range of 30–63 K by the following formula:

$$C_p = C_p(\text{lattice}) + C_{mag}(\text{HT}) = \sum_{i=3}^{m} a_i T^i + bT^{-2} \tag{3}$$

Fitting Equation (3) with $m = 6$ to the experimental heat capacity values yielded $a_3 = 6.07(7) \times 10^{-2}$ J·K^{-4} mol^{-1}, $a_4 = -2.08(4) \times 10^{-3}$ J·K^{-5} mol^{-1}, $a_5 = 2.76(9) \times 10^{-5}$ J·K^{-6} mol^{-1}, $a_6 = -1.33(6) \times 10^{-7}$ J·K^{-7} mol^{-1}, and $b = 45,355.9(1)$ J·K·mol^{-1}. The baseline is defined by C_p(lattice) thus determined and extrapolated down to 0 K (green solid curve in Figure 3). The subtraction of the baseline from the detected heat capacity values yields the magnetic excess heat capacity ΔC_p, see Figure 4.

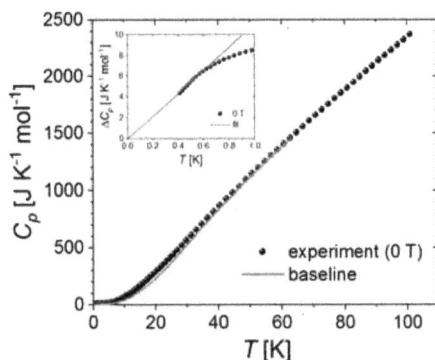

Figure 3. Heat capacity of **1** (symbols) together with the baseline (solid line). Inset: Zero-field excess heat capacity in the low temperature range (symbols) with the algebraic fit fT^μ (solid line).

To calculate the entropy content associated with the excess heat capacity ΔC_p in zero field the low temperature behaviour of ΔC_p was additionally analyzed. A function fT^μ was fitted to the magnetic excess heat capacity in the narrow range of 0.41−0.62 K (10 experimental points). The best fit yielded $f = 11.1(2)$ J·K$^{-(1+\mu)}$mol^{-1} and $\mu = 1.04(2)$ (see Inset of Figure 3). The low-temperature algebraic best-fit function was next used to estimate the entropy contribution in the temperature range 0–0.62 K. The entropic contribution in the temperature range 0.62–63 K was obtained by numerical integration of the area under the excess heat capacity signal $\Delta S = \int \Delta C_p d \ln T$. Finally, the high temperature excess heat capacity bT^{-2} was used to estimate the entropic contribution above 63 K. The total entropy content amounts to 116.97 J·K^{-1}·mol^{-1} and is slightly higher than the value of $S_{max,1} = 116.79$ J·K^{-1}·mol^{-1}. Such a good level of agreement of the calculated magnetic entropy with the theoretically predicted value is not accidental. The particular choice of the temperature interval used to determine the baseline (30–63 K) was made so as to reproduce the total magnetic entropy associated with the reported spin cluster. The above procedure was automatized within a specially designed notebook of the Mathematica 8.0 environment.

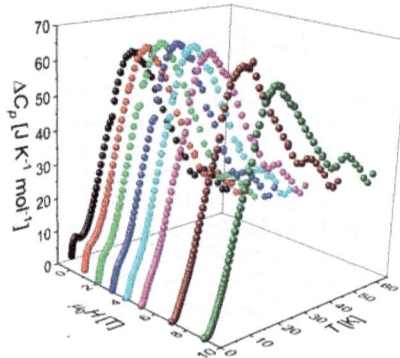

Figure 4. Temperature dependence of the magnetic excess heat capacity of **1**.

Equations (1) and (2) were used to determine temperature dependences of ΔS_m and ΔT_{ad}. The entropy thermodynamic function $S(T,H)$ was calculated upon the algebraic extrapolation of the magnetic excess heat capacity ΔC_p down to 0 K using the data in the interval 0.40–0.63 K (10 experimental points) for each field value. The resultant extrapolating functions were next appended with the baseline to yield the final forms of $C_p(T,H)$.

Temperature dependence of ΔS_m for several indicated field change values ($\mu_0\Delta H = \mu_0(H\text{-}0) = \mu_0 H$) is depicted in Figure 5. The $-\Delta S_m$ vs. T curves display maxima placed in the range 2.4–6.5 K in addition to becoming broader and higher with increasing field change values. The peak values of $|\Delta S_m|$ and positions are given in Table 1. $|\Delta S_m^{peak}|$ for $\mu_0\Delta H = 5$ T of other magnetic molecules was reported to range from 25 to 45 J·K^{-1}·kg^{-1} [19] which is considerably higher than $|\Delta S_m^{peak}| = $ 3.36 J·K^{-1}·kg^{-1} (18.38 J·K^{-1}·mol^{-1}) recorded for **1**. This is most probably due to the relatively higher molecular weight of **1**. On the other hand, the observed $|\Delta S_m^{peak}|$ values lie close to the low temperature physical threshold defined by $R\ln(2S+1) \approx 26.76$ J·K^{-1}·mol^{-1} with $S = 12$. In the whole experimental range, the $-\Delta S_M$ values remain positive exceeding 3 J·K^{-1}·mol^{-1}, although it is apparent they steeply drop below their peaks.

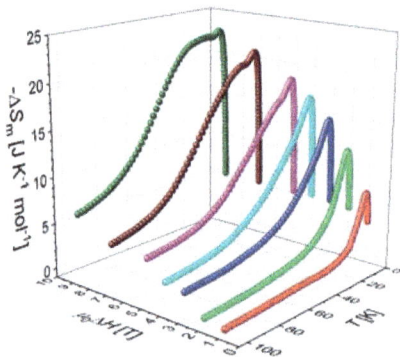

Figure 5. Isothermal entropy change ΔS_m as a function of temperature for **1** (symbols) inferred from the calorimetric data

Table 1. Peak values of ΔS_m and ΔT_{ad} of **1**.

$\mu_0 \Delta H$ (T)	T_{peak} (K)	$|\Delta S_m^{peak}|$ (J·K^{-1}·mol^{-1})	T_{peak} (K)	ΔT_{ad}^{peak} (K)
1	2.4	7.18	2.9	1.6
2	3.0	11.42	2.6	2.8
3	3.5	14.48	2.4	3.6
4	3.9	16.62	2.3	4.2
5	4.3	18.38	2.2	4.6
7	5.3	20.86	2.0	5.2
9	6.5	22.77	2.0	5.6

One usually looks at the parameter n (= d ln $|\Delta S_m|$/d ln H) conveniently quantifying the local sensitivity of the isothermal entropy change ΔS_m to the external field amplitude $H = \Delta H = H_f$ ($H_i = 0$). The value of n means that in the vicinity of a given thermodynamic point (T,H) the entropy change is approximately given by the power function H^n. At high temperatures, where the magnetization is directly proportional to the field (the Curie–Weiss law), the integral version of the Maxwell relation implies a quadratic field dependence of ΔS_m, giving rise to $n = 2$. Figure 6 shows the temperature dependence of the field-averaged value of parameter n for **1**. The field-variation of n differs in intensity at different temperatures, which is reflected by the size of the error bars. The behaviour of n is consistent with the above high-temperature paradigm, showing a steady increase with temperature and approaching the limiting value of 2. At the experimental low-temperature threshold parameter n attains a value as low as 0.16.

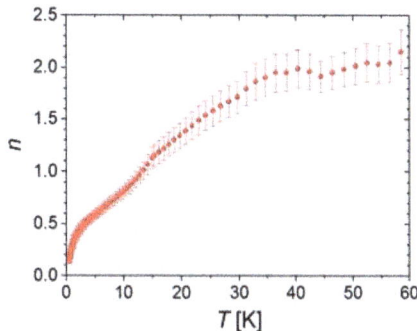

Figure 6. Temperature dependence of exponent n determined for **1**.

In Figure 7 the adiabatic temperature change ΔT_{ad} is depicted. All the $\Delta T_{ad}(T)$ curves can be seen to display well-defined peaks. The peak amplitudes of ΔT_{ad} and the corresponding peak positions are listed in Table 1. The value of ΔT_{ad}^{peak} for $\mu_0 \Delta H = 1, 3, 5, 7$ and 9 T amounts to 1.6, 3.6, 4.6, 5.2, and 5.6 K, respectively. The record beating Gd^{3+} dimer with $\Delta T_{ad}^{peak} \approx 3.5, 9.0, 12.7$ K for $\mu_0 \Delta H = 1, 3$, and 7 T, respectively [40], exceeds the ΔT_{ad}^{peak} values of **1** more than twice. However, they fall closer to those reported for the Mn$_{32}$ cluster ($\Delta T_{ad}^{peak} \approx 2.2, 4.5, 6.7$ K for $\mu_0 \Delta H = 1, 3$, and 7 T, respectively) [41]. The values of ΔT_{ad}^{peak} reported for the compound {[Ni^{II}(pyrazole)$_4$]$_2$ [Nb^{IV}(CN)$_8$]· 4H$_2$O}$_n$, 2.0 K for $\mu_0 \Delta H = 5$ T, and 2.9 K for $\mu_0 \Delta H = 9$ T [42], are significantly lower than for **1**. The still lower value was recorded for molecular magnet Mn$_2$-pyridazine-[Nb(CN)$_8$] (1.5 K for $\mu_0 \Delta H = 5$ T), but this is probably due to the relatively higher transition temperature of 43 K [43].

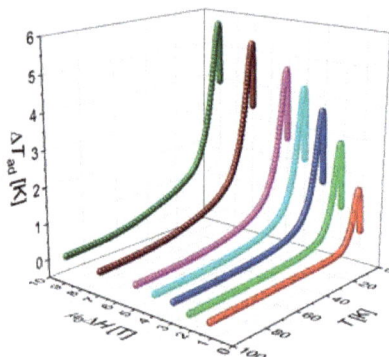

Figure 7. Temperature dependence of the adiabatic temperature change ΔT_{ad} for **1**.

2.3. Coordination Polymer $\{[Mn^{II}(pyrazole)_4]_2[Nb^{IV}(CN)_8]\cdot 4H_2O\}_n$

Compound **2** crystallizes in the tetragonal space group I4$_1$/a [37]. Its unique structure consists of a 3D skeleton, where each Mn^{II} center is bridged to only two Nb^{IV} ions through the cyanido linkages Mn^{II}-NC-Nb^{IV}, while each Nb^{IV} centre is linked to four Mn^{II} ions. The remaining part of the pseudooctahedral coordination sphere of Mn^{II} is substituted with pyrazole molecules, while four further terminal cyanide ligands are linked with the Nb^{IV} ion. It is an interesting and unique structural feature that a three-dimensional (3D) extended network should emerge from such low connectivity indices. Figure 8 shows the crystal structure of the compound for instant reference.

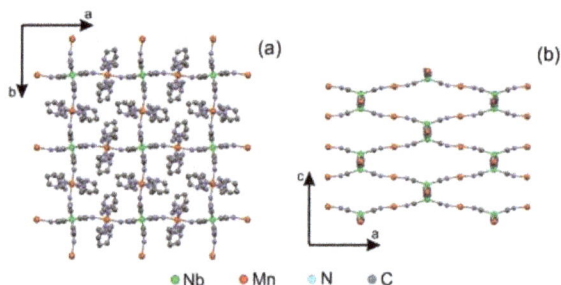

Figure 8. Structure of **2**. (a) View of the structure along the *c* (**a**) and *b* (**b**) crystallographic axes. For clarity the water molecules, the hydrogen atoms, and all nonbridging cyanido ligands have been removed.

The magnetic data imply that at $T_c \approx 23.8$ K compound **1** undergoes a phase transition to the long-range magnetically ordered state [37]. The mean-field approach was employed to analyze the isothermal magnetization and the dc susceptibility, revealing the antiferromagnetic character of the coupling between the Mn^{II} and Nb^{IV} centers, which gives rise to an overall ferromagnetic behaviour. The coupling constant was estimated to amount to -6.8 cm^{-1} (-9.8K) [37].

The relaxation calorimetry technique implemented in the PPMS Quantum Design instrument was used to measure the heat capacity of **1**. The zero-field measurements carried out in the temperature range of 1.9-101 K revealed a well-defined λ-shaped anomaly at 22.8 K (slightly lower than the magnetometric counterpart) providing evidence of the presence of a magnetic continuous transition, see Figure 9. Additionally, the sample was measured in the applied field of $\mu_0 H = 0.1, 0.2, 0.5,$ 1, 2, 3, 4, 5, 7, and 9T in the range of 1.9–40.4 K in the cooling direction. To estimate the normal heat capacity (baseline), the observed C_p values within the specially selected range 31–55 K above the transition temperature (justified in what follows) were considered to involve two independent contributions. The first contribution originates from the lattice degrees of freedom C_p(lattice) and the second contribution is magnetic and due to the short-range order C_{mag} (short-range order). While the latter is introduced by the term proportional to T^{-2}, a polynomial approximation with cubic and higher order terms is used to model the former. Thus the experimental values of the heat capacity within the range of 31–55 K are assumed to be given by the formula:

$$C_p = C_p(\text{lattice}) + C_{mag}(\text{short} - \text{range order}) = \sum_{i=3}^{n} a_i T^i + bT^{-2} \tag{4}$$

Fitting Equation (4) with $n = 6$ to the experimental heat capacity values yielded $a_3 = 1.8674 \times 10^{-2}$ J·K^{-4}·mol^{-1}, $a_4 = -7.0801 \times 10^{-4}$ J·K^{-5}·mol^{-1}, $a_5 = 1.0546 \times 10^{-5}$ J·K^{-6}·mol^{-1}, $a_6 = -5.7808 \times 10^{-8}$ J·K^{-7}·mol^{-1}, and $b = 8345.94$ J·K mol^{-1}. The lattice heat capacity C_p(lattice) thus determined extrapolated down to 0 K defines the baseline (solid curve in Figure 9).

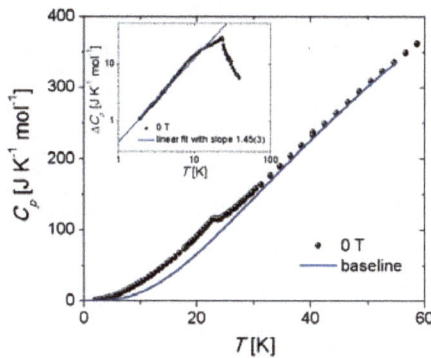

Figure 9. Zero-field total heat capacity of **2** (symbols) with the baseline (solid line). Inset: Zero-field magnetic excess heat capacity in the log-log plot (symbols) with the linear fit in the low temperature regime (solid line).

The subtraction of the lattice heat capacity from the detected heat capacity values yields the magnetic excess heat capacity ΔC_p, see Figure 10. Increasing magnetic field is apparent to suppress the anomaly peak. Moreover, consistently with a system with dominating antiferromagnetic coupling it slightly moves the anomaly toward lower temperatures. The calculation of the entropy associated with zero-field ΔC_p proceeded in three separate steps. Firstly, the contribution to the entropy above 55 K was estimated by considering the high temperature excess heat capacity bT^{-2}. Secondly, the contribution in the temperature range 1.9–55 K was calculated using the formula $\Delta S = \int \Delta C_p \mathrm{d} \ln T$. And finally, to obtain the low temperature contribution function fT^μ was fitted to ΔC_p in the range of 1.9–3.0 K, yielding $f = 0.45(1)$ J·K$^{-(1+\mu)}$ mol^{-1} and $\mu = 1.45(3)$ (see Inset of Figure 9). Next, the fitted function was logarithmically integrated in the interval 0–1.9 K. The total magnetic entropy is received by summing the above three contributions giving 35.31 J·K^{-1} mol^{-1}. This value compares perfectly well with $S_{max,2}$ = 35.56 J·K^{-1}·mol^{-1}, which is attributable to the specific choice of the temperature interval (31–55 K) used to determine the baseline.

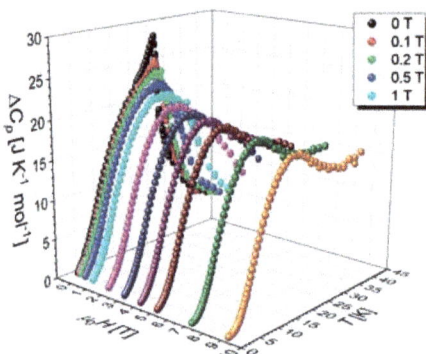

Figure 10. Magnetic excess heat capacity of **2**.

Similarly as for **1** using Equations (1) and (2), the isothermal entropy change ΔS_m and the adiabatic temperature change ΔT_{ad} were determined. The entropy thermodynamic function $S(T,H)$ was calculated using the baseline and the algebraic extrapolation of ΔC_p based on the data in the temperature range of 1.9–3.0 K. Figure 11 shows the temperature dependence of ΔS_m for $\mu_0\Delta H = 0.1$, 0.2, 0.5, 1, 2, 3, 4, 5, 7 and 9T ($H_i = H$, $H_f = 0$). It is apparent that the corresponding curves display a peak. In Table 2 the peak values of ΔS_m together with the peak positions are provided. In addition to the ΔS_m data inferred from the calorimetric measurements Figure 11 shows also the ΔS_m values obtained by using the Maxwell relation from the magnetometric data [42]. Except for the two highest field change values (4 and 5 T), where the magnetometric data are slightly scattered around the smoother calorimetric data, both sets agree strikingly well. The lack of smoothness is probably a consequence of the problems of the particular instrument with temperature stabilization.

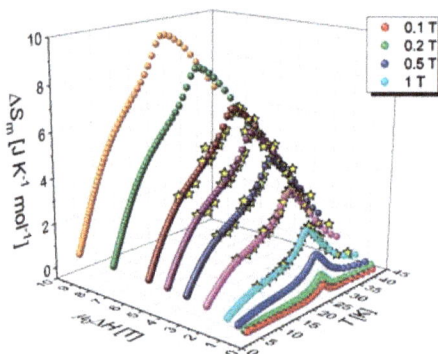

Figure 11. Temperature dependence of the isothermal entropy change ΔS_m of **2** inferred from the heat capacity measurements (spheres) or from the isothermal magnetization measurements (stars).

The value of $\Delta S_m^{peak} = 6.83$ J·K^{-1}·mol^{-1} detected for $\mu_0\Delta H = 5$T slightly exceeds that reported in [42] (6.7 J·K^{-1}·mol^{-1}). The isostructural compound {Ni^{II}(pyrazole)$_4$]$_2$[Nb^{IV}(CN)$_8$]·4H$_2$O}$_n$ reveals at the same time the lower entropy change of 6.1 J·K^{-1}·mol^{-1} [42], which most probably can be attributed to the smaller spin value of the Ni^{II} ion ($S_{Ni} = 1$). Figure 12 shows the thermal dependence of the field-averaged exponent n for **2**. On heating parameter n it is apparent that it smoothly decreases down to the minimal value of 0.62 attained at 23.6 K (slightly above the transition temperature), and subsequently increases toward the high temperature value of 2. At the transition temperature $T_N = 22.8$ K parameter n takes on the value of 0.64. It can be compared to that calculated using the relationship: $n|_{T=T_c} = 1 + (\beta - 1)/(\beta + \gamma)$ [8,44]. The value of 0.6424(4), obtained with the above equation and the theoretical estimates of $\beta = 0.3689(3)$ and $\gamma = 1.3960(9)$ for the 3D Heisenberg universality class [45], is very close to that obtained for **2**.

Table 2. Peak values of ΔS_m and ΔT_{ad} of **2**.

$\mu_0 \Delta H$ (T)	T_{peak} (K)	$|\Delta S_m^{peak}|$ (J·K^{-1}·mol^{-1})	T_{peak} (K)	ΔT_{ad}^{peak} (K)
0.1	23.3	0.29	23.3	0.06
0.2	23.8	0.68	23.3	0.14
0.5	23.8	1.50	23.8	0.30
1	23.8	2.44	23.8	0.50
2	24.3	3.85	23.8	0.80
3	24.3	4.99	23.8	1.04
4	24.3	5.97	23.8	1.24
5	24.3	6.83	23.8	1.42
7	24.3	8.30	23.8	1.73
9	25.5	9.49	23.8	1.97

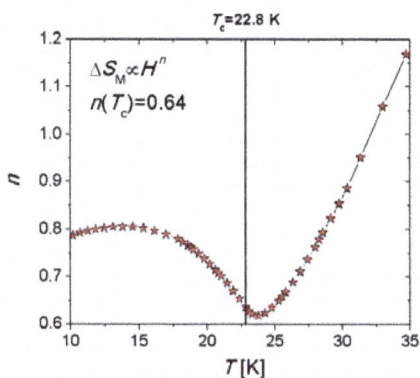

Figure 12. Temperature dependence of exponent n for **2**.

The $\Delta T_{ad}(T)$ curves are shown in Figure 13. A two-peak structure of the curves is apparent. In addition to the expected peak located near the transition temperature they reveal the second peak at the lowest experimentally accessible temperatures. In Table 2 the primary peak values of ΔT_{ad} are provided. The positions of the primary peaks are practically independent of the applied field change. Moreover, there is their slight shift off the transition temperature toward higher temperatures. By contrast, the increasing magnetic field change seems to move the secondary peaks toward lower temperatures. The isostructural compound {Ni^{II}(pyrazole)$_4$]$_2$[Nb^{IV}(CN)$_8$]·4H$_2$O}$_n$ revealed higher values of ΔT_{ad}, i.e., 2.0 K for $\mu_0 \Delta H = 5$ T, and 2.9 K for $\mu_0 \Delta H = 9$ T [42]. The amplitudes of ΔT_{ad} reported for Mn$_2$-pyridazine-[Nb(CN)$_8$] (1.5 K for $\mu_0 \Delta H = 5$ T) [43] and hexacyanochromate Prussian blue analogues (1.2 K for $\mu_0 \Delta H = 7$ T) [46] are at the same time comparable to those for **2**.

The occurrence of the secondary peaks in the $\Delta T_{ad}(T)$ signal around 2 K points to the possibility that beyond the experimentally accessible temperature range an additional magnetic transition is concealed. This conjecture is neither confirmed by the shape of the $\Delta S_m(T)$ signal nor it was confirmed by additional heat capacity measurements in the temperature range 0.8–10 K [38]. Therefore it seems plausible that the reason for the presence of the secondary peaks is merely the proximity effect to the natural boundary of the temperature scale ($T = 0$ K). In order to shed some light onto that point a hypothetical paramagnetic medium was considered which consists of entities carrying the effective spins of $S = 9/2$ ($2 S_{Mn} - S_{Nb} = 5/2 + 5/2 - 1/2$). ΔT_{ad} of this medium may be found solving the following differential equation implied by the Maxwell thermodynamic relation:

$$\frac{dT}{dH} = -\frac{\left(\frac{\partial M(T,H)}{\partial T}\right)_H}{C_L(T) + C_H(T,H)} \tag{5}$$

where $M(H,T)$ denotes the molar magnetization, C_L stands for the lattice contribution to the total heat capacity, C_H is the magnetic contribution. The baseline determined for **2** may serve as C_L. The paramagnetic Hamiltonian corresponding to an isolated spin S subject to applied magnetic field provides the basis to calculate C_H and $\partial M(T,H)/\partial T$. Let us note that in [38] an integral version of Equation (5) was used, which is incorrect due to the explicit T-dependence of its right-hand side. Solving Equation (5) with initial conditions $T(H_i \neq 0) = T_i$ in the field interval $[H_i, H_f = 0]$ for several arbitrary values of the initial temperature T_i yields the system of adiabats shown in Figure 14 in the (H,T) plane.

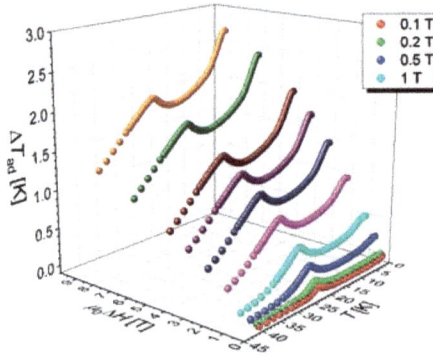

Figure 13. Temperature dependence of ΔT_{ad} of **2**.

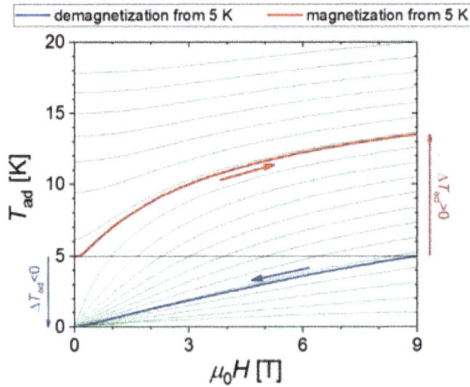

Figure 14. Adiabats obtained by solving Equation (5) with different initial conditions. Two adiabats are highlighted: the adiabat corresponding to the demagnetization process starting from 5 K and 9 T (blue), the adiabat corresponding to the magnetization process starting at 5 K and 0 T continued up to 9 T (red). The arrows show the amplitudes of the corresponding adiabatic temperature changes ΔT_{ad}.

Using Equation (5) the ΔT_{ad} signal was calculated for the magnetization processes $H_i = 0 \rightarrow H_f$, with H_f assuming the integer field values, within the temperature interval $0 - 55$ K. The result is shown in Figure 15 together with the ΔT_{ad} signal of **2** (cf. Figure 13). Two points may be raised here.

Firstly, the $\Delta T_{ad}(T)$ curves of the hypothetical medium are peaked at 0 K, which is the direct vicinity of the secondary peaks in the ΔT_{ad} signal of **2**. Secondly, the magnitude of ΔT_{ad} of **2** is about four times smaller than that of the hypothetical medium. Strikingly enough, it may be concluded that the presence of a magnetic phase transition at some finite temperature has an adverse effect on MCE.

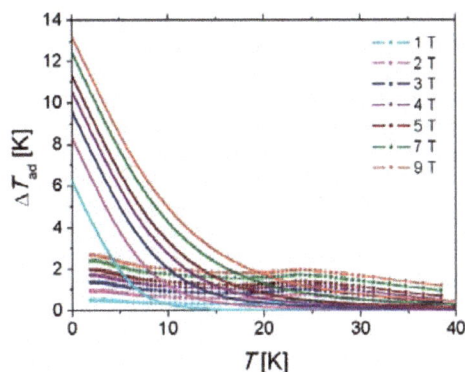

Figure 15. ΔT_{ad} of the hypothetical paramagnetic medium sharing with **2** the same spin content and the lattice heat capacity (solid lines). For the sake of comparison ΔT_{ad} of the studied compound has been also shown (symbols).

2.4. Coordination Polymer $\{[Fe^{II}(pyrazole)_4]_2[Nb^{IV}(CN)_8] \cdot 4H_2O\}_n$

Compounds **2** and **3** share the crystallographic structure. The graphical representation of the crystal structure of **3** is shown in Figure 8 with the Mn^{II} ions replaced by the Fe^{II} ions. The magnetic data implied that the compound undergoes a transition to a long-range magnetically ordered state at $T_c \approx 7.8$ K [3,16]. Moreover, the muon spin rotation spectroscopy (μSR) strongly suggested that the compound may be assigned to the universality class of the 3D Heisenberg model [40]. The calculations based on the mean field-model and the magnetometric data pointed to the antiferromagnetic character of the superexchange coupling between the Fe^{II} and Nb^{IV} centres [37].

The relaxation calorimetry technique implemented in the PPMS Quantum Design instrument was employed to measure the heat capacity of **3**. The measurements were carried out both without applied field and in nonzero field in the range 0.36–20.2 K in the cooling direction. The temperature dependences of ΔS_m and ΔT_{ad} were determined by employing Equations (1) and (2). In this case the entropy thermodynamic function $S(T,H)$ was calculated without determining the normal heat capacity. However, the calculation required the extrapolation of C_p down to 0 K. Two independent contributions were considered in the extrapolation scheme involving two field-dependent parameters: $C_{p,LT}(T,H) = A(H)T^3 + B(H)T^{3/2}$. The first contribution accounts for the lattice degrees of freedom, whereas the second one is due to the magnonic excitations (the Bloch law).

Figure 16 shows the temperature dependence of ΔS_m for $\mu_0\Delta H = 0.1, 0.2, 0.5, 1, 2, 5$, and 9 T ($H_i \neq 0$, $H_f = 0$). Table 3 lists the values of ΔS_m attained at the maximum. The isostructural compounds $\{[M^{II}(pyrazole)_4]_2[Nb^{IV}(CN)_8] \cdot 4H_2O\}_n$ with M = Ni and M = Mn reveal ΔS_m^{peak} for $\mu_0\Delta H = 5$ T equal to 6.1 J·K^{-1}·mol^{-1} and 6.7 J·K^{-1}·mol^{-1}, respectively, which is higher than the value of 4.9 J·K^{-1}·mol^{-1} detected for **3** [42]. It is surprising that it should be so for the Ni congener as the Ni^{II} ion carries lower spin ($S_{Ni} = 1 < S_{Fe} = 2$). However, it is consistent with the ferromagnetic superexchange present in the Ni compound [37] and the relatively stronger anisotropy of the Fe^{II} centre [13].

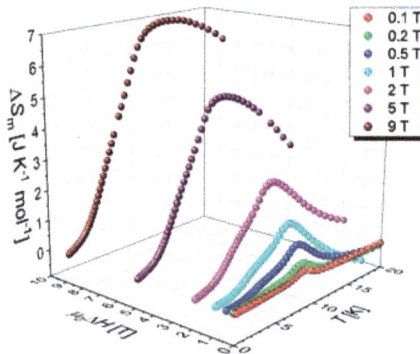

Figure 16. Temperature dependence of the isothermal entropy change ΔS_m.

Table 3. Peak values of ΔS_m and ΔT_{ad} of **3**.

$\mu_0 \Delta H$ (T)	T_{peak} (K)	$\mid \Delta S_m^{peak} \mid$ ($J \cdot K^{-1} \cdot mol^{-1}$)	T_{peak} (K)	ΔT_{ad}^{peak} (K)
0.1	8.9	0.3	8.8	0.1
0.2	8.9	0.5	8.8	0.2
0.5	8.9	1.0	8.9	0.4
1	9.3	1.6	8.8	0.6
2	9.3	2.7	8.8	1.1
5	10.3	4.9	8.9	2.0
9	6.9	6.9	8.8	2.8

The field-averaged exponent n of **3** vs. temperature is depicted in Figure 17. Its estimation drew on the data in Figure 16. The n vs. T curve reveals a minimum of 0.63 slightly above T_c = 8.3 K (implied by the peak in the zero-field heat capacity). On further increase of temperature, it increases toward the value of 2, consistent with the fact that in the paramagnetic phase the susceptibility is independent of field (the Curie-Weiss law). At T_c parameter n takes on the value of 0.64, which, like in the case of **2**, indicates that **3** belongs to the universality class of the 3D Heisenberg model.

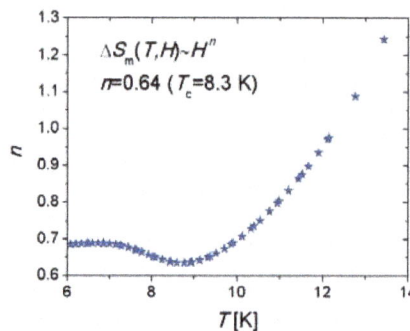

Figure 17. Temperature dependence of exponent n for **3**.

Temperature dependence of ΔT_{ad} of **3** calculated using Equation (2) is shown in Figure 18. All the ΔT_{ad} vs .T curves reveal peaks in the vicinity of T_c. The peak values together with the peak positions are listed in Table 3 which indicates that the peak positions are slightly moved off T_c toward higher temperatures and are practically independent of the applied field change value. The presence of a field induced effect is suggested by the occurrence of an inflection point at about 4 K (most apparent for higher fields) followed by a sharp drop. The values of ΔT_{ad} are comparable to those reported for its

Ni congener (2.0 K for $\mu_0\Delta H = 5$ T, and 2.9 K for $\mu_0\Delta H = 9$ T) [42], and larger than those found for Mn$_2$-pyridazine-[Nb(CN)$_8$] (1.5 K for $\mu_0\Delta H = 5$ T) [43], hexacyanochromate Prussian blue analogues (1.2 K for $\mu_0\Delta H = 7$ T) [46], and the Mn congener (1.42 K for $\mu_0\Delta H = 5$ T, and 1.97 K for $\mu_0\Delta H = 9$ T) [38]. The last case might seem surprising as the Mn ion carries the maximal spin value ($S_{Mn} = 5/2$) among the 3d ions. However, the Mn congener (**2**) displays the transition to the magnetically ordered phase at the temperature (22.8 K) which is more than twice as high than for **3**, and the ratio of the magnetic excess heat capacity to the lattice heat capacity is expected to decrease with temperature. Thus the magnetic entropy for **2** may be admittedly higher than for **3**, but it is faced with a relatively higher heat capacity at the transition temperature, which suppresses the magnitude of ΔT_{ad}.

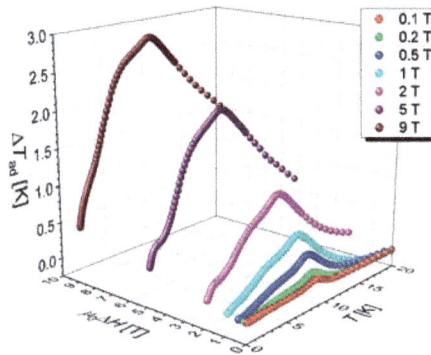

Figure 18. Temperature dependence of ΔT_{ad} for **3**.

2.5. Final Remarks of Section 2

In this Section we have demonstrated how one can employ the calorimetric data to determine two main characteristics of the magnetocaloric effect, i.e., the isothermal entropy change ΔS_m and the adiabatic temperature change ΔT_{ad}. The pivotal step in the corresponding procedure is the calculation of the entropy thermodynamic function $S(T,H)$ in the temperature interval starting from 0 K and for the given applied field values. Due to the experimental limitations, where the heat capacity measurements go down to a possibly small but finite temperature T_L never reaching the limit of 0 K, the procedure is out of necessity only approximate. The approximation consists in the extrapolation of the low-temperature $C_p(T,H)$ values from T_L down to 0 K by using the heuristic algebraic function. Two approaches were considered, the simplified approach obviating the need to determine the normal heat capacity (baseline) and the more comprehensive approach, where the baseline is constructed using the polynomial approximation. Both approaches were exemplified by three instances of molecular magnets, the cluster compound involving 15 ferromagnetically coupled spins (**1**) and two isostructural 3D coordination polymers **2** and **3** including Mn^{II} and Fe^{II} ions, respectively. While **2** and **3** display the transitions to the magnetically long-range ordered phases, compound **1** does not order magnetically above 0.4 K. For the sake of comparison of the MCE signals of the studied compounds Figure 19 shows the temperature dependences of the isothermal entropy change ΔS_m and the adiabatic temperature change ΔT_{ad} corresponding to the field change $\mu_0\Delta H = 5$ T. It is apparent that at the lowest temperatures the cluster compound **1** substantially exceeds **2** and **3** in terms of the ΔS_m and ΔT_{ad} signals. The peak values of the ΔS_m quantity seem to monotonically depend on the constituent spin value with that for **1** being the highest ($S = 12$), that for **2** being intermediate ($S_{Mn} = 5/2$) and that for **3** being the smallest ($S_{Fe} = 2$). This hierarchy breaks down for the ΔT_{ad} peak values with the peak value of **3** exceeding that of **2**, which can be rationalized by remembering that higher temperatures involve higher heat capacity values and a concomitantly lower heating or cooling effect. Compounds **2** and **3** can compete with **1** in terms of the ΔT_{ad} signal only in the intermediate temperature range. One can thus conclude that in order to optimize the MCE parameters one should strive to organize separate spin carriers in ferromagnetically coupled clusters rather than in 3D extended networks.

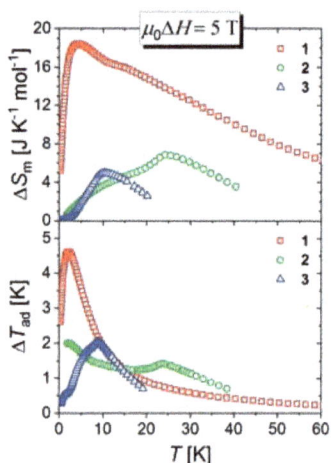

Figure 19. Temperature dependences of the isothermal entropy change ΔS_m and the adiabatic temperature change ΔT_{ad} corresponding to the applied field change of $\mu_0 \Delta H = 5$ T for the three studied compounds.

3. Magnetocaloric Properties and Critical Behaviour in Magnetically Ordered Compounds Investigated by Magnetometry

3.1. $\{[M^{II}(H_2O)_2]_2[Nb^{IV}(CN)_8] \cdot 4H_2O\}_n(M = Fe, Mn)$ Molecular Compounds

$\{[Mn^{II}(H_2O)_2]_2[Nb^{IV}(CN)_8] \cdot 4H_2O\}_n$ **(4)** and $\{[Fe^{II}(H_2O)_2]_2[Nb_{IV}(CN)_8] \cdot 4H_2O\}_n$ **(5)** are isostructural molecular magnets, which crystallize in tetragonal space group I4/m. Despite the same structure these compounds show different types of magnetic moments order. Mn-based compound **(4)** is a ferrimagnet with the coercive field equal to 0 Oe and critical temperature T_c equal to 50 K. Fe-based analogue **(5)** is an example of a ferromagnet with the coercive field equal to 145 Oe and T_c equal to 43 K [47,48]. For both compounds the magnetocaloric effect was determined using an indirect method involving the measurements of a series of isothermal magnetization curves as a function of external magnetic field $M(H)$ in a wide temperature range: above and below the critical temperature (Figure 20a). The magnetic entropy change $|\Delta S|$ was calculated based on the integrated Maxwell's relation given by the equation:

$$\Delta S_M(T, \Delta H) = \int_{H_1}^{H_2} \left(\frac{\partial M(T, H)}{\partial T}\right)_H dH \tag{6}$$

As expected, a maximum ΔS value occurs at temperatures corresponding to T_c, and $|\Delta S|$ increases with ΔH. The replacement of Fe by Mn in isostructural $\{[M^{II}(H_2O)_2]_2[Nb^{IV}(CN)_8] \cdot 4H_2O\}_n$ network causes an increase of $|\Delta S|^{max}$ as well as the temperature at which the maximum value of the entropy change occurs [49]. The maximum values of $|\Delta S|$ are equal to 4.82 J mol^{-1}K^{-1} (8.65 J kg^{-1}K^{-1}) and 5.07 J mol^{-1}K^{-1} (9.09 J kg^{-1}K^{-1}) for the field change of 0–5 T for **5** and **4**, respectively. Slightly higher values of the magnetic entropy change observed for **4** may be related to the spin value of the Mn^{II} ion ($S_{Mn} = 5/2$), being higher than that of the Fe^{II} ion ($S_{Fe} = 2$) and the typically soft magnetic character.

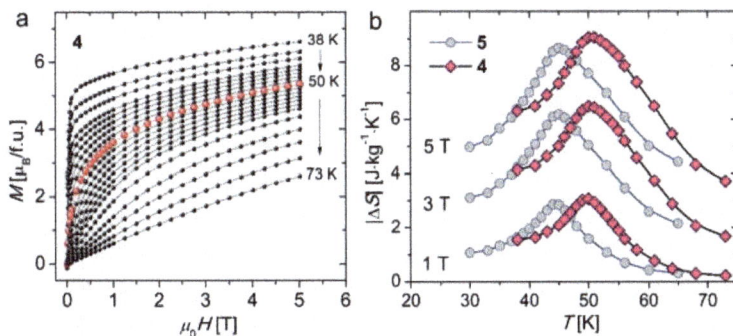

Figure 20. (a) Magnetization vs field isotherms measured in the vicinity of T_c for $\{[Mn^{II}(H_2O)_2]_2$ $[Nb^{IV}(CN)_8]\cdot4H_2O\}_n$ (4) and (b) temperature dependence of magnetic entropy change recorded in various magnetic fields for (4) and $\{[Fe^{II}(H_2O)_2]_2[Nb^{IV}(CN)_8]\cdot4H_2O\}_n$(5).

A useful parameter describing the efficiency of a magnetocaloric material is the relative cooling power (RCP) defined as $RCP = |\Delta S|^{max}\delta_{FWHM}$, where δ_{FWHM} is the full width at half maximum of the magnetic entropy change curve. The RCP measured for an applied magnetic field of 5 T is equal to 118.40 and 125.43 J mol^{-1} (212.61 and 225.59 J kg^{-1}) for 4 and 5, respectively. These values of RCP make about 50% of that of pure gadolinium-prototype magnetocaloric material.

An important aspect of the analysis of MCE is the construction of a universal curve of magnetic entropy change, a so-called master curve. Franco et al. [44,50] proposed that such universal entropy curve can be successfully used for determination of phase transition order. For the substances undergoing the second order phase transition, the temperature dependences of the magnetic entropy changes obtained at different applied magnetic fields may overlap after rescaling. The phenomenological universal curve can be constructed by normalizing all the $\Delta S(T)$ curves using their maximum value ΔS^{max} and, subsequently, re-scaling the temperature axis according to the following expression:

$$\theta = \frac{-(T - T_c)}{(T_{r1} - T_c)}, \; T \leq T_c; \; \theta = \frac{(T - T_c)}{(T_{r2} - T_c)}, \; T > T_c \tag{7}$$

The values of reference temperatures T_{r1}, T_{r2} were calculated based on $\Delta S(T_r)/\Delta S^{max} = h$, where h is a reference level with value from the range 0–1. For both studied materials: 4 and 5, the rescaled $\Delta S(T)$curves, create one master curve in a wide temperature range, and thus the type of the magnetic order does not affect their construction. Figure 21 shows the comparison of universal entropy curves obtained for samples 4 and 5.

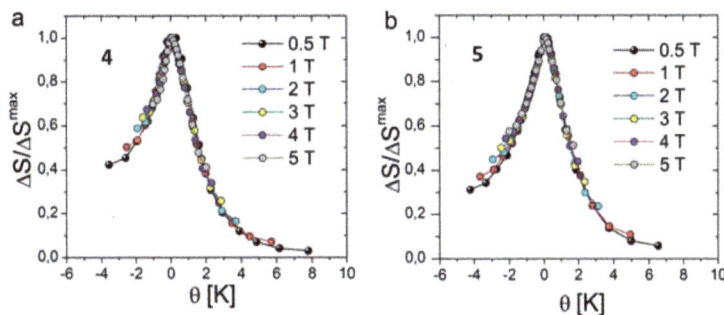

Figure 21. Universal curves of ferrimagnetic $\{[Mn^{II}(H_2O)_2]_2[Nb^{IV}(CN)_8]\cdot4H_2O\}_n$ (4) (a) and ferromagnetic $\{[Fe^{II}(H_2O)_2]_2[Nb^{IV}(CN)_8]\cdot4H_2O\}_n$ (6) (b) compounds.

3.2. {[M^{II}(pyrazole)$_4$]$_2$[Nb^{IV}(CN)$_8$]$_3$·4H$_2$O}$_n$(M = Ni, Mn) Molecular Compounds

{[Mn^{II}(pyrazole)$_4$]$_2$[Nb^{IV}(CN)$_8$]$_3$·4H$_2$O}$_n$ (6) and [Ni^{II}(pyrazole)$_4$]$_2$[Nb^{IV}(CN)$_8$]$_3$·4H$_2$O}$_n$ (7) are subsequent examples of isostructural bimetallic compounds, where pyrazole is the five membered C$_3$H$_4$N$_2$ ring. MCE results for 6 determined by calorimetry, including both the entropy change and the adiabatic temperature change, have been already discussed in paragraph 2.3. Below, we proceed the study to compare the effect in the Mn-based compound with that in the Ni-based compound. Both substances show a sharp phase transition, from a high- temperature paramagnetic to a low-temperature ordered state and while 6 is a ferrimagnet with critical temperature equal to 23.8 K, 7 is a ferromagnet with T_c = 13.4 K [37]. The higher values of the magnetic entropy change were observed for sample 6, despite the ferrimagnetic coupling between magnetic ions (see Figure 22) [42]. The higher value of magnetic entropy change observed for 6 may be related to the spin value of the Mn^{II} ion (S_{Mn}=5/2) being higher than that of the Ni^{II} ion (S_{Ni} = 1). For Ni-based compound, the magnetic entropy change was also determined by means of the heat capacity measurements. Obtained result is in good agreement with the previous estimation inferred from $M(H, T)$ data. The similar reasonable agreement of the magnetically and thermally determined values of the magnetic entropy change was previously observed for some inorganic alloys [51]. Based on the heat capacity data for 7, we estimate the maximum value of the adiabatic temperature change ΔT_{ad} upon the applied magnetic field change of 9 T was equal to 2.9 K [42].

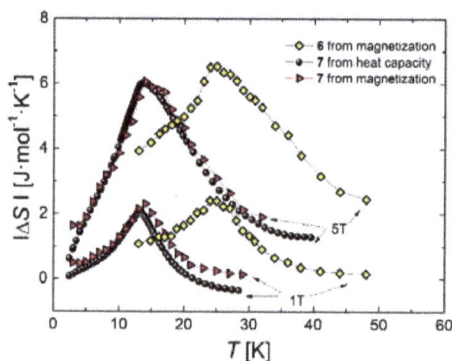

Figure 22. The plot of magnetic entropy change vs temperature corresponding to $\mu_0 \Delta H$ = 1, 5 T measured for {[Mn^{II}(pyrazole)$_4$]$_2$[Nb^{IV}(CN)$_8$]$_3$·4H$_2$O}$_n$ and [Fe^{II}(pyrazole)$_4$]$_2$[Nb^{IV}(CN)$_8$]$_3$·4H$_2$O}$_n$.

3.3. {Mn^{II}$_2$(imH)$_2$(H$_2$O)$_4$[Nb^{IV}(CN)$_8$]·4H$_2$O}$_n$Molecular Magnetic Sponge

{Mn^{II}$_2$(imH)$_2$(H$_2$O)$_4$[Nb^{IV}(CN)$_8$]·4H$_2$O}$_n$ (8) is a molecular magnet, where imidazole is a bridging ligand. This compound is an example of magnetic sponge, because it allows an easy and reproducible control of the amount of water molecules absorbed in this material. During one-step dehydration, one new CN$^-$bridge is formed in this material. The changes in the compound structure generated by the loss of water molecules cause the shift of the magnetic ordering temperature from 25 K (8) to 68 K (8$_{deh}$) [52]. The dehydration process is responsible for the change of magnetocaloric properties of the system: the maximum value of magnetic entropy change determined for 8$_{deh}$ is 40% lower than for the as-synthesized compound 8 [53] (see Figure 23). Using the molecular field approximation (MFA) an attempt was made to explain the origin of the magnetic entropy changes in the system, for both forms 8 and 8$_{deh}$. As shown in Figure 23, in comparison to the experimental MCE results, the calculated entropy change is underestimated above the transition temperature, which can be attributed to the fact that MFA does not account for the short-range correlations. However, below the critical temperature, the calculated entropy change is overestimated in relation to the experimental data [53]. This furthermore, may be attributed to the lack of thermal and quantum fluctuations in the mean field model. These fluctuations impede the magnetic moments reorientation caused by the change of magnetic field, thus the magnetic susceptibility is higher than in normal paramagnets.

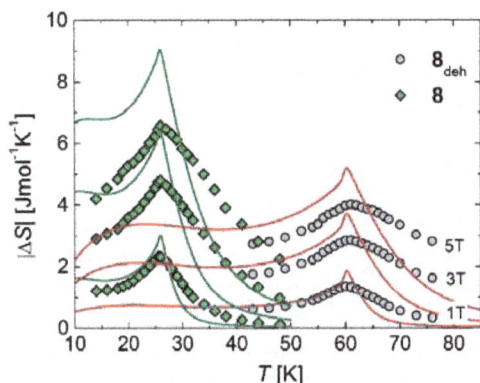

Figure 23. Isothermal entropy change extracted from the experiment (symbols) and calculated within the framework of the molecular field approximation (solid lines) for $\{Mn^{II}_2(imH)_2 (H_2O)_4[Nb^{IV}(CN)_8]\cdot4H_2O\}_n$ and its dehydrated form $\{Mn^{II}_2(imH)_2[Nb^{IV}(CN)_8]\}_n$.

3.4. $[\{[Mn^{II}(pydz)(H_2O)_2][Mn^{II}(H_2O)_2][Nb^{IV}(CN)_8]\}\cdot2H_2O]_n$ Two-Step Molecular Magnetic Sponge

$[\{[Mn^{II}(pydz)(H_2O)_2][Mn^{II}(H_2O)_2][Nb^{IV}(CN)_8]\}\cdot2H_2O]_n$ (**9**) is an example of multifunctional molecular magnet, where a pyridazine,$C_4H_4N_2$, was chosen as a bridging ligand. One of the unique properties of this compound is related to its two-step switchable magnetic sponge behaviour [33]. Within 6 h of dehydration of **9**, the loss of four water molecules is observed, and **9** is transformed into dehydrated form—**9$_{deh}$**. This process is responsible for destabilization of hydrogen bonds and causes intraskeletal molecular rearrangement: formation of a new Nb–CN–Mn bond as well as migration of the pyridazine ligand between two Mn centres. The structural changes occurring during the dehydration induce a significant increase of critical temperature from 43 K, observed for **9**, to 68 K recorded for **9$_{deh}$**. In the second step of dehydration (of about 24 h) two remaining water molecules are removed and new stable, anhydrous phase **9$_{anh}$** is formed. During this process the material structure does not change but some bonds are shortened. The critical temperature of anhydrous sample is equal to 98 K. The transformation from the initial, as synthesized **9** to the anhydrous form **9$_{anh}$** is reversible [33].

As depicted in Figure 24, for all three forms of the compound: **9**, **9$_{deh}$**, and **9$_{anh}$**, a maximum ΔS value occurs at temperatures corresponding to consecutive T_c, and $|\Delta S|$ increases with ΔH. The highest value of magnetic entropy change is revealed for **9**. For dehydrated and anhydrous samples values of magnetic entropy change are similar—lower by 40% than that observed for the as-synthetized material. This result is perfectly correlated with AC susceptibility data: the decrease in ΔS by a factor of 0.6 for **9$_{deh}$** and **9$_{anh}$** relative to **9** corresponds to the same reduction in the $\chi'(T)$ magnitude of **9$_{deh}$** and **9$_{anh}$** [43].

For the as-synthesized sample **9** the heat capacity measurements were performed, first at zero applied field and then at 0.2, 0.5, and 1 T. The data obtained from this experiment were used for the evaluation of magnetocaloric effect based on the magnetothermal method. The values of the magnetic entropy change calculated based on the heat capacity and magnetic data are in very good agreement. Furthermore, it was also possible to estimate, from the heat capacity data, the adiabatic entropy change associated with the change of magnetic entropy. The maximum value of the adiabatic temperature change determined for **9**, corresponding to the change of magnetic field from 0 to 1 T, is equal to 1.5 K [43].

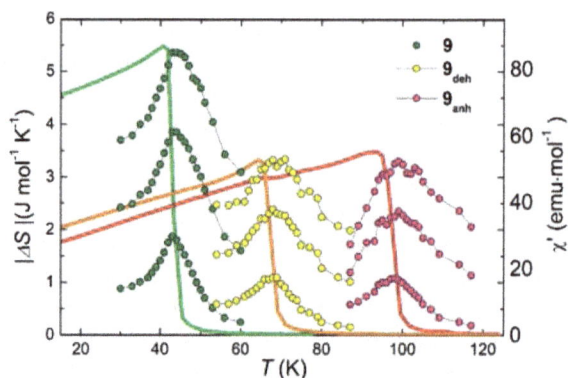

Figure 24. The comparison of temperature dependences of the magnetic entropy change measured for $\mu_0 \Delta H = 1, 3$, and 5 T (symbols) and real component of AC susceptibility for **9**, **9$_{deh}$**, and **9$_{anh}$** (lines).

3.5. [{[Mn^{II}(pydz)(H$_2$O)$_2$][Mn^{II}(H$_2$O)$_2$][Nb^{IV}(CN)$_8$]}·2H$_2$O]$_n$ Molecular Compound under Pressure

The uniqueness of compound **9** is also related to the possibility of the change of the magnetic properties by the application of hydrostatic pressure [54]. The application of hydrostatic pressure of 1.8 GPa results in the compression of the unit cell (ca. 7.6% in total volume) and a noticeable bending of the Mn-NC-Nb linkages. These structural modifications are accompanied by the change in magnetic properties of examined compound. The increase of the applied pressure increases the ordering temperature up to 48 K, 50.5 K and 52.5 K for pressure of 0.27 GPa, 0.57 GPa and 1.19 GPa, respectively. The shift of T_c towards higher temperatures due to application of mechanical stress is the result of the strengthening of Mn- and Nb-sublattice coupling due to the enhanced overlap of magnetic orbitals. The analysis of MCE for sample under pressure of 1.19 GPa (**9$_{HP}$**) showed the reduction of $|\Delta S|^{max}$ value in relation to as-synthesized compound **9** by 13.6% (see Figure 25) [55]. The summary of the magnetocaloric properties of all forms of molecular magnet **9** as well as results obtained for other octacyanoniobate- based compounds are presented in Table 4.

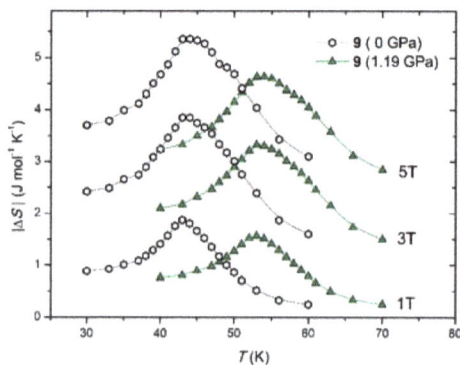

Figure 25. Temperature dependence of the magnetic entropy change of as-synthetized sample **9** ($p = 0$ GPa) (empty symbols) and **9** under pressure $p = 1.19$ GPa (filled symbols) determined for $\mu_0 \Delta H = 1, 3$ and 5 T.

Table 4. Comparison of MCE data obtained for samples discussed in Section 3. $|\Delta S|^{max}$ and RCP were determined for $\mu_0 \Delta H = 5$ T; MCE: magnetocaloric effect; RCP: relative cooling power

Sample	T_c (K)	$\|\Delta S\|^{max}$ (J mol^{-1} K^{-1})	$\|\Delta S\|^{max}$ (J kg^{-1} K^{-1})	RCP (J mol^{-1} K^{-1})	RCP (J kg^{-1} K^{-1})
4	50.0	5.07	9.09	118.40	212.61
5	43.0	4.82	8.65	125.43	225.59
6	23.8	6.70	6.50	136.9	132.9
7	13.4	6.10	5.90	75.6	73.1
8	25.0	6.70	8.95	186.2	248.9
8_{deh}	60.0	4.02	7.73	152.8	293.8
9	43.0	5.36	8.95	160.8	268.0
9_{deh}	68	3.33	5.82	109.9	192.1
9_{anh}	98	3.38	6.88	101.4	206.5
9_{HP}	52.5	4.63	7.73	138.9	231.8

3.6. $T_c^{-2/3}$ Dependence of the Maximum Entropy Change

Oesterreicher and Parker predicted that the maximum value of magnetic entropy change is proportional to $T_c^{-2/3}$ [7]. This relation has been proved for the series of intermetallic samples [9]. Taking into account that in the compounds based on manganese and octacyanoniobate, the structural changes caused by the external stimuli or selection of a bridging ligand did not affect magnetic moment of Mn and Nb sublattices, while the critical temperature was changed, it was possible to verify the relation: $|\Delta S|^{max} \sim T_c^{-2/3}$.

Figure 26a shows the comparison of the temperature dependences of magnetic entropy upon changing the magnetic field from 0 to 3 T, measured for samples: **9**, 9_{deh}, 9_{anh}, 9_{HP}, **8**, 8_{deh}, **4** and **6**. The clear reduction of the height of the $\Delta S(T)$ peak with the increase of the critical temperature is observed. The slight deviation from this behaviour occurs for 9_{deh}, because the maximum value of magnetic entropy change is smaller than expected. It can be explained by instability of partially dehydrated sample. Figure 26b presents the values of magnetic entropy change as a function of $T_c^{-2/3}$ in all Mn-Nb based compounds under study obtained at the change of magnetic field from 0 to 1 T, 3 T and 5 T. The linear character of all determined relations confirms the proportionality of $|\Delta S|^{max}$ to $T_c^{-2/3}$.

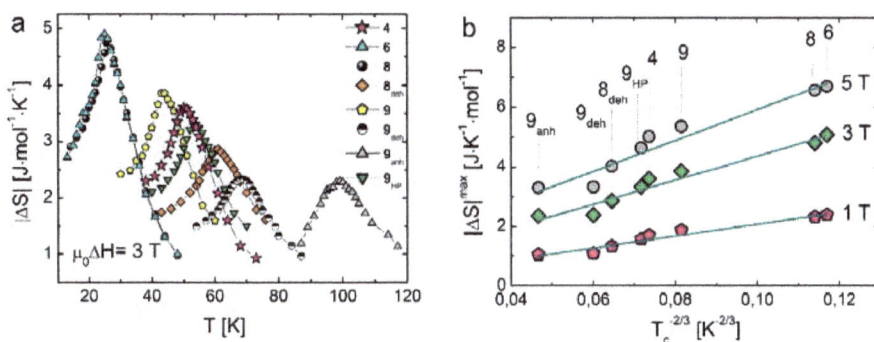

Figure 26. (a) The comparison of temperature dependences of magnetic entropy change determined for the manganese octacyanoniobate-based samples with different ligands: pyridazine for samples **9**, 9_{deh}, 9_{anh}, and 9_{HP}, imidazole for **8** and 8_{deh}, pyrazole for **4** and ligand-free **6**. (b) The maximum values of magnetic entropy change $|\Delta S|^{max}$ obtained for **9**, 9_{deh}, 9_{anh}, 9_{HP}, **8**, 8_{deh}, **4**, and **6** at the change of magnetic field $\mu_0 \Delta H = 1, 3$ and 5 T as a function of $T_c^{-2/3}$.

3.7. The Critical Behaviour of the 3 D Octacyanoniobate- Based Compounds

The MCE is an intrinsic thermodynamic characteristic and in the vicinity of the ferro- and ferrimagnetic transition reflects the critical behaviour of the system. Thus, the critical exponents can be extracted by use of the isothermal magnetization curves measured with the aim to determine the magnetic entropy change. According to the scaling hypothesis, the spontaneous magnetization M_S below T_c, the inverse initial susceptibility χ_0 above T_c, and the magnetization dependence on magnetic field $M(H)$ measured at temperature T_c are characterized by the set of critical exponents β, γ and δ. These exponents are defined by the following relations:

$$M_S(T) = M_0(\varepsilon)^\beta, T < T_c \tag{8}$$

$$\chi_0^{-1}(T) = (h_0/M_0)(\varepsilon)^\gamma, \ T > T_c \tag{9}$$

$$M = DH^{\frac{1}{\delta}}, T = T_c \tag{10}$$

where ε is a reduced temperature defined as

$$\varepsilon = (T - T_c)/T_c$$

According to the mean field theory, which predicts in the vicinity of critical temperature T_c, $\beta = 0.5$ and $\gamma = 1$, the isotherms $M^2(H/M)$, so called Arrot plots, should consist of series of parallel straight lines, and the isotherm measured at T_c should pass through the origin of coordinate system. Arrot plots can be used for the determination of the order of phase transition occurring in the material under study. According to the Banerjee criterion [56], a positive slope of the $M^2(H/M)$ indicates a magnetic phase transition of the second order, while a negative slope corresponds to the first order phase transition. For all the octacyanoniobate-based samples the positive slope of $M^2(H/M)$ plots was observed, indicating the second order of the magnetic phase transition. The non-linear shape of $M^2(H/M)$ Arrot for all discussed samples plots indicates that mean field approach does not describes properly the critical properties of these materials.

The critical exponents β and γ were determined by using the Kouvel–Fisher method [57]. In the Kouvel–Fisher method two new variables X and Y are defined as

$$X(T) = \chi_0^{-1}\left(\frac{\mathrm{d}\chi_0^{-1}}{\mathrm{d}T}\right)^{-1} = \frac{T - T_c}{\gamma}, \tag{11}$$

$$Y(T) = M_0\left(\frac{\mathrm{d}M_0}{\mathrm{d}T}\right)^{-1} = \frac{T - T_c}{\beta}. \tag{12}$$

$X(T)$ and $Y(T)$ are the linear functions of temperature with slope of $1/\gamma$ and $1/\beta$, respectively. The intercepts of $X(T)$ and $Y(T)$ with the temperature axis correspond to the critical temperatures. Figure 27a shows an exemplary result obtained for sample **4** with the Kouvel–Fisher method, which next was used for the determination of the critical exponents β and γ for samples **9**, **9**$_{deh}$, **9**$_{anh}$, **9**$_{HP}$, **8**, **8**$_{deh}$, **4**, and **6** [42,43,53,55,58]. The obtained results are presented in Table 5.

Table 5. Comparison of the critical exponents determined for samples discussed in Section 3 and for the mean field and 3D Heisenberg, 3D Ising, and 3D XY model. n_{MCE}—the critical exponent n determined based on the MCE data, n_{theor}—the critical exponents n determined based on Equation (13).

Sample	β	γ	δ	n_{MCE}	n_{theor}
4	0.41	1.32	4.39	0.69	0.66
5	0.37	1.33	4.37	0.67	0.63
6				0.64	
7				0.59	
8	0.37	1.35	4.48	0.65	0.63
8_{deh}	0.37	1.40	4.95	0.67	0.64
9	0.38	1.35	4.69	0.66	0.64
9_{deh}	0.43	1.38	4.23	0.68	0.69
9_{anh}	0.39	1.37	4.49	0.69	0.65
9_{HP}	0.37	1.40	4.48	0.67	0.64
Mean field model	0.500	1	3		0.66
Heisenberg model	0.365	1.385	4.8		0.64
Ising model	0.325	1.24	4.82		0.61
XY model	0.346	1.316	4.81		0.57

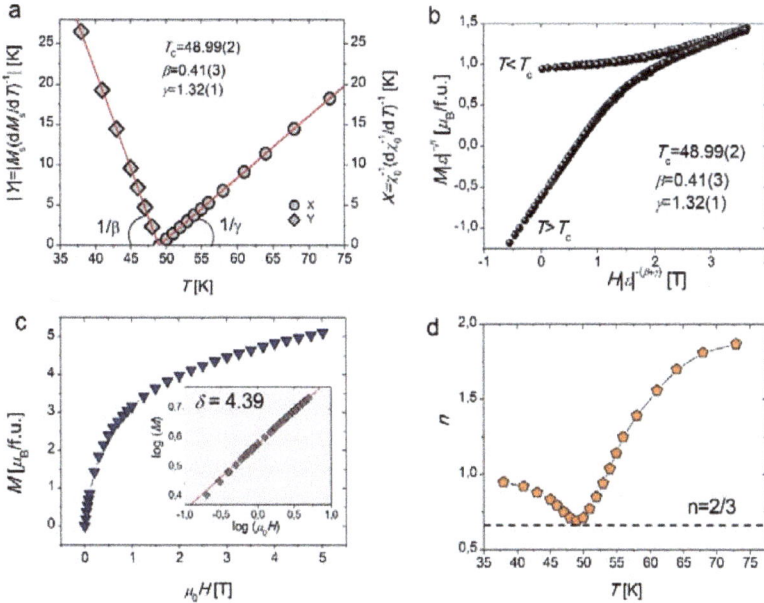

Figure 27. (**a**) Determination of the critical exponents and the critical temperature for **4** using the Kouvel–Fisher method, (**b**) logarithmic scaling plot of $M |\varepsilon|^{-\beta}$ vs. $H \cdot |\varepsilon|^{-(\beta + \gamma)}$ for **4** evidences the validity Kouvel–Fisher method, (**c**) magnetization vs. magnetic field measured for $M(H)$ obtained for **4** at temperature $T = 49$ K; inset shows the same $M(H)$ plots in the log–log scale, which was used for determination of critical exponent δ, (**d**) temperature dependence of exponent n obtained for **4**.

The value of critical exponent δ was estimated using the $M(H)$ isotherm measured at temperature T_c. The magnetization curve measured at T_c presented in log–log plot is expected to be a straight line with a slope $1/\delta$. The example result of the analysis of the critical isotherm measured for **4** is shown in Figure 27c.

In the materials undergoing the second order phase transition, the magnetic entropy change is related to the change of the external magnetic field by the relation: $|\Delta S|(T, H) \propto H^n$ where n is a successive critical exponent. The mean field theory predicts the exponent $n = 1$ for the temperatures well below T_c, and $n = 2$ well above T_c. At critical temperature n is equal to 2/3. The analysis of temperature dependence of exponent n shows that for all octacyanoniobate-based compounds, shapes of the $n(T)$ plot are very similar and the values of n at T_c are very close to 2/3. The example of dependence $n(T)$ obtained for **4** is presented in Figure 27d.

At temperature T_c, critical exponents β and γ are related to exponent n by the formula [8]:

$$n(T_c) = 1 + \frac{\beta - 1}{\beta + \gamma} \tag{13}$$

Thus, having the values of the critical exponents β and γ obtained with the Kouvel–Fisher method, the value of the critical exponent n can be estimated. For all octacyanoniobate-based samples we obtained a very good agreement between the values of critical exponent n determined using Equation (13) (in Table 5 denoted as n_{theor}) with the experimentally determined values based on MCE data (in Table 5 denoted as n_{MCE}).

The analysis of the critical exponents determined for the series of octacyanoniobate-based samples showed that except sample (**7**), all the examined compounds can be assigned to the universality class of the 3D Heisenberg model. It means that neither structural change due to the insertion of bridging ligands into the structure nor the influence of the external stimuli significantly affect the critical behaviour of the octacyanometallates-based materials. The result obtained for sample **7** is more consistent with XY model.

We have also compared the described above values of the critical exponents with the values predicted from the scaling hypothesis. In critical region, the magnetic equation of state can be written as:

$$M(H, \varepsilon) = \varepsilon^\beta f_\pm \left(\frac{H}{\varepsilon^{\beta + \gamma}} \right) \tag{14}$$

where $f_\pm$ is the scaling function: f_+ for $T > T_c$, f_- for $T < T_c$. Equation (14) suggests, that $M |\varepsilon|^{-\beta}$ plotted as a function of $H \cdot |\varepsilon|^{-(\beta + \gamma)}$ should give two different curves: one corresponding to the temperatures below the ordering temperature, and the other for temperatures above T_c. For all the studied samples for which the critical exponent β and γ were determined with Kouvel–Fisher method, $M |\varepsilon|^{-\beta}$ vs. $H \cdot |\varepsilon|^{-(\beta + \gamma)}$ plots exhibit two independent branches, indicating that the values of the critical exponents are reasonably accurate. Figure 27b presents the example of logarithmic scaling plot of $M |\varepsilon|^{-\beta}$ vs. $H \cdot |\varepsilon|^{-(\beta + \gamma)}$ obtained for sample **4**.

3.8. Final Remarks of Section 3

In this section we have investigated the magnetocaloric effect and critical behaviour in molecular magnets showing long-range magnetic order. The most important issue of this study was the test of the influence of bridging ligands and external stimuli on magnetocaloric properties of examined samples.

We have demonstrated that for the manganese octacyanoniobate-based samples with different brigding ligand—pyridazine, pyrazole, imidazole, for which the magnetocaloric effect was determined for as-synthetized samples and modified by external stimuli—the maximum value of magnetic entropy change ΔS^{max} is proportional to $T_c^{-2/3}$.

We have also proved, that 3D Heisenberg model is the most adequate for the description of critical behaviour of octacyanoniobate-based molecular magnets. The analysis of critical exponents β, γ, δ, and n showed that neither structural changes due to the insertion of bridging ligands into the structure nor the influence of external stimuli such as hydrostatic pressure and dehydration/hydration process do not significantly affect the critical behaviour of octacyanometallate- based materials. Moreover, the value of critical exponent n describing the field dependence of MCE according to $|\Delta S|(T, H) \propto H^n$ is consistent with other critical exponents obtained from the magnetization data.

4. Rotating Magnetocaloric Effect in Anisotropic Two-Dimensional Molecular Magnets

The rotating magnetocaloric effect (RMCE) is a new issue in the magnetic cooling research. In contrast to conventional magnetocaloric effect, in RMCE the change of entropy is obtained not by changing the external magnetic field, but with rotation of a single crystal in a constant magnetic field [59–62]. If the compound reveals a substantial magnetic anisotropy, then the magnetic entropy will depend on the crystal orientation in the magnetic field. The rotation is changing the crystal orientation with respect to the applied field direction, therefore a change of magnetic entropy is observed and can be used in cooling cycle. Recently, Balli et al. [63,64] have introduced a realization of refrigerator based on RMCE. This approach has several advantages: simple construction, high efficiency [63–65] (cycles in higher frequency than the conventional MCE), or working in constant field (lower power consumption, possibility of use permanent magnets).

Most of the research concerning rotating magnetocaloric effect deals with inorganic materials [59–61,63,64,66–68] and there are only few examples related to molecular magnets [69]. In our research we have focused on two-dimensional molecular compounds which reveal magnetic anisotropy and transition to long-range ordered phase. Single crystal studies allowed us to explore the anisotropy of MCE. The dependence of MCE on the orientation was used to study the RMCE in case of low (Mn^{II}(R-mpm)$_2$]$_2$[Nb^{IV}(CN)$_8$]}·4H$_2$O) [70] and high ({(tetren)H$_5$)$_{0.8}$Cu^{II}$_4$[W^V(CN)$_8$]$_4$·7.2H$_2$O}$_n$) [71] magnetic anisotropy. In particular, we have shown that inverse magnetocaloric effect can be used to enhance the RMCE up to 51% in respect to conventional MCE.

4.1. Low Anisotropy Case: {Mn^{II}(R-mpm)$_2$]$_2$[Nb^{IV}(CN)$_8$]}·4H$_2$O Crystal

{Mn^{II}(R-mpm)$_2$]$_2$[Nb^{IV}(CN)$_8$]}·4H$_2$O (**10**), where mpm = α-methyl-2-pyridinemethanol, is a two-dimensional coordination ferrimagnet [70]. The separation between layers of square grid topology is 7.5 Å. Nevertheless, the compound reveals a phase transition to 3D long-range magnetic ordered state at T_c = 23.5 K due to intermolecular dipole-dipole interactions. The magnetic single crystal measurements of **10** showed an easy-plane type anisotropy within the layers, whereas the perpendicular direction was a hard axis [72]. The observed anisotropy was not significant, since above 0.35 T applied field both orientations was magnetically undistinguishable. The magnetocaloric effect was obtained by the indirect method from magnetization measurements in two orientations: $bc \parallel H$ (easy plane) and $a* \parallel H$ (hard axis).

In high magnetic fields (above 1.0 T) the difference between magnetic entropy change in both orientations is modest (Figure 28a), as a consequence of low magnetic anisotropy. In lower fields the value of $-\Delta S_m$ is small, but the relative change between easy plane and hard direction is more significant (Figure 28b). Moreover, in the hard axis orientation an inverse magnetocaloric effect can be noticed. The magnetic entropy change related to rotation (ΔS_{RMCE}) by 90° from hard direction ($a* \parallel H$, hard axis) to easy axis ($bc \parallel H$) can be calculated by

$$- \Delta S_{RMCE} = -(\Delta S_{easy} - \Delta S_{hard}) \qquad (15)$$

where ΔS_{easy} and ΔS_{hard} stands for magnetic entropy change in easy plane and hard axis orientations, respectively. Figures 28 and 29 show the obtained values of $-\Delta S_{RMCE}$. In high magnetic fields (Figure 28a) both conventional MCE, for hard axis and easy plane, are greater than the RMCE in whole temperature range. The situation is changing in lower fields $\mu_0 H < 0.2$ T (Figure 28b), for which the RMCE can have higher output even than the MCE for easy plane. This excess is a consequence of rotation from hard axis (higher entropy) to easy plane (lower entropy) orientation and the inverse MCE in hard axis. Figure 30 shows the ratio between the entropy change from RMCE and conventional MCE. Depending on the applied field and the temperature, the RMCE can be more efficient than MCE in an easy plane up to 51%.

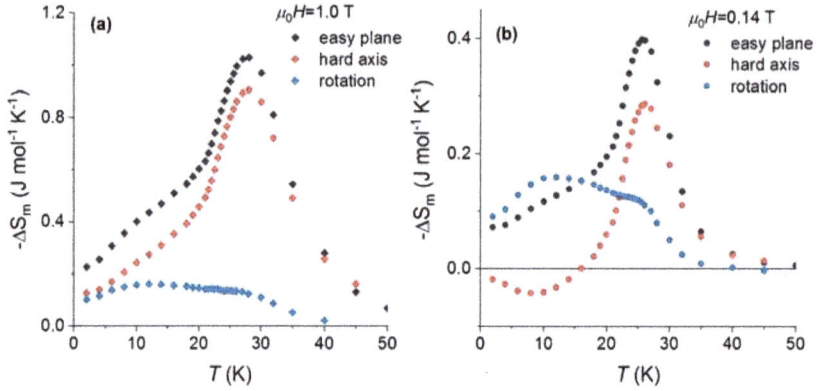

Figure 28. The magnetic entropy change for **10** due to application of the magnetic field 0-$\mu_0 H$ (black-easy plane, red -hard axis) and by rotating the crystals in constant field $\mu_0 H$. (**a**) $\mu_0 H = 1.0$ T, (**b**) $\mu_0 H = 0.14$ T.

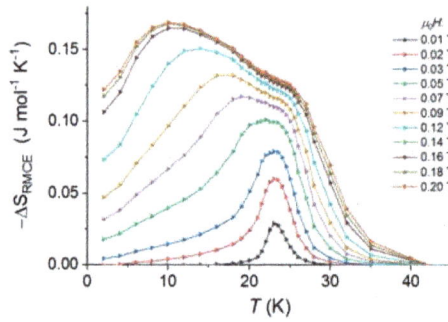

Figure 29. The magnetic entropy change with RMCE for **10**. Solid lines are guides for the eyes. Adapted with permission from Reference [72]. Copyright 2017 American Chemical Society.

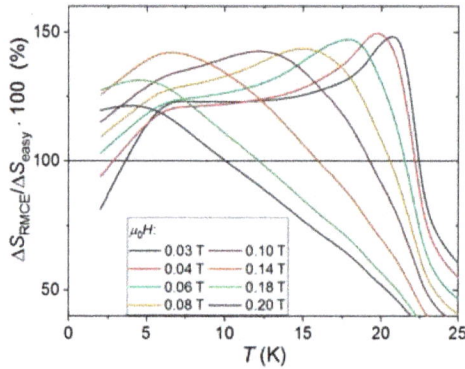

Figure 30. Temperature dependence of percentage ratio between ΔS_{RMCE} and ΔS_{easy} (easy plane) for **10** for different fields. Adapted with permission from Reference [72]. Copyright 2017 American Chemical Society.

4.2. High Anisotropy Case: (tetren)Cu4[W(CN)8]4 Crystal

The high anisotropy case was studied [71] with (tetren)Cu$_4$[W(CN))$_8$]$_4$ (**11**)(full formula: {((tetren)H$_5$)$_{0.8}$Cu$^{II}_4$[W^V(CN)$_8$]$_4$·7.2H$_2$O}$_n$, tetren = tetraethylenepentamine), a 2D cyanido-bridged network with significant 2D XY magnetic anisotropy (*ac* is the easy plane, *b* is the hard axis) and the Berezinskii-Kosterlitz-Thouless [73] topological phase transition at $T_{BKT} \approx 33$ K [74,75]. The anisotropy for **11** is so high, that even $\mu_0H = 7.0$ T magnetic field is too weak to merge magnetization curves of two orientations at 2.0 K [71]. The conventional and rotating magnetocaloric effects were studied in a similar procedure to the low anisotropy case (**10**). However, in this case the difference between magnetic entropy change for hard and easy orientations was relevant up to the highest measured field $\mu_0H = 7.0$ T (Figure 31). Therefore, the absolute values of $-\Delta S_{RMCE}$ for the RMCE were one order of magnitude higher than for **10** (Figure 32). The temperature dependences of ΔS_{RMCE} for **11** have peculiar shapes with two peaks for fields below 7.0 T and a single peak for the highest field. The double peaks are related to different dependences of ΔS maxima for easy and hard orientation. The study also showed that it is possible to obtain an inverse RMCE (Figures 31 and 32), as a consequence of temperature region where the $-\Delta S_m$ for hard axis is higher than for the easy plane.

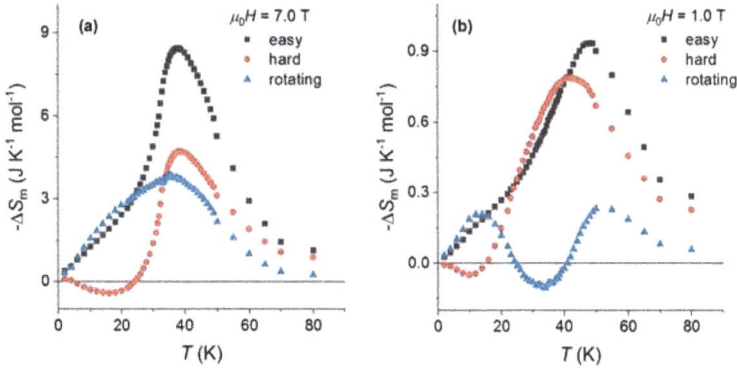

Figure 31. The magnetic entropy change for **11** due to application of the magnetic field 0- μ_0H (black-easy plane, red -hard axis) and by rotating the crystals in constant field μ_0H. (a) $\mu_0H = 7.0$ T, (b) $\mu_0H = 1.0$ T.

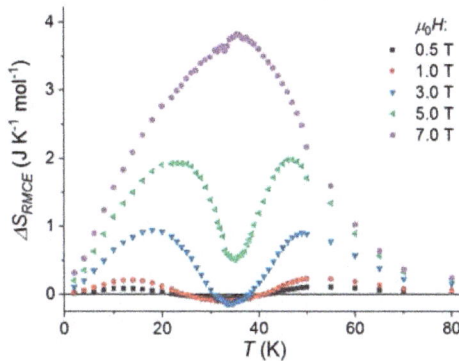

Figure 32. The temperature dependence of magnetic entropy change due to single crystals rotation in the constant field for **11**. Adapted with permission from Reference [71] Copyright 2017 American Chemical Society.

Figure 33 shows the ratio between the conventional and rotating MCE for **11**. Similar to the low anisotropy case of **10**, there is a region where ΔS_{RMCE} is more efficient than the MCE for easy plane. However, the $\Delta S_{RMCE}/\Delta S_{easy}$ ratio shows also negative values for lower fields ($\mu_0 H < 3.0$ T) and temperatures around 30 K. These negative values correspond to the conditions, where the sign of the ΔS_{RMCE} is negative, in other words, where the inverse RMCE occurs.

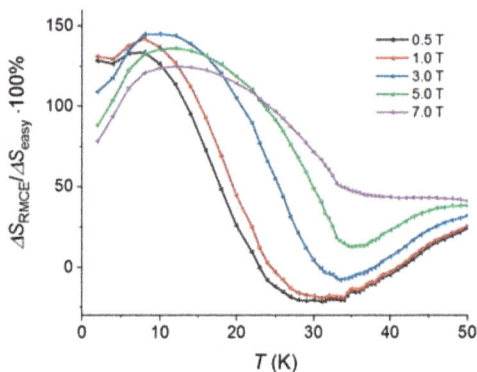

Figure 33. Percentage ratio between ΔS_{RMCE} and ΔS_{easy} (easy plane) for **11** as a function of temperature for different fields. Adapted with permission from Reference [71] Copyright 2017 American Chemical Society.

4.3. Final Remarks of Section 4

The rotational magnetocaloric effect is an alternative method for magnetic cooling. In our works we were studying 2D molecular magnets with low and high anisotropy. We have proven that the RMCE can be, in specific circumstances, more efficient than the conventional MCE and that the inverse RMCE is also possible. In our opinion, the rotating magnetocaloric effect in molecular magnets has potential application as cryogenic refrigerator. However, to compete with other types of low temperature refrigerators (e.g., He-3 type), the good candidate for a molecular RMCE refrigerant should be characterized by the following features [71]: (1) strong magnetic anisotropy, (2) T_c around 2 K, (3) large magnetic moments with ferromagnetic interactions, (4) easy plane anisotropy to achieve the RMCE enhancement due to the inverse MCE.

5. Final Conclusions

In this review, the magneto-thermal properties of octacyanometallate-based molecular magnets showing the different types of crystal architecture have been discussed. The investigation of magnetocaloric effect has been performed by two experimental methods: calorimetry and magnetometry. The highest value of magnetic entropy change and the change of adiabatic temperature was observed for the high-spin dodecanuclear cluster $Ni_9[W(CN)_8]_6$ making this compound a suitable candidate for application in cryogenic magnetic cooling. The systematic study of magnetocaloric effect in the ligand tunable coordination compounds is based on manganese and octacyanoniobate, showing the wide temperature range in which T_c occurs, allowing us to confirm the $\Delta S_m \sim T_c^{-2/3}$ relation stemming from the molecular field theory. The set of critical exponents obtained for these compounds series belong to the 3D Heisenberg universality class. Finally, we have presented the study of rotating magnetocaloric effect (RMCE) in two 2D molecular magnets: ferromagnetic {Mn^{II}(R-mpm)$_2$]$_2$[Nb^{IV}(CN)$_8$]}·4H$_2$O and strongly anisotropic (tetren)Cu$_4$[W(CN)$_8$]$_4$ bilayered magnet showing the topological Berezinskii-Kosterlitz-Thouless transition.

Author Contributions: M.F. wrote Section 3, reviewed and edited the manuscript, R.P. wrote Section 2, P.K. wrote Section 4, M.B. wrote Section 1 and reviewed the manuscript together with M.F.

Funding: This research was funded by the Polish National Science Centre within the frame of Project No. DEC-2013/11/N/ST8/01267 and UMO-2011/03/D/ST5/05400.

Conflicts of Interest: The authors declare no conflict of interest.

References

1. Kahn, O. *Molecular Magnetism*; VCH Weinheim: Weinheim, Germany, 1993.
2. Gatteschi, D.; Sessoli, R.; Villain, J. *Molecular Nanomagnets*; Oxford University Press: Oxford, UK, 2003.
3. Bartolomé, J.; Luis, F.; Fernández, J.F. *Molecular Magnets—Physics and Applications*; Springer: Berlin/Heidelberg, Germany, 2014.
4. Sieklucka, B.; Pinkowicz, D. *Molecular Magnetic Materials—Concepts and Applications*; VCH Weinheim: Weinheim, Germany, 2017.
5. Zhang, Q.; Li, B.; Zhao, X.G.; Zhang, Z.D. Magnetic and reversible magnetocaloric properties of $(Gd_{1-x}Dy_x)_4Co_3$ ferrimagnets. *J. Appl. Phys.* **2009**, *105*, 053902. [CrossRef]
6. Samanta, T.; Das, I.; Banerjee, S. Giant magnetocaloric effect in antiferromagnetic $ErRu_2Si_2$ compound. *Appl. Phys. Lett.* **2007**, *91*, 152506. [CrossRef]
7. Oesterreicher, H.; Parker, F.T. Magnetic cooling near Curie temperatures above 300 K. *J. Appl. Phys.* **1984**, *55*, 4334. [CrossRef]
8. Franco, V.; Blázquez, J.S.; Conde, A. Field dependence of the magnetocaloric effect in materials with a second order phase transition: A muster curve for the magnetic entropy change. *Appl. Phys. Lett.* **2006**, *89*, 222512. [CrossRef]
9. Belo, J.H.; Amaral, J.S.; Pereira, A.M.; Amaral, V.S.; Araújo, J.P. On the Curie temperature dependency of the magnetocaloric effect. *Appl. Phys. Lett.* **2012**, *100*, 242407. [CrossRef]
10. Von Ranke, P.J.; Gama, S.; Coelho, A.A.; De Campos, A.; Magnus, A.; Carvalho, G.; Gandra, F.C.; de Oliveira, N.A. Theoretical description of the colossal entropic magnetocaloric effect: Application to MnAs. *Phys. Rev. B* **2006**, *73*, 014415. [CrossRef]
11. Gutfleisch, O.; Willard, M.A.; Brück, E.; Chen, C.H.; Sankar, S.G.; Liu, J.P. Magnetic Materials and Devices for the 21st Century: Stronger, Lighter, and More Energy Efficient. *Adv. Mater.* **2011**, *23*, 821. [CrossRef]
12. Sessoli, R. Chilling with Magnetic Molecules. *Angew. Chem. Int. Ed.* **2012**, *51*, 43. [CrossRef]
13. Evangelisti, M.; Brechin, E.K. Recipes for enhanced molecular cooling. *Dalton Trans.* **2010**, *39*, 4672–4676. [CrossRef]
14. Sibille, R.; Mazet, T.; Malaman, B.; François, M. A Metal–Organic Framework as Attractive Cryogenic Magnetorefrigerant. *Chem. Eur. J.* **2012**, *18*, 12970. [CrossRef]
15. Evangelisti, M.; Roubeau, O.; Palacios, E.; Camón, A.; Hooper, T.N.; Brechin, E.K.; Alonso, J.J. Cryogenic magnetocaloric effect in a ferromagnetic molecular dimer. *Angew. Chem. Int. Ed.* **2011**, *50*, 6606–6609. [CrossRef] [PubMed]
16. Peng, J.B.; Zhang, Q.C.; Kong, X.J.; Zheng, Y.Z.; Ren, Y.P.; Long, L.S.; Huang, R.B.; Zheng, L.S.; Zheng, Z. High-Nuclearity 3d–4f Clusters as Enhanced Magnetic Coolers and Molecular Magnets. *J. Am. Chem. Soc.* **2012**, *134*, 3314–3317. [CrossRef] [PubMed]
17. Chang, L.X.; Xiong, G.; Wang, L.; Cheng, P.; Zhao, B.A. 24-Gd nanocapsule with a large magnetocaloric effect. *Chem. Commun.* **2013**, *49*, 1055. [CrossRef] [PubMed]
18. Chen, Y.C.; Guo, F.S.; Liu, J.-L.; Leng, J.-D.; Vrábel, P.; Orendáč, M.; Prokleška, J.; Sechovský, V.; Tong, M.L. Switching of the Magnetocaloric Effect of Mn^{II} Glycolate by Water Molecules. *Chem. Eur. J.* **2014**, *20*, 3029. [CrossRef]
19. Liu, J.-L.; Chen, Y.-C.; Guo, F.-S.; Tong, M.-L. Recent advances in the design of magnetic molecules for use as cryogenic magnetic coolants. *Coord. Chem. Rev.* **2014**, *281*, 26. [CrossRef]
20. Entley, W.R.; Girolami, G.S. High-Temperature Molecular Magnets Based on Cyanovanadate Building Blocks: Spontaneous Magnetization at 230 K. *Science* **1995**, *268*, 397–400. [CrossRef]
21. Verdaguer, M.; Girolami, G. Magnetic Prussian Blue Analogs. In *Magnetism: Molecules to Materials V*; Miller, J.S., Drillon, M., Eds.; Wiley-VCH Verlag GmbH: Weinheim, Germany, 2001; pp. 283–347.

22. Sieklucka, B.; Podgajny, R.; Pinkowicz, D.; Nowicka, B.; Korzeniak, T.; Bałanda, M.; Wasiutyński, T.; Pełka, R.; Makarewicz, M.; Czapla, M.; et al. Towards high Tc octacyanometalate-based networks. *CrystEngComm* **2009**, *11*, 2032. [CrossRef]

23. Sieklucka, B.; Podgajny, R.; Korzeniak, T.; Nowicka, B.; Pinkowicz, D.; Kozieł, M. A decade of octacyanides in polynuclear molecular materials. *Eur. J. Inorg. Chem.* **2011**, *3*, 305. [CrossRef]

24. Nowicka, B.; Korzeniak, T.; Stefańczyk, O.; Pinkowicz, D.; Chorąży, S.; Podgajny, R.; Sieklucka, B. The impact of ligands upon topology and functionality of octacyanidometallate-based assemblies. *Coord.Chem.Rev.* **2012**, *256*, 1946. [CrossRef]

25. Song, Y.; Zhang, P.; Ren, X.-M.; Shen, X.-F.; Li, Y.-Z.; You, X.-Z. Octacyanometalate-based single-molecule magnets: $Co^{II}_9M^V_6$ (M = W, Mo). *J. Am. Chem. Soc.* **2005**, *127*, 3708. [CrossRef]

26. Yoo, H.S.; Ko, H.H.; Ryu, D.W.; Lee, J.W.; Yoon, J.H.; Lee, W.R.; Kim, H.C.; Koh, E.K.; Hong, C.S. Octacyanometalate-Based Ferrimagnetic M^VMn^{III} (M = Mo, W) bimetallic chain racemates with slow magnetic relaxations. *Inorg. Chem.* **2009**, *48*, 5617. [CrossRef] [PubMed]

27. Chorazy, S.; Stanek, J.; Nogas, W.; Majcher, A.M.; Rams, M.; Kozieł, M.; Juszyńska-Gałązka, E.; Nakabayashi, K.; Ohkoshi, S.; Sieklucka, B.; et al. Tuning of charge transfer assisted phase transition and slow magnetic relaxation functionalities in {$Fe_{9-x}Co_x[W(CN)_8]_6$} (x = 0–9) molecular solid solution. *J. Am. Chem. Soc.* **2016**, *138*, 1635. [CrossRef] [PubMed]

28. Nowicka, B.; Rams, M.; Stadnicka, K.; Sieklucka, B. Reversible Guest-Induced Magnetic and Structural Single-Crystal-to-Single-Crystal Transformation in Microporous Coordination Network {[Ni(cyclam)]$_3$ [W(CN)$_8$]$_2$}$_n$. *Inorg. Chem.* **2007**, *46*, 8123. [CrossRef] [PubMed]

29. Nowicka, B.; Reczyński, M.; Rams, M.; Nitek, W.; Kozieł, M.; Sieklucka, B. Larger pores and higher T$_C$: {[Ni(cyclam)]$_3$[W(CN)$_8$]$_2$·solv}$_n$—A new member of the largest family of pseudo-polymorphic isomers among octacyanometallate-based assemblies. *CrystEngComm* **2015**, *17*, 3526. [CrossRef]

30. Ohkoshi, S.; Tokoro, H.; Hozumi, T.; Zhang, Y.; Hashimoto, K.; Mathonière, C.; Bord, I.; Rombaut, G.; Verelst, M.; Moulin, C.C.D.; et al. Photoinduced Magnetization in Copper Octacyanomolybdate. *J. Am. Chem. Soc.* **2006**, *128*, 270. [CrossRef] [PubMed]

31. Ohkoshi, S.; Tokoro, H. Photomagnetism in Cyano-Bridged Bimetal Assemblies. *Acc. Chem. Res.* **2012**, *45*, 1749. [CrossRef] [PubMed]

32. Pinkowicz, D.; Podgajny, R.; Nitek, W.; Rams, M.; Majcher, A.M.; Nuida, T.; Ohkoshi, S.; Sieklucka, B. Multifunctional Magnetic Molecular {[Mn^{II}(urea)$_2$(H$_2$O)]$_2$[Nb^{IV}(CN)$_8$]}$_n$ System: Magnetization-Induced SHG in the Chiral Polymorph. *Chem. Mater.* **2011**, *23*, 21. [CrossRef]

33. Pinkowicz, D.; Podgajny, R.; Gaweł, B.; Nitek, W.; Łasocha, W.; Oszajca, M.; Czapla, M.; Makarewicz, M.; Bałanda, M.; Sieklucka, B. Double Switching of a Magnetic Coordination Framework through Intraskeletal Molecular Rearrangement. *Angew. Chem. Int. Ed.* **2011**, *50*, 3973. [CrossRef]

34. Arczyński, M.; Rams, M.; Stanek, J.; Fitta, M.; Sieklucka, B.; Dunbar, K.R.; Pinkowicz, D. A Family of Octahedral Magnetic Molecules Based on [NbIV(CN)$_8$]$^{4-}$. *Inorg. Chem.* **2017**, *56*, 4021. [CrossRef]

35. Nowicka, B.; Stadnicka, K.; Nitek, W.; Rams, M.; Sieklucka, B. Geometrical isomerism in pentadecanuclear high-spin Ni$_9$W$_6$ clusters with symmetrical bidentate ligands detected. *CrystEngComm* **2012**, *14*, 6559. [CrossRef]

36. Gajewski, M.; Pełka, R.; Fitta, M.; Miyazaki, Y.; Nakazawa, Y.; Bałanda, M.; Reczyński, M.; Nowicka, B.; Sieklucka, B. Magnetocaloric effect of high spin cluster with Ni$_9$W$_6$ core. *J. Magn. Magn. Mater.* **2016**, *414*, 25. [CrossRef]

37. Pinkowicz, D.; Pełka, R.; Drath, O.; Nitek, W.; Bałanda, M.; Majcher, A.M.; Poneti, G.; Sieklucka, B. Nature of magnetic interactions in 3D {[M^{II}(pyrazole)$_4$]$_2$[Nb^{IV}(CN)$_8$]·4H$_2$O}$_n$ (M = Mn, Fe, Co, Ni) molecular magnets. *Inorg. Chem.* **2010**, *49*, 7565. [CrossRef]

38. Pełka, R.; Gajewski, M.; Miyazaki, Y.; Yamashita, S.; Nakazawa, Y.; Fitta, M.; Pinkowicz, D.; Sieklucka, B. Magnetocaloric effect in Mn$_2$-pyrazole-[Nb(CN)$_8$] molecular magnet by relaxation calorimetry. *J. Magn. Magn. Mater.* **2016**, *419*, 435. [CrossRef]

39. Pełka, R.; Konieczny, P.; Zieliński, P.M.; Wasiutyński, T.; Miyazaki, Y.; Inaba, A.; Pinkowicz, D.; Sieklcuka, B. Magnetocaloric effect in {[Fe(pyrazole)$_4$]$_2$[Nb(CN)$_8$]·4H$_2$O}n molecular magnet. *J. Magn. Magn. Mater.* **2014**, *354*, 359. [CrossRef]

40. Konieczny, P.; Pełka, R.; Zieliński, P.M.; Pratt, F.L.; Pinkowicz, D.; Sieklucka, B.; Wasiutyński, T. Scaling analysis of [Fe(pyrazole)$_4$]$_2$[Nb(CN)$_8$] molecular magnet. *J. Magn. Magn. Mater.* **2013**, *344*, 105. [CrossRef]

41. Evangelisti, M.; Candini, A.; Affronte, M.; Pasca, E.; de Jongh, L.J.; Scott, R.T.W.; Brechin, E.K. Magnetocaloric effect in spin degenerated molecular nanomagnets. *Phys. Rev. B* **2009**, *79*, 104414. [CrossRef]

42. Fitta, M.; Bałanda, M.; Mihalik, M.; Pełka, R.; Pinkowicz, D.; Sieklucka, B.; Zentková, M. Magnetocaloric effect in M-pyrazole-[Nb(CN)$_8$] molecular compounds. *J. Phys.Condens. Matter.* **2012**, *24*, 506002. [CrossRef] [PubMed]

43. Fitta, M.; Pełka, R.; Bałanda, M.; Czapla, M.; Mihalik, M.; Pinkowicz, D.; Sieklucka, B.; Wasiutyński, T.; Zentková, M. Magnetocaloric effect in Mn$_2$-pyridazine-[Nb(CN)$_8$] molecular magnetic sponge. *Eur. J. Inorg. Chem.* **2012**, *2012*, 3830. [CrossRef]

44. Franco, V.; Conde, A.; Romero-Enrique, J.M.; Blázquez, J.S. A universal curve for the magnetocaloric effect: An analysis based on scaling relations. *J. Phys. Condens. Matter* **2008**, *20*, 285207. [CrossRef]

45. Campostrini, M.; Hasenbusch, M.; Palissetto, A.; Rossi, P.; Vicari, E. Critical exponents and equation of state of the three-dimensional Heisenberg universality class. *Phys. Rev. B* **2002**, *65*, 144520. [CrossRef]

46. Manuel, E.; Evangelisti, M.; Affronte, M.; Okubo, M.; Train, C.; Verdaguer, M. Magnetocaloric effect in hexacyanochromate Prussian blue analogs. *Phys. Rev. B* **2006**, *73*, 172406. [CrossRef]

47. Herrera, J.M.; Franz, P.; Podgajny, R.; Pilkington, M.; Biner, M.; Decurtins, S.; Stoeckli-Evans, H.; Neels, A.; Garde, R.; Dromzee, Y.; et al. Three-dimensional bimetallic octacyanidometalates [M^{IV}{(μ-CN)$_4$Mn^{II}(H$_2$O)$_2$}$_2$·4H$_2$O]$_n$ (M= Nb, Mo, W): Synthesis, single-crystal X-ray diffraction and magnetism. *C. R. Chim.* **2008**, *11*, 1192. [CrossRef]

48. Pinkowicz, D.; Podgajny, R.; Pełka, R.; Nitek, W.; Bałanda, M.; Makarewicz, M.; Czapla, M.; Zukrowski, J.; Kapusta, C.; Zajac, D.; et al. Iron(II)-octacyanoniobate(IV) ferromagnet with T_C 43 K. *Dalton Trans.* **2009**, 7771. [CrossRef] [PubMed]

49. Fitta, M.; Pełka, R.; Sas, W.; Pinkowicz, D.; Sieklucka, B. Dinuclear molecular magnets with unblocked magnetic connectivity: Magnetocaloric effect. *RSC Adv.* **2018**, *8*, 14640. [CrossRef]

50. Franco, V.; Caballero-Flores, R.; Conde, A.; Knipling, K.E.; Willard, M.A. Magnetocaloric effect and critical exponents of Fe$_{77}$Co$_{5.5}$Ni$_{5.5}$Zr$_7$B$_4$Cu$_1$: A detailed study. *J. Appl. Phys.* **2011**, *109*, 07A905. [CrossRef]

51. Foldeaki, M.; Schnelle, W.; Gmelin, E.; Benard, P.; Koszegi, B.; Giguere, A.; Chahine, R.; Boseet, T.K. Comparison of magnetocaloric properties from magnetic and thermal measurements. *J. Appl. Phys.* **1997**, *82*, 309. [CrossRef]

52. Pinkowicz, D.; Podgajny, R.; Bałanda, M.; Makarewicz, M.; Gaweł, B.; Łasocha, W.; Sieklucka, B. Magnetic Spongelike Behavior of 3D Ferrimagnetic {[Mn^{II}(imH)]$_2$[Nb^{IV}(CN)$_8$]}$_n$ with Tc = 62 K. *Inorg. Chem.* **2008**, *47*, 9745. [CrossRef] [PubMed]

53. Fitta, M.; Pełka, R.; Gajewski, M.; Mihalik, M.; Zentkova, M.; Pinkowicz, D.; Sieklucka, B.; Bałanda, M. Magnetocaloric effect and critical behavior in Mn$_2$-imidazole-[Nb(CN)$_8$] molecular magnetic sponge. *J. Magn. Magn. Mater.* **2015**, *396*. [CrossRef]

54. Pinkowicz, D.; Kurpiewska, K.; Lewiński, K.; Bałanda, M.; Mihalik, M.; Zentkova, M.; Sieklucka, B. High-pressure single-crystal XRD and magnetic study of aoctacyanoniobate-based magnetic sponge. *CrystEngComm* **2012**, *14*, 5224. [CrossRef]

55. Fitta, M.; Bałanda, M.; Pełka, R.; Konieczny, P.; Pinkowicz, D.; Sieklucka, B. Magnetocaloric effect and critical behaviour in Mn$_2$-pyridazine-[Nb(CN)$_8$] molecular compound under pressure. *J. Phys. Condens. Matter* **2013**, *25*, 496012. [CrossRef]

56. Banerjee, S.K. On a generalised approach to first and second order magnetic transitions. *Phys. Lett.* **1964**, *12*, 16. [CrossRef]

57. Kouvel, J.S.; Fisher, M.E. Detailed Magnetic Behavior of Nickel Near its Curie Point. *Phys. Rev.* **1964**, *136*, A1626. [CrossRef]

58. Pełka, R.; Pinkowicz, D.; Sieklucka, B.; Fitta, M. Molecular realizations of 3D Heisenberg magnet: Critical scaling. *J. Alloys Compd.* **2018**, *765*, 520. [CrossRef]

59. Nikitin, S.A.; Skokov, K.P.; Koshkid'ko, Y.S.; Pastushenkov, Y.G.; Ivanova, T.I. Giant Rotating Magnetocaloric Effect in the Region of Spin-Reorientation Transition in the NdCo$_5$ Single Crystal. *Phys. Rev. Lett.* **2010**, *105*, 137205. [CrossRef] [PubMed]

60. Orendáč, M.; Gabáni, S.; Gažo, E.; Pristáš, G.; Shitsevalova, N.; Siemensmeyer, K.; Flachbart, K. Rotating Magnetocaloric Effect and Unusual Magnetic Features in Metallic Strongly Anisotropic Geometrically Frustrated TmB$_4$. *Sci. Rep.* **2018**, *8*, 10933. [CrossRef] [PubMed]

61. Zhang, H.; Li, Y.; Liu, E.; Ke, Y.; Jin, J.; Long, Y.; Shen, B. Giant Rotating Magnetocaloric Effect Induced by Highly Texturing in Polycrystalline DyNiSi Compound. *Sci. Rep.* **2015**, *5*. [CrossRef]
62. Jin, J.-L.; Zhang, X.-Q.; Ge, H.; Cheng, Z.-H. Rotating Field Entropy Change in Hexagonal $TmMnO_3$ Single Crystal with Anisotropic Paramagnetic Response. *Phys. Rev. B* **2012**, *85*, 214426. [CrossRef]
63. Balli, M.; Jandl, S.; Fournier, P.; Dimitrov, D.Z. Giant Rotating Magnetocaloric Effect at Low Magnetic Fields in Multiferroic $TbMn_2O_5$ Single Crystals. *Appl. Phys. Lett.* **2016**, *108*, 102401. [CrossRef]
64. Balli, M.; Jandl, S.; Fournier, P.; Gospodinov, M.M. Anisotropy-Enhanced Giant Reversible Rotating Magnetocaloric Effect in $HoMn_2O_5$ Single Crystals. *Appl. Phys. Lett.* **2014**, *104*. [CrossRef]
65. Engelbrecht, K.; Eriksen, D.; Bahl, C.R.H.; Bjørk, R.; Geyti, J.; Lozano, J.A.; Nielsen, K.K.; Saxild, F.; Smith, A.; Pryds, N. Experimental Results for a Novel Rotary Active Magnetic Regenerator. *Int. J. Refrig.* **2012**, *35*, 1498. [CrossRef]
66. Tkáč, V.; Orendáčová, A.; Čižmár, E.; Orendáč, M.; Feher, A.; Anders, A.G. Giant Reversible Rotating Cryomagnetocaloric Effect in $KEr(MoO_4)_2$ Induced by a Crystal-Field Anisotropy. *Phys. Rev. B* **2015**, *92*, 24406. [CrossRef]
67. Caro Patiño, J.; de Oliveira, N.A. Rotating Magnetocaloric Effect in $HoAl_2$ Single Crystal. *Intermetallics* **2015**, *64*, 59. [CrossRef]
68. Balli, M.; Mansouri, S.; Jandl, S.; Fournier, P.; Dimitrov, D.Z. Large Rotating Magnetocaloric Effect in the Orthorhombic $DyMnO_3$ Single Crystal. *Solid State Commun.* **2016**, *239*, 9. [CrossRef]
69. Lorusso, G.; Roubeau, O.; Evangelisti, M. Rotating Magnetocaloric Effect in an Anisotropic Molecular Dimer. *Angew. Chem. Int. Ed.* **2016**, *55*, 3360–3363. [CrossRef] [PubMed]
70. Chorazy, S.; Podgajny, R.; Nitek, W.; Fic, T.; Görlich, E.; Rams, M.; Sieklucka, B. Natural and Magnetic Optical Activity of 2-D Chiral Cyanido-Bridged Mn^{II}–Nb^{IV} Molecular Ferrimagnets. *Chem. Commun.* **2013**, *49*, 6731. [CrossRef] [PubMed]
71. Konieczny, P.; Pełka, R.; Czernia, D.; Podgajny, R. Rotating Magnetocaloric Effect in an Anisotropic Two-Dimensional $Cu^{II}[W^V(CN)_8]^{3-}$ Molecular Magnet with Topological Phase Transition: Experiment and Theory. *Inorg. Chem.* **2017**, *56*, 11971. [CrossRef] [PubMed]
72. Konieczny, P.; Michalski, Ł.; Podgajny, R.; Chorazy, S.; Pełka, R.; Czernia, D.; Buda, S.; Mlynarski, J.; Sieklucka, B.; Wasiutyński, T. Self-Enhancement of Rotating Magnetocaloric Effect in Anisotropic Two-Dimensional (2D) Cyanido-Bridged Mn^{II}–Nb^{IV} Molecular Ferrimagnet. *Inorg. Chem.* **2017**, *56*, 2777. [CrossRef] [PubMed]
73. Baranová, L.; Orendáčová, A.; Čižmár, E.; Tarasenko, R.; Tkáč, V.; Orendáč, M.; Feher, A. Fingerprints of Field-Induced Berezinskii–Kosterlitz–Thouless Transition in Quasi-Two-Dimensional S = 1/2 Heisenberg Magnets $Cu(en)(H_2O)_2SO_4$ and $Cu(tn)Cl_2$. *J. Magn. Magn. Mater.* **2016**, *404*, 53. [CrossRef]
74. Bałanda, M.; Pełka, R.; Wasiutyński, T.; Rams, M.; Nakazawa, Y.; Miyazaki, Y.; Sorai, M.; Podgajny, R.; Korzeniak, T.; Sieklucka, B. Magnetic Ordering in the Double-Layered Molecular Magnet $Cu(tetren)[W(CN)_8]$: Single-Crystal Study. *Phys. Rev. B* **2008**, *78*, 174409. [CrossRef]
75. Czapla, M.; Pełka, R.; Zieliński, P.M.; Budziak, A.; Bałanda, M.; Makarewicz, M.; Pacyna, A.; Wasiutyński, T.; Miyazaki, Y.; Nakazawa, Y.; et al. Critical Behavior of Two Molecular Magnets Probed by Complementary Experiments. *Phys. Rev. B* **2010**, *82*, 94446. [CrossRef]

crystals

MDPI

Article

Ferromagnetic Oxime-Based Manganese(III) Single-Molecule Magnets with Dimethylformamide and Pyridine as Terminal Ligands

Carlos Rojas-Dotti, Nicolás Moliner, Francesc Lloret and José Martínez-Lillo *

Instituto de Ciencia Molecular (ICMol), Universitat de València, C/Catedrático José Beltrán 2, 46980 Paterna, Valencia, Spain; carlos.rojas@uv.es (C.R.-D.); fernando.moliner@uv.es (N.M.); francisco.lloret@uv.es (F.L.)
* Correspondence: f.jose.martinez@uv.es; Tel.: + 34-963-544-460

Received: 1 December 2018; Accepted: 24 December 2018; Published: 31 December 2018

Abstract: Two new members of the [Mn_6] family of single-molecule magnets (SMMs) of formulae [$Mn_6(\mu_3\text{-}O)_2(H_2N\text{-sao})_6(dmf)_8$]($ClO_4$)$_2$ (**1**) and [$Mn_6(\mu_3\text{-}O)_2(H_2N\text{-sao})_6(py)_6(EtOH)_2$][$ReO_4$]$_2\cdot$4EtOH (**2**), (dmf = N,N'-dimethylformamide, py = pyridine, $H_2N\text{-saoH}_2$ = salicylamidoxime) have been synthesized and characterized structurally and magnetically. Both compounds were straightforwardly prepared from the deprotonation of the $H_2N\text{-saoH}_2$ ligand in the presence of the desired manganese salt and solvent (dmf (**1**) vs. py (**2**)). Compound **1** crystallizes in the triclinic system with space group $P\bar{1}$ and **2** crystallizes in the monoclinic system with space group $P2_1/n$. In the crystal packing of **1** and **2**, the (ClO_4)$^-$ (**1**) and [ReO_4]$^-$ (**2**) anions sit between the cationic [Mn_6]$^{2+}$ units, which are H-bonded to $-NH_2$ groups from the salicylamidoxime ligands. The study of the magnetic properties of **1** and **2** revealed ferromagnetic coupling between the Mn^{III} metal ions and the occurrence of slow relaxation of the magnetization, which is a typical feature of single-molecule magnet behavior. The cationic nature of these [Mn_6]$^{2+}$ species suggests that they could be used as suitable building blocks for preparing new magnetic materials exhibiting additional functionalities.

Keywords: manganese(III); salicylamidoxime; molecular magnetism; single-molecule magnets

1. Introduction

Single-molecule magnets (SMMs) have attracted a great deal of attention during the last two decades [1], because of their potential applications in quantum information processing [2], low-temperature cooling [3], and molecular spintronics [4,5]. Most of the reported SMMs are based on paramagnetic 3D metal ions, the Mn^{III} ion being one of the more explored in this multidisciplinary research [6].

In this context, the combination of phenolic oximes and Mn^{III} has proven to be particularly successful in the preparation of SMMs [7]. Thus, a large family of hexanuclear [Mn^{III}_6] complexes based on salicylaldoxime and salicylamidoxime ligands (Scheme 1), along with their derivatives, has been investigated [8–20]. All the family members display the SMM phenomenon, with remarkably different magnetic behavior, antiferromagnetic or ferromagnetic, that is strongly affected by the structural distortion of the Mn–N–O–Mn torsion angles. As a result, it established a semi-quantitative magnetostructural correlation that enables the prediction of the magnetic behavior of new [Mn^{III}_6] systems [8–20].

A search on the Cambridge Structural Database (CSD) revealed more than 100 hits of discrete [Mn^{III}_6] molecules based on salicylaldoxime and salicylamidoxime ligands. However, only six of them were cationic [Mn^{III}_6]$^{2+}$ systems, the rest being neutral complexes [17–19]. This singular type of SMMs suggests that they could be used as suitable building blocks for preparing new magnetic materials, just by replacing the anion by another anionic species exhibiting an additional functionality [17–19].

For that reason, we are motivated to investigate the crystal structure and magnetic properties of cationic [Mn$^{III}_6$]$^{2+}$ SMMs.

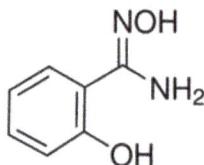

Scheme 1. Structure of the salicylamidoxime ligand (H$_2$N-saoH$_2$).

Herein we report two novel cationic [Mn$_6$]$^{2+}$ complexes with the formulae [Mn$_6$(μ_3-O)$_2$(H$_2$N-sao)$_6$(dmf)$_8$](ClO$_4$)$_2$ (**1**) and [Mn$_6$(μ_3-O)$_2$(H$_2$N-sao)$_6$(py)$_6$(EtOH)$_2$][ReO$_4$]$_2$·4EtOH (**2**) (dmf = *N,N'*-dimethylformamide, py = pyridine, H$_2$N-saoH$_2$ = salicylamidoxime), which have been characterized structurally and magnetically. Both **1** and **2** behave as SMMs.

2. Materials and Methods

2.1. Reagents and Instruments

All manipulations were performed under aerobic conditions, using materials as received (reagent grade). Although no problems were encountered in this work, care should be taken when using the potentially explosive perchlorate anion. The salicylamidoxime ligand was prepared following the synthetic method described in the literature [21].

Elemental analyses (C, H, N) were performed with a CE Instruments CHNS 1100 Elemental Analyzer (samples of 25 (**1**) and 20 mg (**2**)) by the Central Service for the Support to Experimental Research (SCSIE) at the University of Valencia. Infrared spectra of **1** and **2** were recorded with a PerkinElmer Spectrum 65 FT-IR spectrometer in the 4000–400 cm^{-1} region. Variable-temperature, solid-state direct current (DC) magnetic susceptibility data down to 2.0 K were collected on a Quantum Design MPMS-XL SQUID magnetometer equipped (Quantum Design, Inc., San Diego, CA, USA) with a 7 T DC magnet. The experimental magnetic data were corrected for the diamagnetic contributions of the constituent atoms (-990.3×10^{-6} (**1**) and -1219.8×10^{-6} emu mol^{-1} (**2**)) and also for the sample holder (-3.58×10^{-6} and -3.45×10^{-6} emu g^{-1} for **1** and **2**, respectively).

2.2. Single-Crystal X-Ray Diffraction

X-ray diffraction data of single crystals of dimensions $0.26 \times 0.16 \times 0.04$ (**1**) and $0.39 \times 0.31 \times 0.24$ mm^3 (**2**) were collected on a Bruker D8 Venture diffractometer with PHOTON II detector and by using monochromatized Mo-K$_\alpha$ radiation ($\lambda = 0.71073$ Å). Crystal parameters and refinement results for **1** and **2** are summarized in Table 1.

The structures were solved by standard direct methods and subsequently completed by Fourier recycling by using the SHELXTL software packages. The obtained models were refined with version 2013/4 of SHELXL against F^2 on all data by full-matrix least squares [22]. In both systems, all non-hydrogen atoms were refined anisotropically, and the hydrogen atoms were set in calculated positions and refined isotropically by using the riding model. The highest difference Fourier map peaks were 2.262 (**1**) and 1.282 eÅ^{-3} (**2**), which are located at 0.935 Å of Cl(1) and at 1.029 Å of Re(1), respectively. The graphical manipulations were performed with the DIAMOND program [23].

CCDC numbers for **1** and **2** are 1882221 and 1882222, respectively. These data can be obtained free of charge from the Cambridge Crystallographic Data Center on the web (http://www.ccdc.cam.ac.uk/data_request/cif).

Table 1. Summary of the crystal data and structure refinement for **1** and **2**.

Compound	1	2
CCDC	1882221	1882222
Formula	$C_{66}H_{92}O_{30}N_{20}Cl_2Mn_6$	$C_{84}H_{94}O_{28}N_{18}Mn_6Re_2$
$Mr/g\cdot mol^{-1}$	2046.13	2505.81
Crystal system	triclinic	monoclinic
Space group	$P\bar{1}$	$P2_1/n$
$a/\text{Å}$	12.603(8)	13.446(2)
$b/\text{Å}$	13.256(8)	23.254(4)
$c/\text{Å}$	14.501(9)	16.458(3)
$\alpha/^\circ$	114.71(2)	90
$\beta/^\circ$	98.09(2)	105.06(2)
$\gamma/^\circ$	100.59(2)	90
$V/\text{Å}^3$	2098.2(2)	4649.1(1)
Z	1	2
$D_c/g\cdot cm^{-3}$	1.619	1.675
$\mu(MoK_\alpha)/mm^{-1}$	1.032	3.244
$F(000)$	1052	2496
Crystal size	$0.26 \times 0.16 \times 0.04$	$0.39 \times 0.31 \times 0.24$
Goodness-of-fit on F^2	1.079	1.047
$R_1 [I > 2\sigma(I)]$	0.0623	0.0537
$wR_2 [I > 2\sigma(I)]$	0.1731	0.1596
$\Delta\rho_{max,\,min}/e\,\text{Å}^{-3}$	2.262, $-$1.549	1.282, $-$2.359

2.3. Preparation of the Compounds

2.3.1. Synthesis of $[Mn_6(\mu_3\text{-O})_2(H_2N\text{-sao})_6(dmf)_8](ClO_4)_2$ (1)

$Mn(ClO_4)_2\cdot 6H_2O$ (0.249 g, 0.688 mmol) was dissolved with continuous stirring in dmf (10 mL); then, $H_2N\text{-saoH}_2$ (0.103 g, 0.670 mmol) and NEt$_3$ (0.5 mL, 3.6 mmol) were added. The resulting dark green mixture was stirred for 1 h, filtered and layered with Et$_2$O (10 mL). Dark green crystals suitable for X-ray diffraction were obtained in 4 days. Yield: 80%. Elemental analysis calculated (found) for $C_{66}H_{92}O_{30}N_{20}Cl_2Mn_6$ (1): C, 39.1 (39.7); H, 5.2 (5.0); N, 13.8 (13.9)%. Selected IR data (in KBr/cm^{-1}): 3332 (m), 2925 (w), 1653 (vs), 1610 (vs), 1533 (m), 1438 (m), 1384 (m), 1317 (m), 1253 (m), 1150 (m), 1121 (s), 1109 (s), 1022 (m), 881 (m), 762 (w), 685 (s), 649 (m), 578 (w).

2.3.2. Synthesis of $[Mn_6(\mu_3\text{-O})_2(H_2N\text{-sao})_6(py)_6(EtOH)_2][ReO_4]_2\cdot 4EtOH$ (2)

$Mn(NO_3)_2\cdot 4H_2O$ (0.173 g, 0.688 mmol) was dissolved with continuous stirring in EtOH (20 mL), then $H_2N\text{-saoH}_2$ (0.103 g, 0.670 mmol) was added, followed by pyridine (1 mL, 12.4 mmol) and NEt$_3$ (0.1 mL, 0.72 mmol). Next, $(NH_4)[ReO_4]$ (0.184 g, 0.688 mmol) was added to the dark green solution, which was stirred for 1 h. The final dark brown solution was left to evaporate in a fume hood at room temperature. Crystals of **2** were obtained in 3 days and were suitable for X-ray diffraction. Yield: 70%. Elemental analysis calculated (found) for $C_{84}H_{94}O_{28}N_{18}Mn_6Re_2$ (2): C, 40.3 (40.8); H, 3.8 (4.0); N, 10.1 (9.9)%. Selected IR data (in KBr/cm^{-1}): 3426 (vs), 3314 (vs), 1614 (vs), 1564 (m), 1529 (s), 1483 (w), 1442 (s), 1384 (w), 1311 (w), 1251 (w), 1147 (vw), 1022 (m), 910 (s), 881 (m), 755 (w), 683 (m), 642 (w), 580 (vw).

3. Results and Discussion

3.1. Synthetic Procedure

By reacting $Mn(ClO_4)_2\cdot 6H_2O$ (1) and $Mn(NO_3)_2\cdot 4H_2O$ (2) with the salicylamidoxime ligand (Scheme 1) in the presence of the coordinating solvent dmf (1) and py (2), along with NEt$_3$ (1 and 2) and $[ReO_4]^-$ (2), we obtained dark green crystals of hexametallic Mn^{III} complexes of the well-known family of [Mn$_6$] systems. Both $(ClO_4)^-$ and $[ReO_4]^-$ anions were chosen because of their diamagnetic character, and also for giving a suitable solubility to the final compounds. The crystallization techniques employed for **1** and **2** were slow diffusion by layering with Et_2O (10 mL) and slow evaporation at room temperature of the resulting solutions, respectively. Both compounds were obtained in satisfactory yields.

It is worth noting that both **1** and **2** are cationic oxime-based $[Mn_6]^{2+}$ complexes, and only six systems of this type exist in literature, all of them being obtained with the salicylamidoxime ligand [17–19]. This fact is in contrast to the results obtained from analogous reactions employing similar phenolic oximes, such as salicylaldoxime and its alkyl derivatives, where the isolated complex of the reported works is always a neutral $[Mn_6]$ or $[Mn_3]$ system.

3.2. Description of the Crystal Structures

The crystal structure and exact chemical composition of **1** and **2** were established by single-crystal X-ray diffraction. While **1** crystallizes in the triclinic crystal system with space group $P\bar{1}$, **2** crystallizes in the monoclinic crystal system with space group $P2_1/n$ (Table 1). The structures of **1** and **2** are made up of $[Mn_6]^{2+}$ cations (**1** and **2**) and $(ClO_4)^-$ (**1**) and $[ReO_4]^-$ (**2**) anions. There are solvent molecules of crystallization in only **2**, these are EtOH molecules.

Each cationic $[Mn_6]^{2+}$ unit contains two symmetry equivalent $\{Mn_3(\mu_3\text{-}O)\}$ triangular moieties, which are linked by two phenolate and two oximate O-atoms and related by an inversion center (Figure 1). Each edge of the triangle is spanned by the $-N-O-$ group of the salicylamidoxime ligand, with the central oxo ion displaced 0.102 (**1**) and 0.184 Å (**2**) above the plane of the $[Mn_3]$ triangle, towards the dmf (**1**) and py (**2**) terminal ligands.

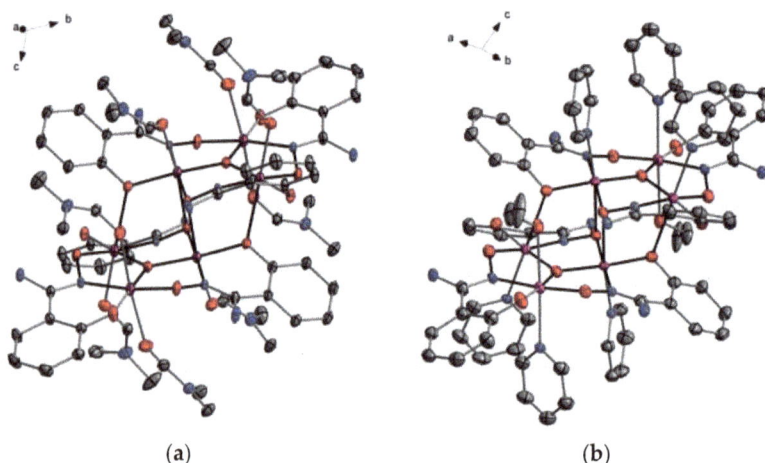

(a)   (b)

Figure 1. (a) Molecular structure of the $[Mn_6(\mu_3\text{-}O)_2(H_2N\text{-}sao)_6(dmf)_8]^{2+}$ cation of **1**. H atoms and $(ClO_4)^-$ anion have been omitted for clarity. Thermal ellipsoids are depicted at 50% probability level. (b) Molecular structure of the $[Mn_6(\mu_3\text{-}O)_2(H_2N\text{-}sao)_6(py)_6(EtOH)_2]^{2+}$ cation of **2**. H atoms, $[ReO_4]^-$ anions and EtOH solvent molecules have been omitted for clarity. Thermal ellipsoids are depicted at 50% probability level. Color code: Pink, Mn; red, O; blue, N; grey, C.

The six Mn^{III} ions in the core of **1** and **2** exhibit coordination environments rather similar to those of previously reported salicylamidoxime-based $[Mn_6]^{2+}$ complexes [17–19], with distorted octahedral geometries and Jahn-Teller axes approximately perpendicular to the $[Mn_3]$ planes. The remaining coordination site on the third Mn ion [Mn(2a)] ((a) = 1 − x, 1 − y, 1 − z for **1** and (a) = 1 − x, 1 − y, −z for **2**) is occupied by a dmf (in **1**) or EtOH (in **2**) molecule. The Mn–N–O–Mn torsion angles of the $[Mn^{III}_3(\mu_3\text{-}O)\text{-}(H_2N\text{-}sao)_3]$ triangular units are 46.5°, 36.3° and 30.3° for **1** and 41.4°, 38.1°, 28.9° for **2** (Table 2).

Table 2. Selected magneto-structural parameters for compounds **1** and **2**.

Compound	Crystal System	Space Group	$\alpha/°$ (Mn-N-O-Mn)	J_1/cm^{-1}	J_2/cm^{-1}	g	τ_0/s^{-1}	$E^{\#}/K$
1	triclinic	$P\bar{1}$	46.5, 36.3, 30.3	+0.90	+0.84	1.99	1.6×10^{-11}	66
2	monoclinic	$P2_1/n$	41.4, 38.1, 28.9	+1.88	+0.72	1.98	8.4×10^{-9}	41

In the crystal packing of **1** and **2**, the $(ClO_4)^-$ (**1**) and $[ReO_4]^-$ (**2**) anions sit between the cationic $[Mn_6]^{2+}$ units, which are H-bonded to $-NH_2$ groups from salicylamidoxime ligands. In **2** the O···N distances are shorter than in **1**, linking the anions and cations into chains [N(4)···O(14b) distance of 2.895(1) Å; (b) = $\frac{1}{2} - x, \frac{1}{2} + y, \frac{1}{2} - z$], as shown in Figure 2. In **1**, the cationic $[Mn_6]^{2+}$ units are somewhat less separated from each other, the shortest intermolecular Mn···Mn distance being 9.831(1) Å [Mn(1)···Mn(2c), (c) = $1 - x, 1 - y, 2 - z$] (Figure 3), whereas the shortest intermolecular Mn···Mn distance in **2** is 10.467(1) Å [Mn(1)···Mn(2b), (b) = $\frac{1}{2} - x, \frac{1}{2} + y, \frac{1}{2} - z$] (Figure 4).

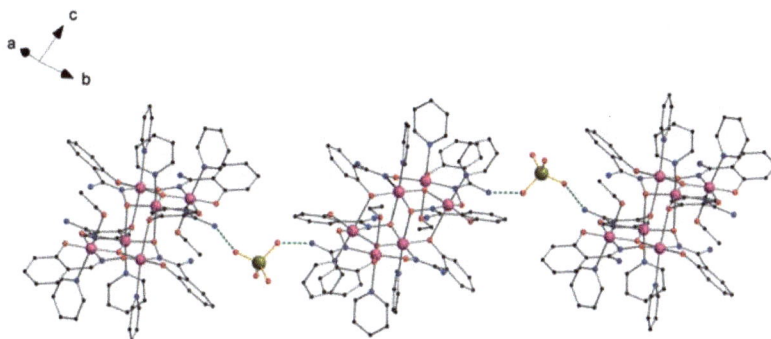

Figure 2. Perspective view of the one-dimensional arrangement of $[Mn_6(\mu_3-O)_2(H_2N\text{-sao})_6(py)_6(EtOH)_2]^{2+}$ cations and $[ReO_4]^-$ anions in the crystal of compound **2** through H-bonding interactions (dashed lines). H atoms and solvent molecules have been omitted for clarity. Color code: Pink, Mn; red, O; blue, N; black, C; green, Re.

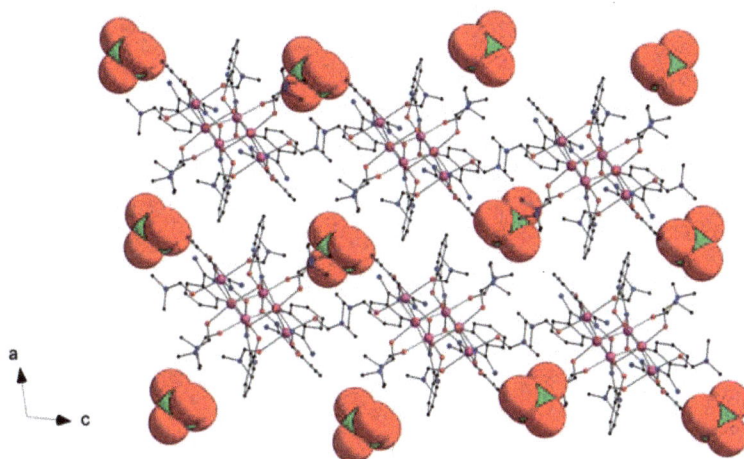

Figure 3. View along the crystallographic *b* axis of a fragment of the packing of **1** showing the arrangement of the $[Mn_6]^{2+}$ cations and $(ClO_4)^-$ anions (space-filling model). H atoms have been omitted for clarity. Colour code: Pink, Mn; red, O; blue, N; black, C; green, Cl.

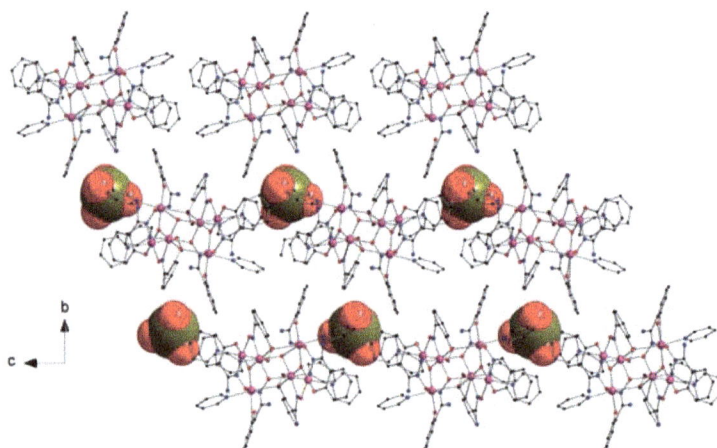

Figure 4. View along the crystallographic *a* axis of a fragment of the packing of **2** showing the arrangement of the [Mn$_6$]$^{2+}$ cations and [ReO$_4$]$^-$ anions (space-filling model). H atoms have been omitted for clarity. Color code: Pink, Mn; red, O; blue, N; black, C; green, Re.

In both compounds, additional weak C···C interactions of different types are also observed. In **1**, there exist π···π off-center parallel stacking interactions of approximately 3.38 Å between aromatic rings of salicylamidoxime ligands of adjacent [Mn$_6$]$^{2+}$ complexes, and also weak C–H···C(O) interactions between dmf molecules of neighboring [Mn$_6$]$^{2+}$ cations (ca. 3.45 Å). In **2**, π···π edge-to-face stacking interactions of ca. 3.49 Å connect aromatic rings of coordinated py molecules and salicylamidoxime ligands of adjacent [Mn$_6$]$^{2+}$ units. All these additional interactions help in stabilizing the supramolecular arrangement in **1** and **2**.

3.3. Magnetic Properties

DC magnetic susceptibility measurements were performed on microcrystalline samples of **1** and **2** in the 2.0–300 K temperature range and under an external magnetic field of 0.1 T. The magnetic properties of **1** and **2** in the form of $\chi_M T$ vs. *T* plot (χ_M being the molar magnetic susceptibility), are shown in Figure 5. The $\chi_M T$ values observed at 300 K are approximately 20.2 and 20.7 cm^3·mol^{-1} K for **1** and **2**, respectively. Although these values are somewhat higher than that expected for six magnetically isolated Mn^{III} ions ($\chi_M T \approx 18.0$ cm^3·mol^{-1} K with g = 1.99), they have been previously observed in ferromagnetically coupled [Mn$_6$] systems [8–20]. Upon cooling, the $\chi_M T$ values rise gradually with decreasing temperature for both compounds, reaching maxima of 38.9 cm^{3}mol^{-1}K at 8.0 K for **1** and 35.8 cm^3·mol^{-1} K at 17.0 K for **2**. In both compounds, $\chi_M T$ values decrease at lower temperatures giving final values of 23.0 (**1**) and 13.0 cm^3·mol^{-1} K (**2**) at 2.0 K, which are observed due to the presence of intermolecular interactions and/or zero-field splitting (ZFS) effects.

The experimental data of the $\chi_M T$ vs. *T* plots of **1** and **2** were treated by using the 2*J* model described by the Hamiltonian of Equation (1), where J_1 and J_2 are the exchange coupling constants for the intramolecular Mn–Mn interactions associated with exchange pathways involving the Mn–N–O–Mn torsion angles of the [Mn$_6$] core, and g is the Landé factor for the Mn^{III} ions. The theoretical parameters thus obtained are summarized in Table 2.

$$\hat{H} = -2J_1 \, (\hat{S}_1 \hat{S}_3 + \hat{S}_1 \hat{S}_{3\prime} + \hat{S}_1 \hat{S}_{1\prime} + \hat{S}_{1\prime} \hat{S}_3 + \hat{S}_{1\prime} \hat{S}_{3\prime})$$
$$-2J_2 \, (\hat{S}_1 \hat{S}_2 + \hat{S}_2 \hat{S}_3 + \hat{S}_{1\prime} \hat{S}_{2\prime} + \hat{S}_{2\prime} \hat{S}_{3\prime}) + \mu_B g H \hat{S} \tag{1}$$

These features reveal an intramolecular ferromagnetic coupling between the Mn^{III} metal ions in both **1** and **2**. In previous studies dealing with DFT calculations on salicylamidoxime-based [Mn$_6$] complexes [16], a critical angle (ca. 27.0°) that is directly correlated to the Mn–N–O–Mn exchange pathway between neighboring Mn^{III} ions was found. Mn–N–O–Mn torsion angles upper than this critical angle switch the magnetic exchange from antiferromagnetic ($J < 0$) to ferromagnetic ($J > 0$). Given that **1** and **2** show Mn–N–O–Mn torsion angles higher than 27.0°, it would be expected to obtain a ferromagnetic coupling as the predominant magnetic interaction for both compounds, as observed experimentally (Figure 5 and Table 2).

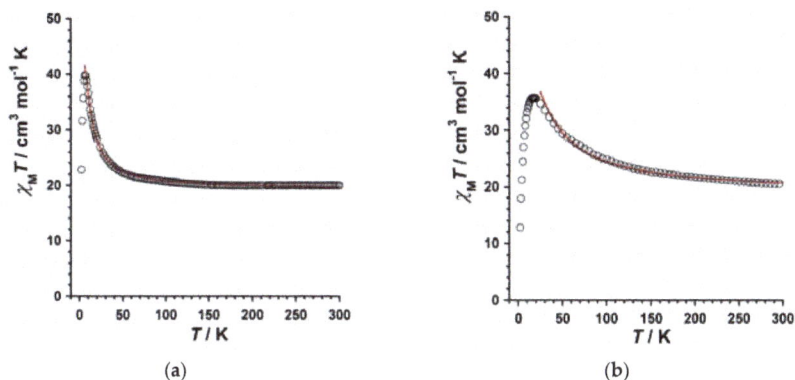

Figure 5. Thermal variation of the $\chi_M T$ product for compounds **1** (**a**) and **2** (**b**). The solid red line represents the best-fit of the experimental data.

The complexes that form the large family of oxime-based [Mn$^{III}_6$] SMMs display ground state spin values that vary from 4 to 12. In general, a spin value of S = 4 is found in antiferromagnetic [Mn$^{III}_6$] systems, whereas ferromagnetic [Mn$^{III}_6$] complexes show a spin value of S = 12. A ground state spin value of S = 12 was obtained for **1** and **2** from the magnetic susceptibility data, hence supporting the ferromagnetic nature for both compounds. Thus, the isotropic simulation of the magnetic susceptibility of **1** and **2** generated the plots of the energy versus total spin shown in Figure 6. The first excited state found in **1** is S = 11 placed at 2.25 cm^{-1}, and the first excited state in **2** is also S = 11, which is located at 1.85 cm^{-1} (Figure 6).

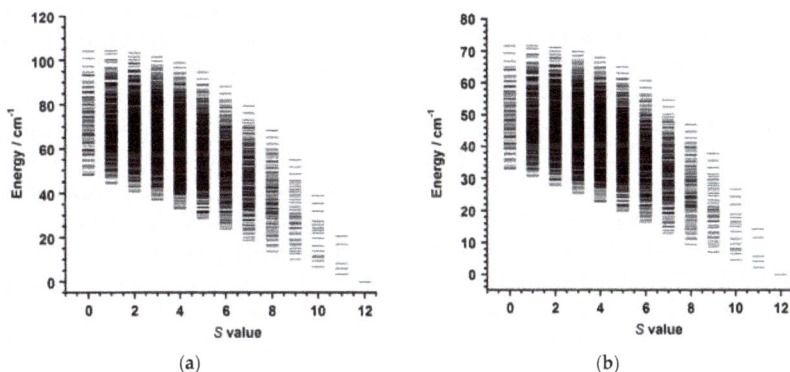

Figure 6. Plot of energy versus total spin state, extracted from the isotropic simulation of the susceptibility data for **1** (**a**) and **2** (**b**).

Additionally, variable temperature-variable field DC magnetization data were measured for **1** and **2** in the 2–7 K temperature and 0.5–7.0 T field ranges. The experimental data are given as reduced magnetization ($M/N\mu_B$ *versus* $\mu_0 H/T$) in Figure 7, which were fitted to a Zeeman plus axial zero-field splitting Hamiltonian ($\hat{H} = D(\hat{S}_z^2 - S(S+1)/3) + \mu_B g H \hat{S}$, where D is the axial anisotropy of the cationic $[Mn_6]^{2+}$ complex, μ_B is the Bohr magneton, $\hat{S}_z$ is the easy-axis spin operator, and H is the applied field) assuming only the ground state is populated. The best fits afforded the parameters $S = 12$, $g = 1.99$ and $D = -0.48$ cm^{-1} for **1** and $S = 12$, $g = 1.98$ and $D = -0.39$ cm^{-1} for **2**, which are in line with those values reported for similar cationic $[Mn_6]^{2+}$ complexes [17–19].

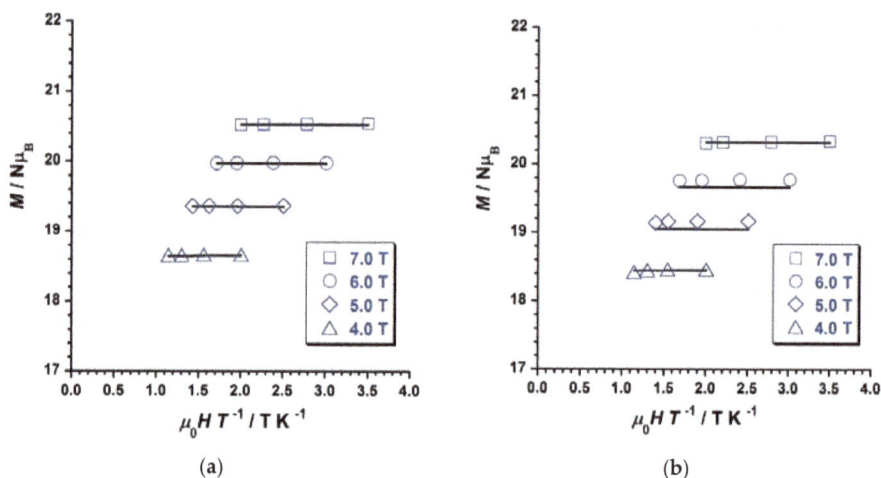

(a)   (b)

Figure 7. Plot of the reduced magnetization ($M/N\mu_B$ *versus* $\mu_0 H/T$) for **1** (a) and **2** (b) at 4, 5, 6 and 7 T fields and temperatures 2–5 K. The solid lines represent the best fit of the experimental data.

AC susceptibility measurements were performed on samples of **1** and **2** in the temperature range 2–10 K, in zero applied DC field, and a 3.9 G AC field oscillating in the 5–1000 Hz range of frequencies. Out-of-phase AC signals (χ''_M) for **1** and **2** are shown in Figure 8. The respective χ''_M versus T plots exhibited frequency dependence of the χ''_M maxima for **1** and **2**. This feature is consistent with SMM behavior. In addition, it was observed that the χ''_M maxima increased with the decreasing frequency for both compounds. These data were fitted to the Arrhenius equation ($\tau = \tau_0 \exp(E^\#/k_B T)$, where τ_0 is the pre-exponential factor, τ is the relaxation time, $E^\#$ is the barrier to relaxation of the magnetization, and k_B is the Boltzmann constant) and plotted in the respective insets of Figure 8. The values obtained for the τ_0 and $E^\#$ parameters are listed in Table 2. The $E^\#$ values for **1** [66.0 K (45.9 cm^{-1})] and **2** [41.0 K (28.5 cm^{-1})] fall into the range for previously reported salicylamidoxime-based $[Mn^{III}_6]$ complexes (24.0 K (16.7 cm^{-1}) < $E^\#$ < 86.0 K (59.8 cm^{-1})). Nevertheless, it is worth pointing out that the $E^\#$ value calculated for **1** is the higher obtained so far for a cationic oxime-based $[Mn^{III}_6]^{2+}$ single-molecule magnet.

This last result is interesting since this type of cationic SMMs can be used as precursors of new multifunctional magnetic materials because the $(ClO_4)^-$ (**1**) and $[ReO_4]^-$ (**2**) anions can be changed through the incorporation of anionic species that bring another physical property or functionality to the final material, for instance, conductivity or luminescence.

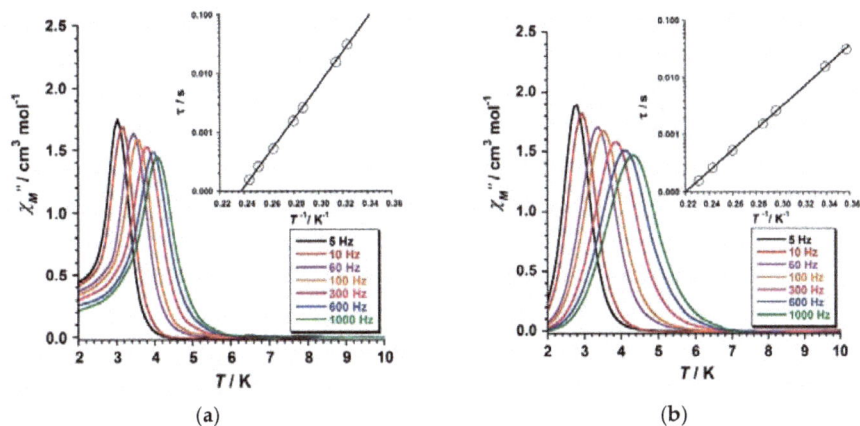

Figure 8. Out-of-phase AC susceptibility (χ''_M) versus T plots for compounds **1** (a) and **2** (b). The insets show the Arrhenius best-fit plot (see text).

4. Conclusions

In summary, two new members of the family of oxime-based [Mn$_6$] complexes have been synthesized and magnetostructurally characterized. Both compounds display a magnetic behavior consistent with the single-molecule magnet (SMM) phenomenon. The barrier value to the relaxation of the magnetization ($E^{\#}$) for compound **1** is the highest reported so far for cationic oxime-based [Mn$_6$]$^{2+}$ systems. Finally, due to their cationic character, these singular SMMs could be used as suitable building blocks for preparing new magnetic materials, just by replacing the anion by another anionic species exhibiting an additional functionality, namely, conductivity or luminescence. This work is in progress.

Author Contributions: J.M.-L. and F.L. obtained funding for the project. C.R.-D. performed the synthesis and the X-ray data collection. N.M. designed and carried out the SQUID measurements. J.M.-L. analyzed the data associated with all the experiments and wrote the manuscript, which all authors discussed and commented on.

Funding: This research was funded by the Spanish Ministry of Science, Innovation and Universities with grant numbers MDM-2015-0538 and CTQ2016-75068P.

Acknowledgments: The Spanish "Ramón y Cajal" Programme is gratefully acknowledged.

Conflicts of Interest: The authors declare no conflict of interest.

References

1. Sessoli, R.; Gatteschi, D. Quantum Tunneling of Magnetization and Related Phenomena in Molecular Materials. *Angew. Chem. Int. Ed.* **2003**, *42*, 268–297. [CrossRef]
2. Timco, G.A.; McInnes, E.J.L.; Winpenny, R.E.P. Physical studies of heterometallic rings: An ideal system for studying magnetically-coupled systems. *Chem. Soc. Rev.* **2013**, *42*, 1796–1806. [CrossRef] [PubMed]
3. Evangelisti, M.; Brechin, E.K. Recipes for enhanced molecular cooling. *Dalton Trans.* **2010**, *39*, 4672–4676. [CrossRef] [PubMed]
4. Bogani, L.; Wernsdorfer, W. Molecular spintronics using single-molecule magnets. *Nat. Mater.* **2008**, *7*, 179–186. [CrossRef] [PubMed]
5. Tyagi, P.; Li, D.; Holmes, S.M.; Hinds, B.J. Molecular Electrodes at the Exposed Edge of Metal/Insulator/Metal Trilayer Structures. *J. Am. Chem. Soc.* **2007**, *129*, 4929–4938. [CrossRef]
6. Ferrando-Soria, J.; Vallejo, J.; Castellano, M.; Martínez-Lillo, J.; Pardo, E.; Cano, J.; Castro, I.; Lloret, F.; Ruiz-García, R.; Julve, M. Molecular magnetism, quo vadis? A historical perspective from a coordination chemist viewpoint. *Coord. Chem. Rev.* **2017**, *339*, 17–103. [CrossRef]

7. Milios, C.J.; Raptopoulou, C.P.; Terzis, A.; Lloret, F.; Vicente, R.; Perlepes, S.P.; Escuer, A. Hexanuclear manganese(III) single-molecule magnets. *Angew. Chem. Int. Ed.* **2004**, *43*, 210–212. [CrossRef]
8. Milios, C.J.; Vinslava, A.; Wood, P.A.; Parsons, S.; Wernsdorfer, W.; Christou, G.; Perlepes, S.P.; Brechin, E.K. A Single-Molecule Magnet with a "Twist". *J. Am. Chem. Soc.* **2007**, *129*, 8–9. [CrossRef]
9. Milios, C.J.; Vinslava, A.; Wernsdorfer, W.; Moggach, S.; Parsons, S.; Perlepes, S.P.; Christou, G.; Brechin, E.K. A Record Anisotropy Barrier for a Single-Molecule Magnet. *J. Am. Chem. Soc.* **2007**, *129*, 2754–2755. [CrossRef]
10. Milios, C.J.; Piligkos, S.; Brechin, E.K. Ground state spin-switching via targeted structural distortion: Twisted single-molecule magnets from derivatised salicylaldoximes. *Dalton Trans.* **2008**, *14*, 1809–1817. [CrossRef]
11. Inglis, R.; Milios, C.J.; Jones, L.F.; Piligkos, S.; Brechin, E.K. Twisted molecular magnets. *Chem. Commun.* **2012**, *48*, 181–190. [CrossRef] [PubMed]
12. Kalofolias, D.A.; Flamourakis, A.G.; Siczek, M.; Lis, T.; Milios, C.J. A bulky oxime for the synthesis of Mn(III) clusters. *J. Coord. Chem.* **2015**, *68*, 1–20. [CrossRef]
13. Tomsa, A.-R.; Martínez-Lillo, J.; Li, Y.; Chamoreau, L.-M.; Boubekeur, K.; Farias, F.; Novak, M.A.; Cremades, E.; Ruiz, E.; Proust, A.; et al. A new family of oxime-based hexanuclear manganese(III) single molecule magnets with high anisotropy energy barriers. *Chem. Commun.* **2010**, *46*, 5106–5108. [CrossRef] [PubMed]
14. An, G.-Y.; Cui, A.-L.; Kou, H.-Z. Assembly of oximate-bridged Mn$_6$ cluster to a one-dimensional chain. *Inorg. Chem. Commun.* **2011**, *14*, 1475–1478. [CrossRef]
15. Martínez-Lillo, J.; Chamoreau, L.-M.; Proust, A.; Verdaguer, M.; Gouzerh, P. Hexanuclear manganese(III) single-molecule magnets from derivatized salicylamidoximes. *C. R. Chim.* **2012**, *15*, 889–894. [CrossRef]
16. Martínez-Lillo, J.; Tomsa, A.-R.; Li, Y.; Chamoreau, L.-M.; Cremades, E.; Ruiz, E.; Barra, A.-L.; Proust, A.; Verdaguer, M.; Gouzerh, P. Synthesis, crystal structure and magnetism of new salicylamidoxime-based hexanuclear manganese(III) single-molecule magnets. *Dalton Trans.* **2012**, *41*, 13668–13681. [CrossRef] [PubMed]
17. Martínez-Lillo, J.; Dolan, N.; Brechin, E.K. A cationic and ferromagnetic hexametallic Mn(III) single-molecule magnet based on the salicylamidoxime ligand. *Dalton Trans.* **2013**, *42*, 12824–12827. [CrossRef]
18. Martínez-Lillo, J.; Dolan, N.; Brechin, E.K. A family of cationic oxime-based hexametallic manganese(III) single-molecule magnets. *Dalton Trans.* **2014**, *43*, 4408–4414. [CrossRef]
19. Martínez-Lillo, J.; Cano, J.; Wernsdorfer, W.; Brechin, E.K. The Effect of Crystal Packing and Re^{IV} Ions on the Magnetisation Relaxation of [Mn$_6$]-Based Molecular Magnets. *Chem. Eur. J.* **2015**, *21*, 8790–8798. [CrossRef]
20. Rojas-Dotti, C.; Martínez-Lillo, J. Thioester-functionalised and oxime-based hexametallic manganese(III) single-molecule magnets. *RSC Adv.* **2017**, *7*, 48841–48847. [CrossRef]
21. Eloy, F.; Lenaers, R. The Chemistry of Amidoximes and Related Compounds. *Chem. Rev.* **1962**, *62*, 155–183. [CrossRef]
22. *SHELXTL-2013/4, Bruker Analytical X-ray Instruments*; Bruker: Madison, WI, USA, 2013.
23. *DIAMOND 4.5.0, Crystal Impact GbR*; Crystal Impact: Bonn, Germany, 2018.

crystals

MDPI

Article

Modification of Structure and Magnetic Properties in Coordination Assemblies Based on [Cu(cyclam)]$^{2+}$ and [W(CN)$_8$]$^{3-}$

Aleksandra Pacanowska, Mateusz Reczyński and Beata Nowicka *

Jagiellonian University in Krakow, Faculty of Chemistry, Gronostajowa 2, 30-387 Kraków, Poland;
o.pacanowska@gmail.com (A.P.); mateusz.reczynski@uj.edu.pl (M.R.)
* Correspondence: beata.nowicka@uj.edu.pl; Tel.: +48-12-686-2475

Received: 14 December 2018; Accepted: 10 January 2019; Published: 16 January 2019

Abstract: The 1D {[Cu^{II}(cyclam)]$_3$[W^V(CN)$_8$]$_2$·5H$_2$O}$_n$ (**1·5H$_2$O**) (cyclam = 1,4,8,11-tetraazacyclotetradecane) coordination polymer of ladder topology can be obtained in water-alcohol solution from [Cu(cyclam)]$^{2+}$ and [W(CN)$_8$]$^{3-}$ building blocks. Upon dehydration, **1·5H$_2$O** undergoes a single-crystal-to-single-crystal structural transformation to the anhydrous {[Cu^{II}(cyclam)]$_3$[W^V(CN)$_8$]$_2$}$_n$ (**1**) form, which retains the same topology, but is characterized by shorter Cu-W distances and significantly more bent CN-bridges. The deformation of the coordination skeleton is reflected in magnetic properties: the predominant intra-chain interactions change from ferromagnetic in **1·5H$_2$O** to antiferromagnetic in **1**. The reaction between the same building blocks in water solution under slow diffusion conditions leads to the formation of a 0D {[Cu^{II}(cyclam)(H$_2$O)]$_2$[Cu^{II}(cyclam)][W^V(CN)$_8$]$_2$}·3H$_2$O pentanuclear assembly (**2·3H$_2$O**).

Keywords: molecular magnetism; octacyanotungstate(V); copper(II); cyclam; cyano bridge; magnetic properties

---

1. Introduction

One of the interesting features that can be achieved in molecular magnetics is the possibility to modify magnetic properties by sorption of small guest molecules. At the turn of the century magnetic sponges was defined as compounds that change structure and magnetic properties upon reversible removal of coordinated or non-coordinated water molecules [1–3]. Later, the term solvatomagnetism was introduced to describe changes in magnetic behavior upon sorption or desorption of different guest molecules.

We have previously shown that the solvatomagnetic effect can be repeatedly observed in CN-bridged networks containing the [Ni(cyclam)]$^{2+/3+}$ cationic building block. By combining it with different polycyanometallates, we obtained two main families of coordination polymers of 1D and 2D connectivity. The 2D {[Ni^{II}(cyclam)]$_3$[M(CN)$_n$]$_2$}$_\infty$ (M = Cr^{III}, Fe^{III}, n = 6; M = W^V, n = 8) networks [4–7] show honeycomb-like topology with microporous channels running across the coordination layers. For these networks, we observed formation of different hydrates as well as the possibility to introduce small organic molecules. For each of the three compounds, four pseudopolymorphic forms differing in structure and magnetic properties have been characterized. The second family of solvatomagnetic polymers are non-porous 1D bimetallic alternating chains. The {[Ni^{III}(cyclam)][M^{III}(CN)$_6$]$_2$·6H$_2$O}$_\infty$ (M^{III} = Cr, Fe) [8] chains can be reversibly dehydrated, which affects their magnetic properties. The {(H$_3$O)[Ni^{III}(cyclam)][M^{II}(CN)$_6$]$_2$·5H$_2$O}$_\infty$ (M^{II} = Fe, [9] Ru, Os [10]) chains, where the charge of the coordination skeleton is compensated by the presence of the H$_3$O$^+$ ions, are rare examples of sorption-driven charge-transfer. The removal of water causes partial electron transfer from M^{II} centers to Ni^{III} ions. The process is reversible for Fe^{II} and irreversible for Ru^{II} and Os^{II}—based compounds.

Cyclam complexes with different metal centers, including Ni^{II} [11,12], Cu^{II} [13–17] and Mn^{III} [18–20], were often used to build CN-bridged coordination polymers. However, apart from our studies, there are no reports on the desolvation effects in this group of compounds. Thus, we decided to examine the potential of cyclam complexes with other metal ions for the construction of solvatomagnetic materials. The manganese(III)-based alternating bimetallic chains {[Ni^{III}(cyclam)][M^{III}(CN)$_6$]$_2$·6H$_2$O}$_\infty$ (M^{III} = Cr, Fe) exhibit loss of crystallization water accompanied by structural and magnetic changes [21]; however, in contrast to their Ni^{III}-based congeners [8], the dehydration process was irreversible. The Cu^{II} cationic building blocks often differ from those of other 3d metals, due to the strong Jahn-Teller effect. The {[Cu^{II}(cyclam)]$_3$[W^{V}(CN)$_8$]$_2$·5H$_2$O}$_n$ compound [22] shows 1D ladder-like topology, unlike the series of Ni^{II}-based 2D honeycomb-like networks of the same stoichiometry described above. It was characterized in terms of structure and magnetic properties, but dehydration effect was not mentioned in the report [22]. Here, we present the post-synthetic modification of structure and magnetic properties of this ladder chain network and the synthesis of a new 0D assembly based on the same building blocks.

2. Materials and Methods

The Na$_3$[W(CN)$_8$]·4H$_2$O precursor complex was prepared according to the published procedure [23]. Other reagents and solvents were commercially available and used as supplied. Measurements of powder X-ray diffraction were carried out on a PANanalitical X'Pert Pro powder diffractometer with the Cu Kα radiation source. Elemental analysis for carbon, nitrogen and hydrogen were carried out on a Vario Micro Cube elemental analyser. IR spectra of single crystals were recorded using a Nicolet iN10 MX FTIR microscope operating in the transmission mode. The dehydration process was characterized by a dynamic vapor sorption method using an SMS DVS Resolution apparatus in the 40–0% RH range at 298 K. Magnetic measurements were performed on a Quantum Design MPMS3 magnetometer. Variable-temperature magnetic susceptibility was measured at an applied field of 1000 Oe in the temperature range of 300 K to 1.8 K and magnetization at 1.8 K up to 70 kOe. The samples were sealed in polyethylene bags to avoid dehydration at low pressure.

2.1. Synthesis of the [Cu(cyclam)](NO$_3$)$_2$·6H$_2$O Precursor Complex

The compound was prepared according to the modified published procedure [24]. Cu(NO$_3$)$_2$·3H$_2$O (1 mmol, 245.4 mg) was dissolved in 4 ml of absolute ethanol at 60 °C. To the above solution, solid cyclam (1 mmol, 200.0 mg) was added and immediate precipitation of violet product was observed. The suspension was left in the freezer for 1h and then filtered. Yield: 396.0 mg (79%).

2.2. Synthesis of {[Cu(cyclam)]$_3$[W(CN)$_8$]$_2$·5H$_2$O}$_n$ (1·5H$_2$O)

The compound was prepared according to the modified published procedure [22]. A solution of [Cu(cyclam)](NO$_3$)$_2$·6H$_2$O (0.13 mmol, 66.4 mg) in a water-ethanol mixture (1:2, 20 mL) was layered over a solution of Na$_3$[W(CN)$_8$]·4H$_2$O (0.07 mmol, 35.6 mg) in a water-ethanol mixture (2:1, 20 mL) in test tubes. Brownish-red needle-shaped crystals were formed after three days. (Found: C, 33.53; H, 4.976; N, 23.44; calc. C$_{46}$H$_{82}$Cu$_3$N$_{28}$O$_5$W$_2$: C, 33.17; H, 4.96; N, 23.55) IR νCN: 2167 cm^{-1}, 2151 cm^{-1}, 2145 cm^{-1}, 2139 cm^{-1}.

2.3. Synthesis of {[Cu(cyclam)]$_3$[W(CN)$_8$]$_2$}$_n$ (1)

The anhydrous compound was obtained by drying **1·5H$_2$O** at 50 °C for 1 h. IR νCN: 2166 cm^{-1}, 2161 cm^{-1}, 2151 cm^{-1}, 2144 cm^{-1}, 2139 cm^{-1}, 2134 cm^{-1}

2.4. Synthesis of {[Cu^{II}(cyclam)(H$_2$O)]$_2$[Cu^{II}(cyclam)][W^{V}(CN)$_8$]$_2$}·3H$_2$O (2·3H$_2$O)

The solutions of [Cu(cyclam)](NO$_3$)$_2$·6H$_2$O (0.05 mmol, 26.8 mg in 8 mL of H$_2$O) and Na$_3$[W(CN)$_8$]·4H$_2$O (0.03 mmol, 14.5 mg in 8 mL of H$_2$O) were placed in separate sections of an

H-tube and water was added to fill the horizontal connecting tube. Brownish-red plate-shaped crystals were formed after 2–3 weeks. (Found: C, 33.42; H, 4.76; N, 23.53; cal. $C_{46}H_{82}Cu_3N_{28}O_5W_2$: C, 33.17; H, 4.96; N, 23.55). IR νCN: 2166 cm^{-1}, 2151 cm^{-1}, 2144 cm^{-1}, 2139 cm^{-1}.

2.5. Structure Determination

Single crystal X-ray diffraction measurements for **1** and **2·3H₂O** were performed on a Bruker D8 QUEST diffractometer. The structures were solved by direct methods using SHELXT [25]. Refinement and further calculations were carried out using SHELXL [25]. The non-H atoms were refined anisotropically, apart from some O atoms of disordered crystallization water, which were refined isotropically. H—atoms were placed in idealized positions and refined using riding model. H—atoms were not considered for disordered water molecules in **2·3H₂O**; however, they were included in the calculation of the molecular weight. The unit cell of **1** was transformed in order to allow direct comparison to the structure of **1·5H₂O**, therefore it does not conform to the requirements of a reduced cell. Due to high absorption and relatively poor quality of the crystals of **1** after solid state transformation, some residual electron density peaks appear around W. The crystallographic data for **1** and **2·3H2O** are presented in Table 1, with structure parameters for **1·5H₂O** [22] added for comparison. Graphics were created with Mercury 3.9.10. The analysis of coordination polyhedra was performed using SHAPE 2.1 [26]. CCDC 1885028 and 1885029 contain the supplementary crystallographic data for this paper. These data can be obtained free of charge via http://www.ccdc.cam.ac.uk/conts/retrieving.html

Table 1. Structure and refinement data for **1** and **2·3H₂O**; data for **1·5H₂O** included for comparison.

	1·5H₂O [1]	**1**	**2·3H₂O**
Empirical formula	$C_{46}H_{82}Cu_3N_{28}O_5W_2$	$C_{46}H_{72}Cu_3N_{28}W_2$	$C_{46}H_{82}Cu_3N_{28}O_5W_2$
Formula weight	1665.72	1575.57	1665.72
Crystal system	triclinic	triclinic	monoclinic
Space group	$P-1$	$P-1$	$P 2_1/n$
a [Å]	8.9255(11)	9.0228(5)	14.0566(7)
b [Å]	10.0938(13)	10.8399(8)	12.8685(7)
c [Å]	19.549(3)	17.8734(13)	36.7198(19)
α [°]	78.342(5)	75.409(6)	90.0
β [°]	83.074(5)	87.810(4)	100.821(2)
γ [°]	69.472(5)	60.092(7)	90.0
V [Å³]	1613.0(4)	1458.5(8)	6524.0(3)
Z	1	1	4
dc [g·cm^{-3}]	1.715	1.794	1.696
μ [mm^{-1}]	4.590	5.064	4.539
F(000)	829	779	3276
θ range [°]	1.07–28.28	2.79–25.4	2.37–29.15
Reflns collected	21,064	5 261	17,526
No. of params	380	356	746
R_{int}	0.0693	0.0532	0.0362
GooF (F²)	-	1.018	1.059
R_1 [I > 2σ(I)]	0.0693	0.0708	0.0446
wR_2 (all data)	0.1662	0.1510	0.0973

[1] Data from reference [22].

3. Results and Discussion

3.1. Synthesis and Post-Synthetic Modification

The reaction between $[Cu(cyclam)]^{2+}$ and $[W(CN)_8]^{3-}$ building blocks in water-alcohol solution leads to the formation of $\{[Cu^{II}(cyclam)]_3[W^V(CN)_8]_2 \cdot 5H_2O\}_n$ (**1·5H₂O**) of the ladder-chain structure reported earlier [22]. The crystals of **1·5H₂O** are stable under ambient conditions, but at slightly elevated temperatures (30–50 °C), they lose crystallization water to result in an anhydrous $\{[Cu^{II}(cyclam)]_3[W^V(CN)_8]_2\}_n$ (**1**) form. The process occurs with retention of crystallinity and thus the structure of **1** could be established from a single crystal XRD measurement. The single-crystal-to-single-crystal structural transformations upon desolvation are relatively rare [5,27,28]. In the family of the coordination polymers based on [Ni(cyclam)]$^{2+/3+}$ ions and polycyanometallates, we observed this phenomenon only for the 2D $\{[Ni^{II}(cyclam)]_3[W^V(CN)_8]_2\}_\infty$ network [5]. The crystals of **1** exposed at 25 °C to humid air (40–60% RH) regain their original composition. However, upon rehydration, the crystals collapse and the PXRD results (Figure S1) show that the rehydration process leads to a mixture of phases, one of which is the original **1·5H₂O** form. We followed the dehydration process using the gravimetric dynamic vapor sorption method. The sample was subjected to decreasing relative humidity from 40% to 0% at a constant temperature of 298K. The sharp decrease in mass starts at about 15% RH (Figure S2), and takes place in two unresolved steps, which correspond to the subsequent loss of 3 and 2 water molecules.

When the reaction between the same building blocks is carried out in a water solution under slow diffusion conditions, a 0D pentanuclear assembly of the formula $\{[Cu^{II}(cyclam)(H_2O)]_2[Cu^{II}(cyclam)][W^V(CN)_8]_2 \cdot 3H_2O\}$ (**2·3H₂O**) is obtained. Interestingly, quick precipitation from the water solution does not lead to the same result. The analysis of the PXRD patterns (Figure S3) suggests that **1·5H₂O** is formed instead. Similar disparity between the course of slow diffusion and quick precipitation reactions was observed for the 3D $\{[Ni^{II}(cyclam)]_2[W^{IV}(CN)_8]\}_\infty$ network [29]. Due to a very low yield and long time taken to grow crystals, the dehydration of **2·3H₂O** was not studied, but it seems likely that both lattice and coordinated water could be removed, possibly with the formation of additional CN-bridges between neighboring molecules.

3.2. Description of Structures

Upon dehydration of **1·5H₂O**, the triclinic system and space group P-1 are retained, but cell parameters change significantly (Table 1). In particular, period *c* shortens by 1.7 Å (8.7%) and the γ angle decreases by 9.38° (13.5%). The cell volume decreases by 155 Å^3 (9.6%) per formula unit, which corresponds quite well to the volume expected for five H-bound water molecules. The connectivity of the coordination network is retained, with Cu and W CN-bridged centers forming a 1D ladder-like chain. Apart from the removed water molecules, the asymmetric unit of **1** (Figure 1) contains the same set of atoms as **1·5H₂O**, including Cu1 and W1 ions located in general positions and Cu2 ions in the inversion center. Each tungsten center (W1) is coordinated by eight cyanide ligands, three of which form bridges to copper centers (two Cu1 and one Cu2). The copper ions are connected with two tungsten centers through cyanide bridges coordinated in axial positions, with equatorial positions blocked by the cyclam ligand. The dehydration process strongly affects the geometry of CN-bridges and coordination polyhedra. The Continuous Shape Measure analysis [26] shows that the geometry of the [W(CN)₈]$^{3-}$ ion in the hydrated form **1·5H₂O** is close to square antiprism, while for the anhydrous form (**1**), it changes to a triangular dodecahedron (Table S1). Both Cu ions in **1·5H₂O** show a highly distorted octahedral geometry (Table S2), due to the significant Jahn-Teller effect causing elongation of the axial bonds. The non-centrosymmetric Cu1 center is more distorted, due to the asymmetry of the axial bonds, one of which (Cu1-N1 of 2.698 Å) can be classified as a semi-coordination bond. Upon dehydration, the Cu2-N5 distances shorten by 0.053 Å, while the Cu2-N distances become more symmetrical (Table 2). As an effect, the octahedral geometry of both copper centers in the anhydrous form **1** is less distorted (Table S2). The angles of cyanide bridges (Cu1-N1-C1, Cu1-N4-C4, Cu2-N5-C5) decrease by 13.74–24.78° (Table 2). Therefore, dehydration leads to significant shortening of the distances between the CN-linked metal centers by 0.219–0.392 Å.

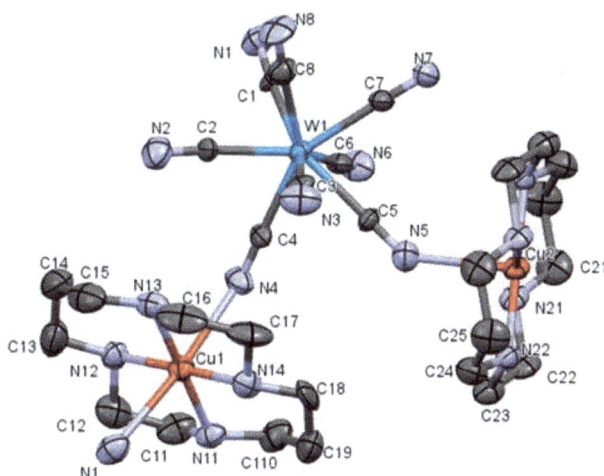

Figure 1. The asymmetric unit of **1** with atom numbering; ellipsoids at 50% probability.

Table 2. Selected bond lengths (Å) and angles (°) in **1·5H₂O** and **1**.

Bond	1·5H₂O [1]	1	Angle	1·5H₂O [1]	1
W1-C1	2.181	2.149	Cu1-N1-C1	143.0	129.3
W1-C2	2.157	2.129	Cu1-N4-C4	157.1	135.3
W1-C3	2.154	2.159	Cu2-N5-C5	157.1	142.1
W1-C4	2.169	2.155	N11-Cu1-N12	93.22	86.1
W1-C5	2.165	2.152	N12-Cu1-N13	86.44	95.5
W1-C6	2.152	2.159	N13-Cu1-N14	94.23	86.6
W1-C7	2.158	2.149	N14-Cu1-N11	85.93	91.9
W1-C8	2.169	2.214	N21-Cu2-N22	86.34	86.2
Cu1-N1	2.698	2.606	W1-C1-N1	176.75	176.3
Cu1-N4	2.510	2.585	W1-C4-N4	177.74	179.4
Cu2-N5	2.556	2.503	W1-C5-N5	174.43	176.5
Cu1-N11	2.004	2.037			
Cu1-N12	2.023	2.027			
Cu1-N13	2.024	2.018			
Cu1-N14	2.014	1.998			
Cu2-N21	2.015	2.011			
Cu2-N22	2.032	2.028			

[1] Data from reference [22].

Projections of structure along the crystallographic axes for **1·5H₂O** and **1** (Figure 2) show the ladder-like topology retained upon dehydration. In both structures, the ladders lie in the (11–1) plane and run along the [1–10] direction. However, the changes in the coordination geometry of the $[W(CN)_8]^{3-}$ ion and different alignment of the cyclam ligand in both forms are noticeable. Likewise, the bending of the CN-bridges can be seen, particularly in the view along the *a* period. Upon dehydration, the distance between the neighboring chains in the [001] direction shortens, which is visible in the projections along the *b* and *c* axes. The distance between the Cu1 and W1 centers from the neighboring chains shortens by 0.725 Å. In the hydrated form **1·5H₂O**, the coordination ladders are bound into a 3D supramolecular network by H-bonds between cyanide ligands, NH groups of the cyclam molecules, and crystallization water [22]. Upon loss of crystallization water, the hydrogen bonds are rearranged (Figure S4). The ladders are bound into stacks along the [110] direction, resulting in the 2D supramolecular network. Conversely, there is no connection in the [001] direction, despite shorter distance between the ladders. Moreover, due to the bending of CN-bridges, intramolecular H-bonds appear. They link the terminal CN ligands with the NH groups of cyclam from the W1 and

Cu1 centers, and thus stabilize strong bending of the Cu1-N1-C1 bridge (Table 2). This may be the reason why the rehydration process leads to the collapse of crystals and is not fully reversible.

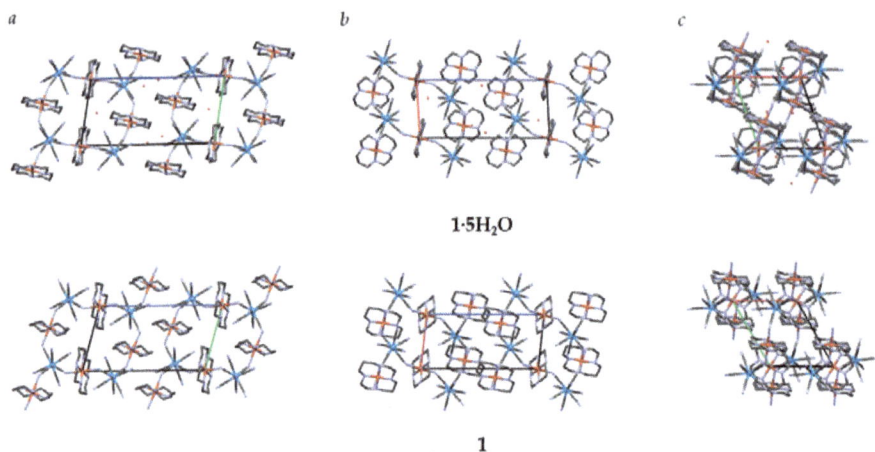

a

b

c

$1 \cdot 5H_2O$

1

Figure 2. Projection of structures of $1 \cdot 5H_2O$ and **1** along (a–c) crystallographic directions.

The 0D $\{[Cu^{II}(cyclam)(H_2O)]_2[Cu^{II}(cyclam)][W^V(CN)_8]_2\} \cdot 3H_2O$ assembly ($2 \cdot 3H_2O$) is obtained from the same building blocks as the $1 \cdot 5H_2O$ ladder chain, but under different synthetic conditions. The compound crystallizes in a monoclinic system, space group P $2_1/n$ (Table 1). Its structure is composed of Z-shaped pentanuclear molecules containing two W and three Cu centers connected by CN-bridges with water molecules coordinated to the terminal Cu ions (Figure 3). There are two symmetrically independent molecules. Each of them is centrosymmetric with the central Cu ions located at the inversion centers. Therefore, the asymmetric unit consist of two tungsten and two copper centers (W1, W2, Cu1, Cu3) in general positions and two copper centers (Cu2 and Cu4) in special positions. Each W atom is coordinated by eight cyanide ligands, two of which form bridges to two Cu centers (W1 with Cu1 and Cu2; W2 with Cu3 and Cu4). The coordination geometry of both W centers is close to ideal square antiprism (Table S1). All Cu ions have cyclam ligand coordinated in equatorial positions. The centrosymmetric Cu2 and Cu4 atoms are linked by two cyanide bridges, whereas the Cu1 and Cu3 centers located at the ends of the Z-shaped molecule are coordinated by one cyanide bridge and one water molecule in axial positions. The octahedral coordination geometry is distorted due to the Jahn-Teller effect (Table S2). There is a large disproportion of the extent of this distortion between the centrosymmetric Cu centers: Cu2 with the very long and bent NC-bridges (Table 3) shows very large distortion, while Cu4 shows the lowest distortion among the Cu centers in **1** and **2**.

The average length of the Cu-N bond in the CN-bridge is 2.543 Å, which is shorter than for $1 \cdot 5H_2O$ (2.588 Å) or **1** (2.565 Å). The average value of the Cu-N-C angles of the CN-bridges (147.87°) is closer to those observed in $1 \cdot 5H_2O$ (153.58°) than those in the anhydrous form **1** (132.31°). The distances between the CN-bridged metal centers are shorter for the central Cu (W1-Cu2, W2-Cu4) than for the terminal ones (W1-C1, W2-Cu3) by about 0.25 Å, due to the stronger bending of the central CN bridges.

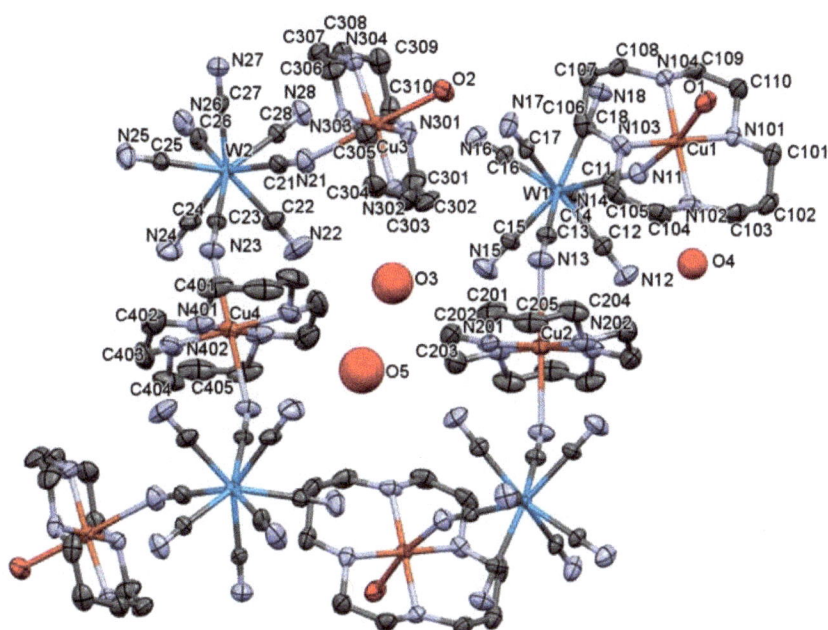

Figure 3. Structure of **2·3H2O** with atom numbering; ellipsoids at 50% probability.

Table 3. Selected bond lengths (Å) and angles (°) in **2·3H$_2$O**.

Cu1-N11	2.523	Cu1-N11-C11	157.9
Cu1-O1	2.518	Cu2-N13-C13	137.2
Cu2-N13	2.596	Cu3-N21-C21	153.9
Cu3-N21	2.594	Cu4-N23-C23	142.8
Cu3-O2	2.422	N101-Cu1-N102	93.3
Cu4-N23	2.457	N102-Cu1-N103	86.0
Cu1-N101	2.013	N103-Cu1-N104	95.0
Cu1-N102	2.037	N104-Cu1-N101	85.7
Cu1-N103	2.011	N201-Cu2-N202	86.0
Cu1-N104	2.023	N301-Cu3-N302	85.8
Cu2-N201	2.008	N302-Cu3-N303	92.8
Cu2-N202	2.019	N303-Cu3-N304	86.4
Cu3-N301	2.021	N304-Cu3-N301	95.0
Cu3-N302	2.029	N401-Cu4-N402	85.6
Cu3-N303	2.008	W1-C11-N11	178.5
Cu3-N304	2.013	W1-C13-N13	177.2
Cu4-N401	2.024	W2-C21-N21	179.6
Cu4-N402	2.016	W2-C23-N23	176.5

Projection along the *a* and *b* crystallographic directions (Figure 4) shows that the symmetrically independent molecules (marked red and blue) are arranged in independent layers in the (10–1) and (20–2) planes. They are bound into a 3D supramolecular network by the net of hydrogen bonds between cyanide ligands, NH groups of cyclam, aqua ligands, and crystallization water (Figure S5). The molecules of the same symmetry are bound within the layers by H-bonds formed by coordinated water, which donates two protons to two CN ligands of two neighboring molecules. The layers are linked by direct bonds between CN ligands and NH groups, as well as by H-bonds between CN-ligands mediated by crystallization water.

a b

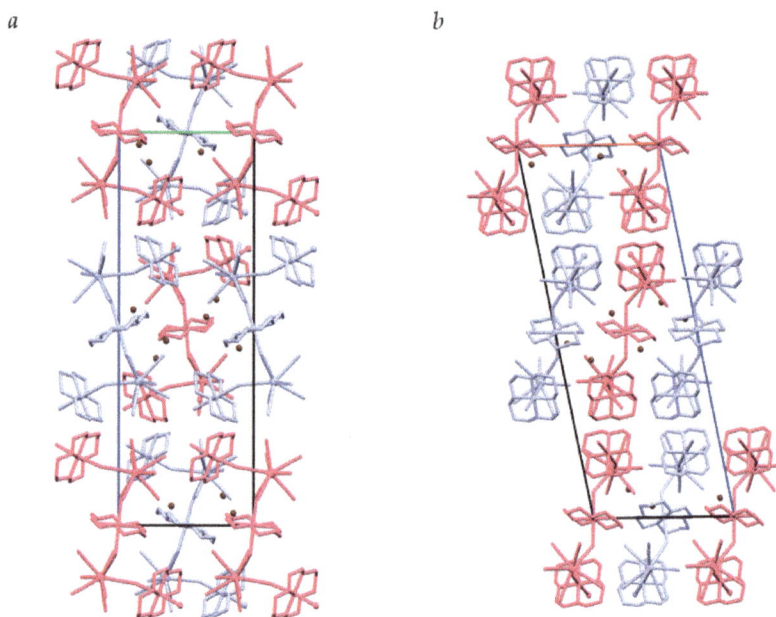

Figure 4. Projection of the structure of **2·3H$_2$O** along (**a**,**b**) crystallographic directions; symmetrically independent molecules marked pink and violet.

3.3. Magnetic Properties

All compounds were characterized by DC susceptibility measurements at the applied field of 1000 Oe in the temperature range 1.8–300 K (Figure 5) and magnetization at 1.8K in the field up to 70 kOe (Figure 6). At the high temperature limit, the value of χT for all compounds is close to the Curie constant of 1.88 cm^{3}K/mol, expected for three Cu^{II} and two W^V ions with s = 1/2 and g = 2, and it remains constant down to about 60 K. Below that temperature, differences in magnetic behavior between the compounds become apparent. For **1·5H$_2$O**, where intra-chain interactions are predominantly ferromagnetic [22], the χT curve rises slightly to reach the maximum of 2.03 cm^{3}K/mol at 7.5 K. For the dehydrated sample **1**, the χT value decreases below 60 K down to 0.87 cm^{3}K/mol at 1.8 K, indicating predominantly antiferromagnetic interaction within the chains. This change in magnetic behavior can be attributed to the bending of the CN-bridges. As was shown before [30–32] for d^9 configuration, the bridge geometry strongly affects the character and strength of magnetic superexchange through the π* orbitals of CN$^-$ ions, which changes from ferromagnetic for a linear bridge to antiferromagnetic for strongly bent bridges. The antiferromagnetic character of intra-chain interactions in **1** is also visible in the magnetization at 1.8 K. The increase of the magnetic moment in an increasing field is much slower for the dehydrated sample **1** than for **1·5H$_2$O** and it reaches only 2.98 Nβ at 70 kOe, which is far from the expected saturation value of 5 Nβ. The magnetic characteristic that we obtained for **1·5H$_2$O** is very similar, but not identical with the published data [22]. Slightly lower values of magnetization and high-temperature susceptibility are most probably caused by the relatively low accuracy of small sample weight assessment. The difference in the χT course in the low temperature range, where we observed a lower maximum at a slightly higher temperature, may be due to partial loss of water and onset of the structural transformation in one of the samples.

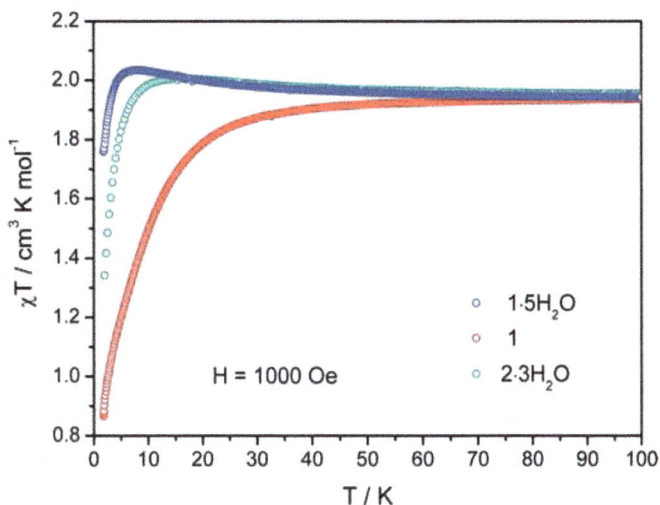

Figure 5. Temperature dependence of the DC magnetic susceptibility at 1000 Oe for **1·5H₂O, 1** and **2·3H₂O**.

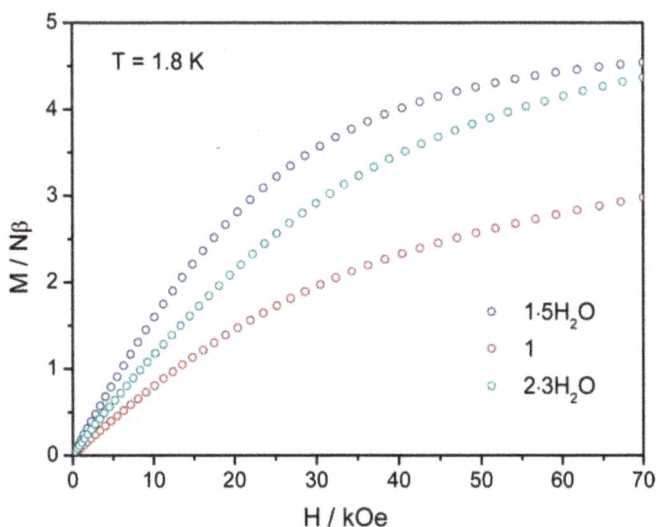

Figure 6. Field dependence of magnetization at 1.8 K for **1·5H₂O, 1** and **2·3H₂O**.

The magnetic behavior of **2·3H₂O** is similar to that of **1·5H₂O**. In the χT dependence on T (Figure 5), a slight increase below 60 K is observed with a broad maximum of 2.01 cm³K/mol at about 14 K and an abrupt drop below 7 K. The magnetization curve lies slightly below the one of **1·5H₂O** and at 70 kOe reaches 4.37 Nβ (Figure 6). These results suggest the presence of weak intramolecular ferromagnetic interactions. This is consistent with the observed geometry of CN-bridges, which in **2·3H₂O** is similar to that in **1·5H₂O**.

4. Conclusions

We have shown that similarly to [Ni(cyclam)]$^{2+/3+}$ complexes, the [Cu(cyclam)]$^{2+}$ ion can be used as a cationic building block in combination with polycyanometallates to construct solvatomagnetic coordination networks. The 1D $\{[Cu^{II}(cyclam)]_3[W^V(CN)_8]_2 \cdot 5H_2O\}_n$ (**1·5H$_2$O**) polymer exhibits a rare single-crystal-to-single-crystal structural transformation to the anhydrous $\{[Cu^{II}(cyclam)]_3[W^V(CN)_8]_2\}_n$ (**1**) form. Although the topology of the compound is retained, the modification of bonds geometry and intermolecular interactions results in a noticeable change of magnetic properties, switching the predominant intra-chain interactions from ferromagnetic in **1·5H$_2$O** to antiferromagnetic in **1**. The dehydration process is not fully reversible, probably due to the formation of intra-chain H-bonds, which stabilize strong bending of the CN-bridges. The rehydration causes collapse of the crystals and partial degradation of the sample. Apart from the post-synthetic modification of **1·5H$_2$O**, we have also shown that the same building block can be arranged in a different structure by slight modification of synthetic conditions. The $\{[Cu^{II}(cyclam)(H_2O)]_2[Cu^{II}(cyclam)][W^V(CN)_8]_2\} \cdot 3H_2O$ pentanuclear Z-shaped molecule (**2·3H$_2$O**), obtained from water under slow diffusion conditions, exhibits weak ferromagnetic interactions through CN-bridges.

Supplementary Materials: The following are available online at http://www.mdpi.com/2073-4352/9/1/45/s1, Figure S1: PXRD pattern for the rehydrated sample of 1 in comparison to 1·5H$_2$O and 1; Figure S2: Mass loss upon dehydration of 1·5H$_2$O monitored by dynamic vapor sorption method; Figure S3: PXRD pattern for sample obtained from [Cu(cyclam)](NO$_3$)$_2$ and Na$_3$[W(CN)$_8$] by quick precipitation from water solution in comparison to 1·5H$_2$O and 2·3H$_2$O; Table S1: Continuous shape measure parameters for octa-coordinated W centers in the structures of 1·5H$_2$O, 1 and 2·3H$_2$O; Table S2: Continuous shape measure parameters for hexa-coordinated Cu centers of 1·5H$_2$O, 1 and 2·3H$_2$O; Figure S4: Inter- and intra-chain H-bonds in 1; Figure S5: Inter- and intra-molecular H-bonds in 2·3H$_2$O.

Author Contributions: Syntheses, original draft preparation: A.P.; crystal structures: A.P. and B.N.; magnetic characterization: M.R.; conceptualization, supervision, funding acquisition and project administration, manuscript review and editing: B.N.

Funding: This research was funded by Polish National Science Centre, grant number 2015/19/B/ST5/00922.

Conflicts of Interest: The authors declare no conflict of interest.

References

1. Kahn, O.; Larionova, J.; Yakhmi, J.V. Molecular Magnetic Sponges. *Chem. Eur. J.* **1999**, *5*, 3443–3449. [CrossRef]
2. Maspoch, D.; Ruiz-Molina, D.; Wurst, K.; Domingo, N.; Cavallini, M.; Biscarini, F.; Tejada, J.; Rovira, C.; Veciana, J. A nanoporous molecular magnet with reversible solvent-induced mechanical and magnetic properties. *Nat. Mater.* **2003**, *2*, 190–195. [CrossRef]
3. Pinkowicz, D.; Podgajny, R.; Gawel, B.; Nitek, W.; Lasocha, W.; Oszajca, M.; Czapla, M.; Makarewicz, M.; Balanda, M.; Sieklucka, B. Double Switching of a Magnetic Coordination Framework through Intraskeletal Molecular Rearrangement. *Angew. Chem. Int. Edit.* **2011**, *50*, 3973–3977. [CrossRef]
4. Nowicka, B.; Balanda, M.; Gawel, B.; Cwiak, G.; Budziak, A.; Lasocha, W.; Sieklucka, B. Microporous $\{[Ni(cyclam)]_3[W(CN)_8]_2\}_n$ affording reversible structural and magnetic conversions. *Dalton Trans.* **2011**, *40*, 3067–3073. [CrossRef]
5. Nowicka, B.; Rams, M.; Stadnicka, K.; Sieklucka, B. Reversible Guest-Induced Magnetic and Structural Single-Crystal-to-Single-Crystal Transformation in Microporous Coordination Network $\{[Ni(cyclam)]_3[W(CN)_8]_2\}_n$. *Inorg. Chem.* **2007**, *46*, 8123–8125. [CrossRef]
6. Nowicka, B.; Reczyński, M.; Bałanda, M.; Fitta, M.; Gaweł, B.; Sieklucka, B. The Rule Rather than the Exception: Structural Flexibility of [Ni(cyclam)]$^{2+}$-Based Cyano-Bridged Magnetic Networks. *Cryst. Growth Des.* **2016**, *16*, 4736–4743. [CrossRef]
7. Nowicka, B.; Reczyński, M.; Rams, M.; Nitek, W.; Kozieł, M.; Sieklucka, B. Larger pores and higher T_c: $\{[Ni(cyclam)]_3[W(CN)_8]_2 \cdot solv\}_n$—A new member of the largest family of pseudo-polymorphic isomers among octacyanometallate-based assemblies. *CrystEngComm* **2015**, *17*, 3526–3532. [CrossRef]
8. Nowicka, B.; Heczko, M.; Reczyński, M.; Rams, M.; Gawel, B.; Nitek, W.; Sieklucka, B. Exploration of a new building block for the construction of cyano-bridged solvatomagnetic assemblies: [Ni(cyclam)]$^{3+}$. *CrystEngComm* **2016**, *18*, 7011–7020. [CrossRef]

9. Nowicka, B.; Reczyński, M.; Rams, M.; Nitek, W.; Zukrowski, J.; Kapusta, C.; Sieklucka, B. Hydration-switchable charge transfer in the first bimetallic assembly based on the [Ni(cyclam)]$^{3+}$—Magnetic CN-bridged chain {(H$_3$O)[Ni^{III}(cyclam)][Fe^{II}(CN)$_6$]·5H$_2$O}$_n$. *Chem. Commun.* **2015**, *51*, 11485–11488. [CrossRef]

10. Reczyński, M.; Nowicka, B.; Näther, C.; Kozieł, M.; Nakabayashi, K.; Ohkoshi, S.-I.; Sieklucka, B. Dehydration-Triggered Charge Transfer and High Proton Conductivity in (H$_3$O)[Ni^{III}(cyclam)][M^{II}(CN)$_6$] (M = Ru, Os) Cyanide-Bridged Chains. *Inorg. Chem.* **2018**, *57*, 13415–13422. [CrossRef]

11. Colacio, E.; Dominguez-Vera, J.M.; Ghazi, M.; Moreno, J.M.; Kivekas, R.; Lloret, F.; Stoeckli-Evans, H. A novel two-dimensional honeycomb-like bimetallic iron(III)-nickel(II) cyanide-bridged magnetic material [Ni(cyclam)]$_3$[Fe(CN)$_6$]$_2$·nH$_2$O (cyclam = 1,4,8,11-tetraazacyclodecane). *Chem. Commun.* **1999**, 987–988. [CrossRef]

12. Ferlay, S.; Mallah, T.; Vaissermann, J.; Bartolome, F.; Veillet, P.; Verdaguer, M. A chromium(III) nickel(II) cyanide-bridged ferromagnetic layered structure with corrugated sheets. *Chem. Commun.* **1996**, *21*, 2481–2482. [CrossRef]

13. Hiroko, T.; Kosuke, N.; Koji, N.; Toshinori, K.; Kazuhito, H.; Shin-ichi, O. Photoreversible Switching of Magnetic Coupling in a Two-dimensional Copper Octacyanomolybdate. *Chem. Lett.* **2009**, *38*, 338–339. [CrossRef]

14. Larionova, J.; Clérac, R.; Donnadieu, B.; Willemin, S.; Guérin, C. Synthesis and Structure of a Two-Dimensional Cyano-Bridged Coordination Polymer [Cu(cyclam)]$_2$[Mo(CN)$_8$]·10.5H$_2$O (Cyclam = 1,4,8,11-Tetraazacyclodecane). *Cryst. Growth Des.* **2003**, *3*, 267–272. [CrossRef]

15. Long, J.; Chamoreau, L.-M.; Marvaud, V. Supramolecular Heterotrimetallic Assembly Based on Octacyanomolybdate, Manganese, and Copper. *Eur. J. Inorg. Chem.* **2011**, *2011*, 4545–4549. [CrossRef]

16. Colacio, E.; Domínguez-Vera, J.M.; Ghazi, M.; Kivekäs, R.; MaríaMoreno, J.; Pajunen, A. Structure and magnetic properties of two cyanide-bridged one-dimensional M–CuII (M = FeIII or FeII) bimetallic assemblies from ferricyanide and CuN4^{2+} (N4 = 1,4,8,11-tetraazacyclotetradecane and *N*,*N*'-bis(2-pyridylmethylene)-1,3-propanediamine) building blocks. *J. Chem. Soc. Dalton Trans.* **2000**, *4*, 505–509. [CrossRef]

17. Salah El Fallah, M.; Ribas, J.; Solans, X.; Font-Bardia, M. A new one-dimensional ferromagnet based on copper(II) and hexacyanochromate(III), with a rope-ladder chain structure. *New J. Chem.* **2003**, *27*, 895–898. [CrossRef]

18. Sen, R.; Bhattacharya, A.; Mal, D.; Bhattacharjee, A.; Gütlich, P.; Mukherjee, A.K.; Solzi, M.; Pernechele, C.; Koner, S. A cyano-bridged bimetallic ferrimagnet: Synthesis, X-ray structure and magnetic study. *Polyhedron* **2010**, *29*, 2762–2768. [CrossRef]

19. Iijima, S.; Honda, Z.; Koner, S.; Mizutani, F. Magnetic and Mössbauer studies of cyanide-bridged bimetallic assembly [Mn(cyclam)][Fe(CN)$_6$]·3H$_2$O. *J. Magn. Magn. Mater.* **2001**, *223*, 16–20. [CrossRef]

20. Bhattacharjee, A.; Miyazaki, Y.; Nakazawa, Y.; Koner, S.; Iijima, S.; Sorai, M. Study of the magnetic phase transition in a cyanide-bridged molecule-based material: [Mn(cyclam)][Fe(CN)$_6$]·3H$_2$O (cyclam=1,4,8,11-tetraazacyclotetradecane). *Phys. B Condens. Matter* **2001**, *305*, 56–64. [CrossRef]

21. Nowicka, B.; Heczko, M.; Rams, M.; Reczyński, M.; Gaweł, B.; Nitek, W.; Sieklucka, B. Solvatomagnetic Studies on Cyano-Bridged Bimetallic Chains Based on [Mn(cyclam)]$^{3+}$ and Hexacyanometallates. *Eur. J. Inorg. Chem.* **2017**, *1*, 99–106. [CrossRef]

22. Lim, J.H.; You, Y.S.; Yoo, H.S.; Yoon, J.H.; Kim, J.I.; Koh, E.K.; Hong, C.S. Bimetallic M^V$_2$Cu^{II}$_3$ (M = Mo, W) Coordination Complexes Based on Octacyanometalates: Structures and Magnetic Variations Tuned by Chelated Tetradentate Macrocyclic Ligands. *Inorg. Chem.* **2007**, *46*, 10578–10586. [CrossRef]

23. Samotus, A. Photochemical Properties of Octacyanotnagstic Acids. 2. Photolisys of Octacyanotungstic(V) Acid. *Pol. J. Chem.* **1973**, *47*, 265–278.

24. Berry, D.E.; Girard, S.; McAuley, A. The Synthesis and Reactions of Nickel(III) Stabilized by a Nitrogen-Donor Macrocycle. *J. Chem. Educ.* **1996**, *73*, 551. [CrossRef]

25. Sheldrick, G. A short history of SHELX. *Acta Crystallogr. A* **2008**, *64*, 112–122. [CrossRef]

26. Llunell, M.; Casanova, D.; Cirera, J.; Alemany, P.; Alvarez, S. *SHAPE v. 2.1*; University of Barcelona: Barcelona, Spain, 2013.

27. Korzeniak, T.; Jankowski, R.; Kozieł, M.; Pinkowicz, D.; Sieklucka, B. Reversible Single-Crystal-to-Single-Crystal Transformation in Photomagnetic Cyanido-Bridged Cd_4M_2 Octahedral Molecules. *Inorg. Chem.* **2017**, *56*, 12914–12919. [CrossRef]

28. Reczyński, M.; Chorazy, S.; Nowicka, B.; Sieklucka, B.; Ohkoshi, S.-I. Dehydration of Octacyanido-Bridged Ni^{II}-W^{IV} Framework toward Negative Thermal Expansion and Magneto-Colorimetric Switching. *Inorg. Chem.* **2017**, *56*, 179–185. [CrossRef]

29. Nowicka, B.; Balanda, M.; Reczyński, M.; Majcher, A.M.; Koziel, M.; Nitek, W.; Lasocha, W.; Sieklucka, B. A water sensitive ferromagnetic [Ni(cyclam)]$_2$[Nb(CN)$_8$] network. *Dalton Trans.* **2013**, *42*, 2616–2621. [CrossRef]

30. Lim, J.H.; Yoon, J.H.; Choi, S.Y.; Ryu, D.W.; Koh, E.K.; Hong, C.S. Cyano-Bridged Pentanuclear and Honeycomblike $M^{III}Cu^{II}$ (M = Fe, Cr) Bimetallic Assemblies: Structural Variations Modulated by Side Groups of Macrocyclic Ligands and Magnetic Properties. *Inorg. Chem.* **2011**, *50*, 1749–1757. [CrossRef]

31. Marvaud, V.; Decroix, C.; Scuiller, A.; Tuyèras, F.; Guyard-Duhayon, C.; Vaissermann, J.; Marrot, J.; Gonnet, F.; Verdaguer, M. Hexacyanometalate Molecular Chemistry: Di-, Tri-, Tetra-, Hexa- and Heptanuclear Heterobimetallic Complexes; Control of Nuclearity and Structural Anisotropy. *Chem. Eur. J.* **2003**, *9*, 1692–1705. [CrossRef]

32. Tuyèras, F.; Scuiller, A.; Duhayon, C.; Hernandez-Molina, M.; de Biani, F.F.; Verdaguer, M.; Mallah, T.; Wernsdorfer, W.; Marvaud, V. Hexacyanidometalate molecular chemistry, part III: Di-, tri-, tetra-, hexa- and hepta-nuclear chromium–nickel complexes: Control of spin, structural anisotropy, intra- and inter-molecular exchange couplings. *Inorg. Chim. Acta* **2008**, *361*, 3505–3518. [CrossRef]

crystals

MDPI

Article

Bis(triphenylphosphine)iminium Salts of Dioxothiadiazole Radical Anions: Preparation, Crystal Structures, and Magnetic Properties

Paweł Pakulski[ID], Mirosław Arczyński * and Dawid Pinkowicz *[ID]

Jagiellonian University, Faculty of Chemistry, Gronostajowa 2, 30-387 Kraków, Poland;
pawel.pakulski@doctoral.uj.edu.pl
* Correspondence: miroslaw.arczynski@doctoral.uj.edu.pl (M.A.); dawid.pinkowicz@uj.edu.pl (D.P.)

Received: 30 November 2018; Accepted: 31 December 2018; Published: 7 January 2019

Abstract: Phenanthroline dioxothiadiazoles are redox active molecules that form stable radical anions suitable for the construction of supramolecular magnetic materials. Herein, the preparation, structures and magnetic properties of bis(triphenylphosphine)iminium (PPN) salts of [1,2,5] thiadiazole[3,4-f][1,10]phenanthroline 1,1-dioxide (**L**), [1,2,5]thiadiazole[3,4-f][4,7]phenanthroline 1,1-dioxide (**4,7-L**), 5-bromo-[1,2,5]thiadiazolo[3,4-f][1,10]phenanthroline 2,2-dioxide (**BrL**), and 5,10-dibromo-[1,2,5]thiadiazolo[3,4-f][1,10]phenanthroline 2,2-dioxide (**diBrL**) are reported. The preparation of new bromo derivatives of the **L**: 5-bromo-[1,2,5]thiadiazolo[3,4-f][1,10]phenanthroline 2,2-dioxide (**BrL**) and 5,10-dibromo-[1,2,5]thiadiazolo[3,4-f][1,10]phenanthroline 2,2-dioxide (**diBrL**)—suitable starting materials for further derivatization—are described starting from a commercially available and cheap 1,10-phenanthroline. All PPN salts show antiferromagnetic interactions between the pairs of radical anions, which in the case of **PPN(diBrL)** are very strong (-116 cm^{-1}; using $\hat{H} = -2JSS$ type of exchange coupling Hamiltonian) due to a different crystal packing of the anion radicals as compared to **PPN(L)**, **PPN(4,7-L)**, and **PPN(BrL)**.

Keywords: radical anion; redox; magnetism; antiferromagnetic coupling; dioxothiadiazole

---

1. Introduction

Purely organic molecular materials that show electric conductivity and non-trivial magnetic properties [1–4] are at the forefront of molecular materials science due to the tremendous flexibility and tunability of organic molecules. Moreover, the potential synergy and interplay between the properties of redox-active organic molecules and metal complexes open new routes to redox-active multistable systems [5,6], single molecule magnets (SMMs) and single chain magnets (SCMs) [7–9], and switchable magnetic conductors [10,11]. Achievement of such advanced properties requires, however, an expansion of the library of easily accessible and electroactive molecules with relatively stable radical forms. The most commonly studied radicals comprise α-diimines [12], dithiolenes [13–15], oxolenes [16,17], nitronyl nitroxides [18,19], tetrathiafulvalene (TTF) derivatives [20], π-conjugated macrocycles [21], TCNE and TCNQ derivatives [22–25], and verdazyl radicals [26].

Dioxothiadiazole-based electroactive molecules [27–33] and their complexes [34], on the other hand, are still underrepresented, despite their obvious advantages: easy reduction to a stable radical form, good coordination abilities (including bridging mode) and chemical tunability presented here for the first time. In 2011, Awaga et al. studied [1,2,5]thiadiazole[3,4-f][1,10]phenanthroline 1,1-dioxide (**L**) in context of its electrochemical properties and synthesis of radical salts [35]. Later on, the same group reported a number of radical salts of **L**, which revealed efficient π-orbitals overlap [36,37] transmitting efficient magnetic exchange interactions varying in strength from ferro- to very strong antiferromagnetic.

Some of us focused on the design of a dioxothiadiazole derivative with the ability to bridge two 3d metals [34]: [1,2,5]thiadiazole[3,4-f][4,7]phenanthroline 1,1-dioxide (**4,7-L**). The bridging potential was tested resulting in the formation of a coordination chain ([CuCl$_2$(**4,7-L**)]}$_n$. Both **L** and **4,7-L** reveal very similar mild reduction potentials to the radical form

Introduction of substituents to tune the redox potential of organic molecules is a well-known concept in electrochemistry [26,38]. We used the same approach to shift the reduction potentials of phenanthroline based dioxothiadiazole derivatives towards less negative values. Herein, we present two new members of the dioxothiadiazole family: 5-bromo-[1,2,5]thiadiazolo[3,4-f][1,10] phenanthroline 2,2-dioxide (**BrL**) and 5,10-dibromo-[1,2,5]thiadiazolo[3,4-f][1,10]phenanthroline 2,2-dioxide (**diBrL**) (Figure 1) and expand the library of potential spin carrying substrates by preparing and investigating bis(triphenylphosphine)iminium (PPN) salts of **L**, **4,7-L**, **BrL** and **diBrL** anion-radicals. The synthetic scheme is presented concisely in Figure 1. Structurally, these compounds reveal alternating cation–anion layers with the exception of the **PPN(diBrL)** which comprises chains of π-conjugated molecules. The influence of the structural differences on the magnetic properties of the organic salts is analyzed.

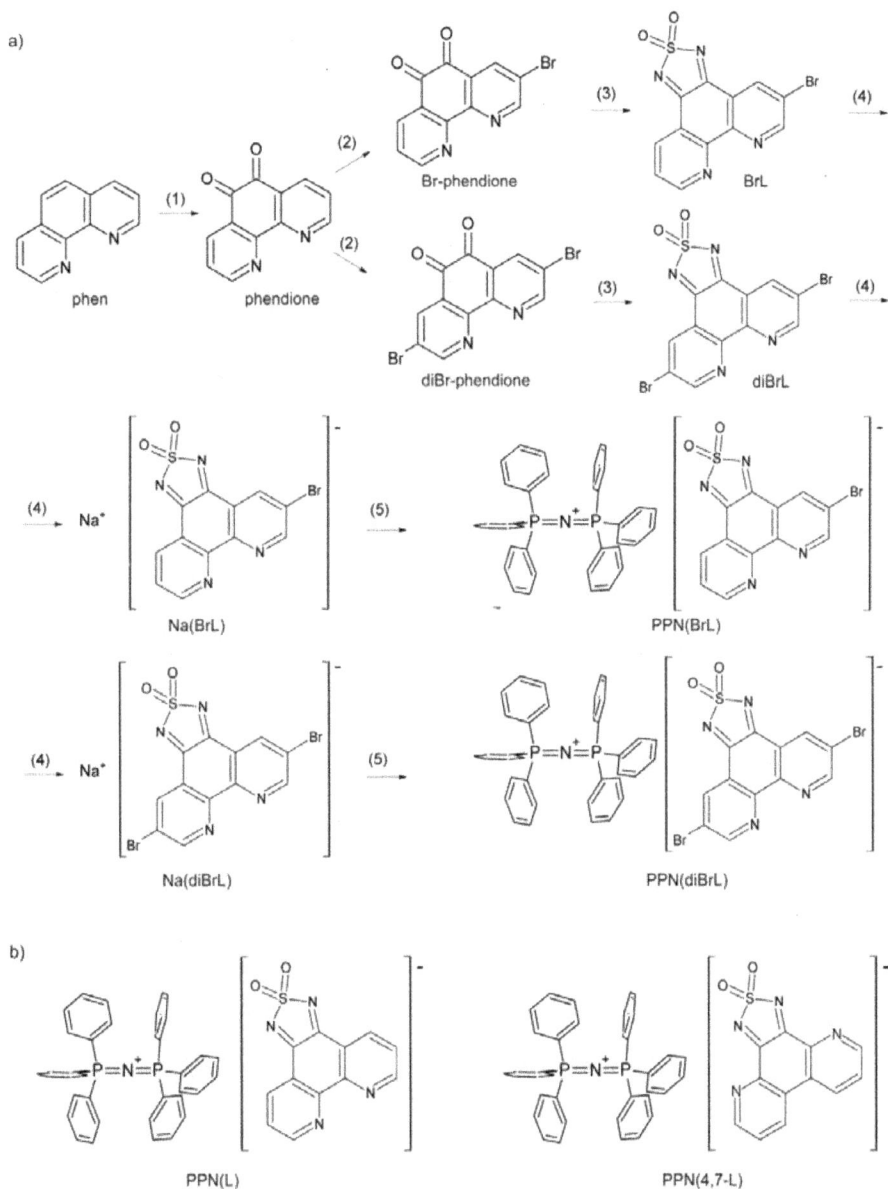

Figure 1. Schematic view of the synthetic route to **BrL**, **diBrL** and their radical salts: **PPN(BrL)**, **PPN(diBrL)** (**a**) and (**b**) structural formulas of **PPN(L)** and **PPN(4,7-L)**. The reaction conditions and yields are as follows: (1) $H_2SO_{4(conc.)}$, KBr, 3 h reflux, yield 95%; (2) $H_2SO_{4(98\%)}$, $HNO_{3(99\%)}$, KBr, 16 h reflux, yield 9% for diBr-phendione and 22% for Br-phendione; (3) Ethanol$_{(anh.)}$, sulfamide (3 portions), 7 d reflux, yield 91% for **BrL**, 35% for **diBrL**; (4) anhydrous acetonitrile, NaI, sonication; (5) THF, PPNCl, yield 11% for **PPN(BrL)**, and 22% for **PPN(diBrL)**.

2. Materials and Methods

Chemicals and reagents were of analytical grade unless otherwise stated. Inert PureSolv-MD-5/7 solvent purification system with alumina filled columns and argon gas was used for the deoxygenation and dehydration of acetonitrile and tetrahydrofuran. 1,10-phenanthroline-5,6-dione (**phendione**) was prepared according to literature procedure [39]. Some of the operations that required inert gas atmosphere were performed using Inert PureLab HE glove box filled with Ar gas.

2.1. Syntheses

2.1.1. 3-Bromo-1,10-phenantroline-5,6-dione (**Br-phendione**) and 3,8-Dibromo-1,10-phenanthroline-5,6-dione (**diBr-phendione**)

Bromo derivatives of phendione were prepared according to the modified literature procedure [40]. 1,10-phenanthroline-5,6-dione (**phendione**) (20 g, 95.2 mmol) was dissolved in a chilled mixture of 120 mL 98% H_2SO_4 and 60 mL 99% HNO_3 in a 500 mL round-bottom flask. KBr (20 g, 168 mmol) was added and the mixture was refluxed at 120 °C overnight. After 16 h, the mixture was allowed to cool down to room temperature and poured onto 500 mL of crushed ice, pH was carefully adjusted to around 5 using 0.25 M NaOH solution. The mixture was extracted with chloroform (6 × 150 mL) and the combined organic extracts were dried using $MgSO_4$ and evaporated to dryness. Column chromatography (SiO_2, eluent: $CHCl_3$:AcOEt:n-hexane, 10:2:1) afforded pure 3-bromo-1,10-phenanthroline-5,6-dione (6.0 g, 22%), 3,8-dibromo-1,10-phenanthroline-5,6-dione (3.0 g, 9%) and the unreacted phendione (0.9 g). 3-bromo-1,10-phenanthroline-5,6-dione ^{1}H NMR ($CDCl_3$) δ [ppm]: 9.14 (d, 1H, J = 2.5 Hz), 9.11 (dd, 1H, J = 4.8, 1.9 Hz), 8.60 (d, 1H, J = 2.5 Hz), 8.50 (dd, 1H, J = 7.8, 1.8 Hz), 7.61 (dd, 1H, J = 7.9, 4.7 Hz); 3,8-dibromo-1,10-phenanthroline-5,6-dione ^{1}H NMR ($CDCl_3$) δ [ppm]: 9.15 (d, 2H, J = 2.3 Hz), 8.62 (d, 2H, J = 2.5 Hz). ^{1}H NMR spectra can be found in the Supplementary Materials.

2.1.2. 5-Bromo-[1,2,5]thiadiazolo[3,4-f][1,10]phenanthroline 2,2-Dioxide (**BrL**)

3-bromo-1,10-phenanthroline-5,6-dione (1.0 g, 3.5 mmol) was suspended in 35 mL of anhydrous ethanol, sulfamide (0.56 g, 5.8 mmol) was added and the mixture refluxed for seven days. After 24 h, another quantity (0.25 g) of sulfamide was added. Additions were repeated daily for one week. Next, the mixture was allowed to cool down to room temperature and filtered. The yellow solid was washed with two aliquots of ethanol (2 × 20 mL) and dried in vacuo to afford 1.1 g (91%) of the desired product. ^{1}H NMR ($CDCl_3$) δ [ppm]: 9.22 (d, 1H, J = 2.5 Hz), 9.19 (dd, 1H, J = 4.8, 1.8 Hz), 8.84 (d, 1H, J = 2.3 Hz), 8.72 (dd, 1H, J = 7.9, 1.8 Hz), 7.67 (dd, 1H, J = 7.9, 4.7 Hz). ^{1}H NMR spectra can be found in the Supplementary Materials.

2.1.3. 5,10-Dibromo-[1,2,5]thiadiazolo[3,4-f][1,10]phenanthroline 2,2-dioxide (**diBrL**)

3,8-dibromo-1,10-phenanthroline-5,6-dione (150 mg, 0.4 mmol) was suspended in 20 mL of anhydrous ethanol, sulfamide (50 mg, 0.5 mmol) was added and the mixture refluxed for three days. After 24 h, another quantity (50 mg) of sulfamide was added. Additions were repeated daily for the duration of the synthesis. Next, the brownish-orange mixture was allowed to cool down to room temperature and filtered to afford a brownish-yellow which was washed with two aliquots of ethanol (2 × 10 mL) and dried under vacuum. The crude product was once more suspended in 20 mL of the ethanolic solution of sulfamide (50 mg) and refluxed for another 24 h. The mixture was allowed to cool down to room temperature and filtered. An orange solid was washed with 20 mL of ethanol, 10 mL of cold diethyl ether and dried under vacuum to afford 198 mg (17%) of the desired product. ^{1}H NMR ($CDCl_3$) δ [ppm]: 9.19 (d, 2H, J = 2.2 Hz), 8.42 (d, 2H, J = 2.3 Hz). ^{1}H NMR spectra can be found in the Supplementary Materials.

2.1.4. Bis(triphenylphosphine)iminium [1,2,5]thiadiazole[3,4-f][1,10]phenanthroline 1,1-Dioxide $H_2O/(CH_3)_2CO$ Solvate (**PPN(L)**)

This compound was obtained in a two-step procedure: the suspension of 310 mg of [1,2,5]thiadiazole[3,4-f][1,10]phenanthroline 1,1-dioxide (1.15 mmol) (**L**) in 90 mL of MeCN was stirred overnight with 4 g of NaI (26.7 mmol) under ambient atmosphere. The resulting dark red precipitate was filtered and washed with 60 mL of MeCN until the colour of the product **Na(L)** changed from dark red to purple. The **Na(L)** was dried in vacuo for a few hours. Anal. calcd. for $C_{12}H_7N_4NaO_{2.5}S$ (Na(1,10-tdapO$_2$)·0.5 H$_2$O): C, 47.68; H, 2.33; N, 18.54; S, 10.61. Found: C, 47.75; H, 2.45; N, 18.55; S, 10.20.

293 mg of **Na(L)**·0.5H$_2$O was dissolved in 210 mL of acetone followed by the addition of 2.0 g (3.6 mmol) of PPNCl and stirring for 40 min. After that time the purple/violet solution was filtered to remove the precipitated NaCl. The filtrate was concentrated in a rotary evaporator to ca. 10 mL, which was left undisturbed for 1 h for crystallization.

The product was collected by decantation, filtered and washed with a single drop of cold acetone. Product was dried in air. Yield 660 mg (75%) based on **Na(L)**. Anal. calcd. for **PPN(L)**·H$_2$O·(CH$_3$)$_2$CO, $C_{51}H_{44}N_5O_4P_2S$ (884.9 g/mol): C, 69.22; H, 5.01; N, 7.91; S, 3.62. Found: C, 68.97; H, 4.89; N, 7.96; S, 4.06.

2.1.5. Bis(triphenylphosphine)iminium [1,2,5]thiadiazole[3,4-f][4,7]phenanthroline 1,1-Dioxide (**PPN(4,7-L)**)

Product was obtained similarly to **PPN(L)**. The suspension of 63 mg (0.23 mmol) of [1,2,5]thiadiazole[3,4-f][4,7]phenanthroline 1,1-dioxide (**4,7-L**) in 35 mL of dry MeCN was stirred overnight under an inert atmosphere with a 1.0 g (6.6 mmol) of NaI. Next day the dark violet precipitate was separated from the mother solution by filtration and washed with dry MeCN (ca. 20 mL). The powder attains violet colour after washing. After vacuum drying for a few hours the 30 mg of crude **Na(4,7-L)** was used to obtain **PPN(4,7-L)**. This was achieved by stirring MeCN (135 mL) suspension with 210 mg of PPNCl for two hours. The NaCl precipitate was filtered off and a clear violet solution was quickly condensed on a rotary evaporator to ca. 6 mL and transferred to the Ar-filled glovebox, where it was left for 2 h for crystallization. Large elongated block crystals were separated from the mother suspension by decantation, filtered and washed with a single drop of cold MeCN. Yield 49 mg (59% based on **Na(4,7-L)**). Anal. calcd. for **PPN(4,7-L)**, $C_{48}H_{36}N_5O_2P_2S$ (808.8 g/mol): C, 71.28; H, 4.4; N, 8.66; S, 3.96. Found: C, 70.67; H, 4.30; N, 8.62; S, 4.12.

2.1.6. Bis(triphenylphosphine)iminium 5-Bromo-[1,2,5]thiadiazole[3,4-f][1,10]phenanthroline 1,1-dioxide THF Solvate (**PPN(BrL)**)

350 mg of **BrL** (1.0 mmol) and 2.0 g of NaI (13.3 mmol) was sonicated for 20 min in anhydrous acetonitrile (50 mL). During sonication the yellow suspension turned dark violet due to the reduction of **BrL** to a radical anion BrL$^{\bullet-}$ by iodide. The mixture was filtered and the dark violet precipitate was washed with three portions of anhydrous acetonitrile (3 mL each). The sodium salt **Na(BrL)** was dried under vacuum for 1 h (300 mg, 81%) and then suspended in dry tetrahydrofuran (50 mL). Solid PPNCl (1.15 g, 2.0 mmol) was added resulting in the color change of the liquid phase to deep violet and the precipitation of white sodium chloride. NaCl was removed by filtration and the violet THF solution was concentrated by rotary evaporation to ca. 15 mL. Large crystals of the THF solvate were obtained by slow vapor diffusion of dry diethyl ether onto the THF mother solution (two days). The violet crystals (ca. 1–3 mm) were separated by hand under the microscope from the smaller colorless crystals of PPNCl. Yield 90 mg (11% based on **BrL**). Anal. calcd. for **PPN(BrL)**·2THF, $C_{56}H_{51}BrN_5O_4P_2S$ (1032.0 g/mol): C: 65.18, H: 4.98, 6.79, S: 3.11. Found: C: 65.49, H: 5.15, N: 6.58, S: 2.71.

2.1.7. Bis(triphenylphosphine)iminium 5,10-Dibromo-[1,2,5]thiadiazole[3,4-f][1,10]phenanthroline 1,1-Dioxide (**PPN(diBrL)**)

150 mg of **diBrL** (0.35 mmol) and 1.0 g of NaI (6.6 mmol) was sonicated for 10 min in anhydrous acetonitrile (20 mL). During the sonication, the yellow suspension turned dark violet due to the reduction of **diBrL** to a radical anion by iodide. The mixture was filtered and the dark violet precipitate was washed with three portions of anhydrous acetonitrile (2 mL each). The sodium salt **Na(diBrL)** was dried under vacuum for 1 h (95 mg, 60%) and then suspended in dry tetrahydrofuran (50 mL). Solid PPNCl (0.4 g, 0.7 mmol) was added in small portions resulting in the color change of the liquid phase to deep violet and the precipitation of sodium chloride. NaCl was removed by filtration and the violet THF solution was evaporated to dryness. The crude **PPN(diBrL)** was dissolved in ca. 15 mL of MeCN. Then the solution was evaporated to ca. 2 mL. The sample was left for an hour for crystallization and then the crystals were separated and purified similar to **PPN(4,7-L)**. Yield 45 mg (22%). Anal. calcd. for **PPN(diBrL)**, $C_{48}H_{34}Br_2N_5O_2P_2S$ (966.6 g/mol): C: 59.64, H: 3.55, N: 7.25, S: 3.32. Found: C: 59.32, H: 3.46, N: 7.15, S: 3.05.

2.2. Other Physical Measurements

IR spectra were collected using Nicolet iN10 MX FT-IR microscope in the transmission mode. Cyclic voltammetry was performed using Mtm-anko M-161C electrochemical analyzer. Glassy carbon electrodes were used in both experiments. ^{1}H NMR spectra were measured using Bruker Avance II 300 MHz spectrometer. Elemental CHNS analysis were done with ELEMENTAR Vario Micro Cube CHNS analyzer.

2.2.1. Magnetic Measurements

The magnetic measurements were carried out using Quantum Design MPMS3 Evercool SQUID magnetometer. The samples were sealed in HDPE foil bags to protect them from the crystallization solvent loss. Corrections for the diamagnetism of the sample holder and the compounds themselves (Pascal constants) were applied [41].

2.2.2. X-ray Diffraction Data Collection/Refinement

Single crystal X-ray diffraction (XRD) data was collected on a Bruker D8 Quest Eco diffractometer equipped with Photon 50^{TM} CMOS detector and Mo-K$_\alpha$ Triumph® monochromator. The data were collected at low temperature using Kryoflex II low-temperature device. Data were integrated using SAINT [42], while multi-scan absorption corrections were applied using SADABS or TWINABS [43,44], all incorporated into APEX3 environment [45]. The structures were solved using SHELXT [46] and refined with SHELXL [47,48] software within the Olex2 package [49]. All hydrogen atoms were refined using riding model, and the non-hydrogen atoms were refined anisotropically using weighted full-matrix least-squares on F^2. Disorder of THF crystalline solvent molecules in **PPN(BrL)** was refined using constraints. Ideal THF geometry was supported by IMGL library [50]. The occupancy factors of THF molecules are changed so that the refinement could be stable. The positions should not be taken as perfect because the model shows just one possibility. One of the phenyl rings of PPN cation is also strongly disordered, apparently over four positions, but it was refined using only two of them, where positive electron densities were the strongest. **PPN(diBrL)** was refined as two component twin with scales 0.6227(6). In case of **PPN(L)** the occupation factor of the water oxygen atom was fixed at low value due to disorder of the solvent molecules.

CCDC 1882326-1882331 contains the supplementary crystallographic data for this paper. These data can be obtained free of charge via http://www.ccdc.cam.ac.uk/conts/retrieving.html (or from the CCDC, 12 Union Road, Cambridge CB2 1EZ, UK; Fax: +44 1223 336033; E-mail: deposit@ccdc.cam.ac.uk).

2.2.3. Calculation Details

Density functional theory (DFT) calculations were done in Gaussian09 program [51]. For each molecule hybrid Becke 3-parameter Lee-Yang-Parr (B3LYP) [52,53] functional with 6-311++G(2d,2p) basis set was used [54]. Single crystal XRD structural models were taken as a starting geometry. No structural constraints were imposed on any atom.

3. Results

3.1. Syntheses

The **Br-phendione** and **diBr-phendione** were synthesized by bromination of **phendione** using KBr in the mixture of concentrated HNO_3 and H_2SO_4 acids (Figure 1). The products were purified by column chromatography. Both were further used in the procedure of attaching the thiadiazole dioxide functional group to the phenanthroline backbone. This includes the reaction of alpha-diketone group with sulfamide in the anhydrous boiling ethanol (Figure 1). In contrast to the preparation of **L** or **4,7-L**, reactions of sulfamide with **Br-phendione** and **diBr-phendione** require longer time and additional quantities of sulfamide added every 24 h of the reaction under reflux to achieve good yields. The synthesis of all PPN radical salts was carried out in two steps. In the first step, the reduction of the respective dioxothiadiazole derivative with sodium iodide led to the quantitative precipitation of poorly soluble sodium salts. In the second step, the metathesis of the obtained sodium salts using PPNCl resulted in the final PPN salts, which are very well soluble in THF, acetonitrile, dichloromethane, and chloroform (Figure 1).

3.2. BrL and diBrL—Crystal Structures and DFT Calculations

Crystal structures of **BrL** and **diBrL** were determined using single crystal X-ray diffraction (Table S1 in the Supplementary Materials). **BrL** crystallizes in the *P*-1 space group while **diBrL** in $P2_1/c$. The asymmetric units of both compounds are presented in Figure 2a,b. **BrL** and **diBrL** molecules are equipped with bromine substituents which change the crystal packing of the molecules as compared to the non-substituted **L**. **BrL** forms π-π stacks along "*a*" crystallographic direction with the parallel off-centered arrangement of the molecules forced by steric hindrance of the SO_2 group (Figure S1a in the Supplementary Materials). The plane of each molecule in a stack is inclined relative to the direction of stack propagation. The structure of **BrL** seems to be similar to that of **L** with the layers of molecules forming a two-dimensional (2-D) network of double N···H-C hydrogen bonds (donor acceptor distance of 3.525 Å on one side of the molecule and 3.642 Å on the other). The presence of the Br substituent in **BrL** disrupts the H-bonding and the molecular packing as compared to **L**. One side of the molecule forms slightly different interactions, namely N···S close contacts (3.331 Å) which replace the N···H-C bonds (Figure S1b in the Supplementary Materials). Also, the donor acceptor distances in N···H-C hydrogen bonds of **BrL** are a little shorter (3.467 Å) compared to those in **L**. The molecules of **BrL** do not lie completely flat, but are slightly tilted directing the bromine atoms slightly below the plane of the neighboring molecule.

The structure of the **diBrL** is completely different. The chains of parallel hydrogen bonded dimers interacting with each other through π-orbitals and bromine atoms are easily distinguished (Figure S2 in the Supplementary Materials). The neighboring chains run in two perpendicular directions, and the source of closest contacts between them are N and O atoms of the dioxothiadiazole groups.

The DFT calculations (B3LYP exchange-correlation functional; 6-311++G(2d,2p) basis set) were performed for **BrL** and **diBrL**. Noteworthy, the inspection of molecular orbitals revealed that low-lying LUMO is not spread over the bromine atoms in contrast to the HOMO (Figure 3). The LUMOs for both **BrL** and **diBrL** molecules exhibit an antibonding character at the N-C bonds of the dioxothiadiazole group (nodal plane) and a bonding character at the dioxothiadiazole's S-N and C-C bonds, which is consistent with the observed shortening and elongation of the respective bonds discussed below.

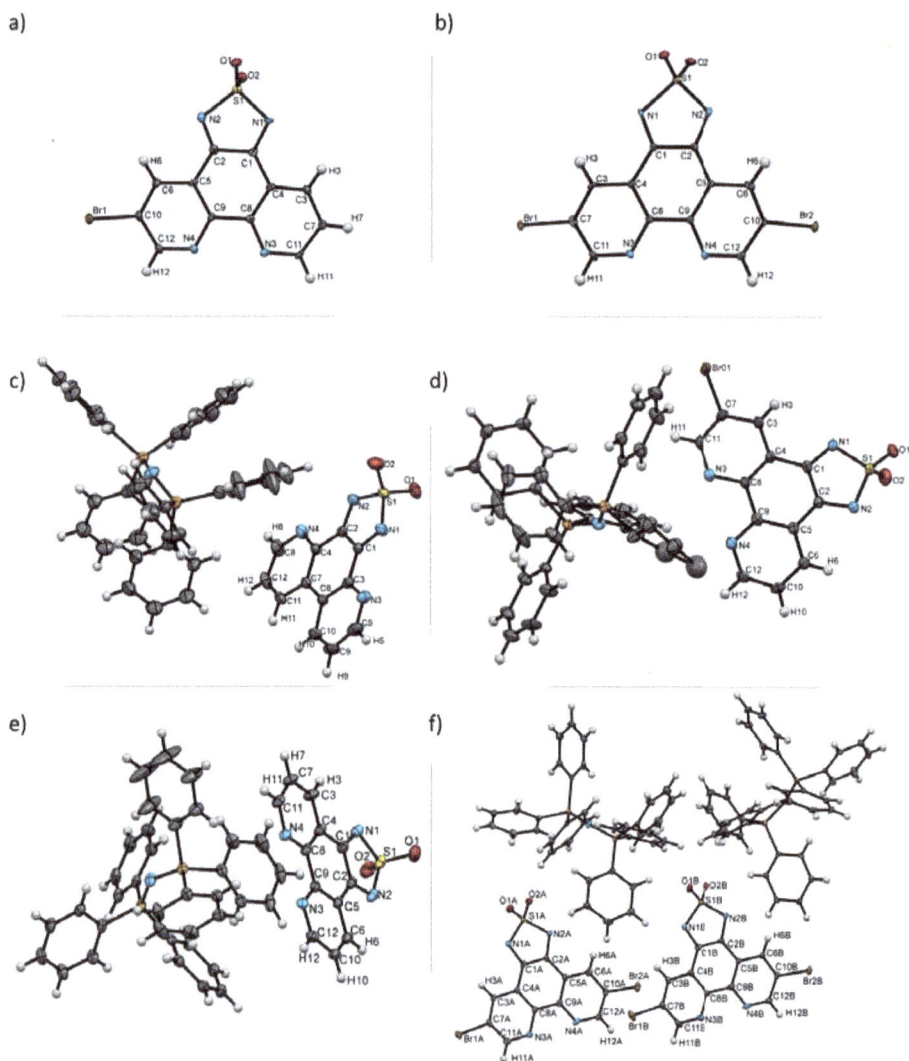

Figure 2. Asymmetric units of (**a**) **BrL**; (**b**) **diBrL**; (**c**) **PPN(4,7-L)**; (**d**) **PPN(BrL)**; (**e**) **PPN(L)**; (**f**) **PPN(diBrL)**; crystallization solvent molecules and disorder treated parts omitted for the sake of clarity.

a)

HOMO

b)

LUMO

c)

HOMO

d)

LUMO

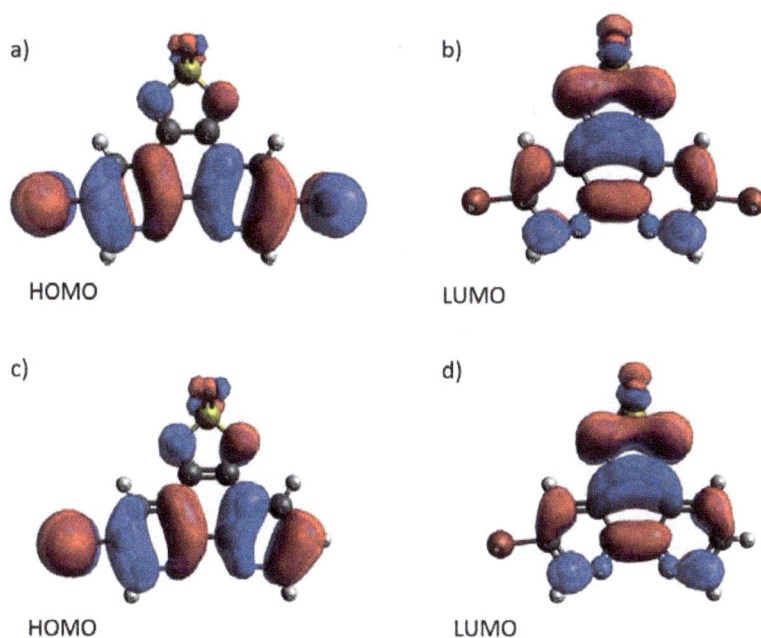

Figure 3. Results of the DFT B3LYP calculations results: HOMO of **diBrL** (**a**), LUMO of **diBrL** (**b**), HOMO of **BrL** (**c**), and LUMO of **BrL** (**d**).

3.3. BrL and diBrL—Cyclic Voltammetry

Electrochemical properties of brominated **BrL** and **diBrL** are similar to the previously reported **L** [35] and **4,7-L** [34] featuring two distinct reduction processes (Figure 4). The first reduction leads to an anion radical and appears at ca. −441 mV vs Fc^+/Fc. The second one results in a diamagnetic dianion and appears ca. 800 mV below the first reduction. The exact values of the reduction potentials are presented in Table 1. The first reduction potential of **diBrL** (−441 mV) and **BrL** (−471 mV) are shifted to less negative values as compared to **L** (−499 mV) and **4,7-L** (−520 mV). This trend is also reflected in the HOMO-LUMO gap, which is much smaller for the brominated species (Table 1).

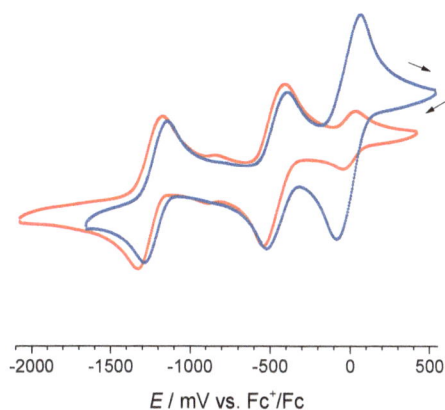

E / mV vs. Fc^+/Fc

Figure 4. Cyclic voltammograms of **BrL** (red line) and **diBrL** (blue line) in 100 mM MeCN solution of n-Bu4NPF6 recorded at 100 mV/s vs. Fc^+/Fc.

Table 1. Reduction half-potentials and HOMO-LUMO gap (Δ) of the dioxothiadiazole-based molecules.

Compound	Reduction Wave No.	E_{ox}/mV	E_{red}/mV	$E_{1/2}$/mV	Δ/eV	Ref.
diBrL	1	-387	-495	-441	3.183	this work
	2	-1139	-1279	-1208		
BrL	1	-405	-538	-471	3.352	this work
	2	-1166	-1324	-1245		
L	1			-499	3.562	[35]
	2			-1320		
4,7-L	1			-520	3.583	[34]
	2			-1229		

3.4. PPN Radical Salts—Crystal Structures

Structural X-ray diffraction data revealed that **PPN(L)**, **PPN(BrL)**, and **PPN(diBrL)** crystallize in *P*-1 while PPN(**4,7-L**) crystallizes in *P*2₁/c space group (Table S1 in the Supplementary Materials).

The asymmetric unit (ASU) of **PPN(diBrL)** contains two anion-radicals and two PPN⁺ cations while ASUs of **PPN(4,7-L)**, **PPN(L)**, and **PPN(BrL)** comprise only one respective radical anion and one counter-cation (Figure 2). **PPN(4,7-L)** and **PPN(diBrL)** crystallize without solvent molecules while **PPN(L)** incorporates one acetone and one water molecule and **PPN(BrL)** crystallizes with tetrahydrofuran molecules.

In terms of crystal packing **PPN(L)**, **PPN(BrL)**, and **PPN(4,7-L)** exhibit alternating anion-cation layered arrangement presented in Figure 5. Due to this particular arrangement the radical anions in these three organic salts exhibit negligible π-orbital overlap and a number of N···H-C hydrogen bonds (Figure 5) that connect neighboring anions in pairs as indicated in Figure 5 by the green ovals. Crystal packing of **PPN(diBrL)** is completely different and comprises infinite π-π stacks of radical anions separated by PPN cations (Figure 5d). The shortest D···A (H-bond) or π-π contacts connecting the dioxothiadiazole groups, where the spin density is the highest, is as follows: 3.403 Å (**PPN(4,7-L)**), 3.520 Å (**PPN(BrL)**), 3.455 Å (**PPN(L)**) and 3.322 Å (**PPN(diBrL)**). Please note, that in the case of (**PPN(4,7-L)**) the separation of dioxothiadiazole groups is the most efficient, despite the shortest D···A distances.

As already mentioned, **L**•⁻ and **BrL**•⁻ anion radicals in their respective PPN salts form almost flat anionic layers (Figure 5b,c) while **4,7-L**•⁻ layer is significantly corrugated (as in corrugated cardboard) (Figure 5a).

Similarly to previously reported dioxothiadiazoles, a one electron reduction results in the contraction of the C-C and S-N bonds (by ca. 0.06 Å and 0.04 Å, respectively), as well as the elongation of C=N and S=O (by ca. 0.05 Å and 0.015 Å, respectively). These results line up with the shape of the LUMO, which is occupied in the radical form [34,35,37]. These particular bond lengths are indicative of the oxidation state of the molecule. Table 2 compares bond lengths of neutral and radical dioxothiadiazole-based compounds presented here.

As can be seen from Table 2, the bond lengths in all presented compounds do not deviate from those of previously reported dioxothiadiazoles. Most notable difference between neutral and anion-radical molecules are found in C-C bond lengths which are shortened due to one electron reduction by more than 0.06 Å as compared to the neutral form (from ca 1.51 Å to ca. 1.44 Å, respectively).

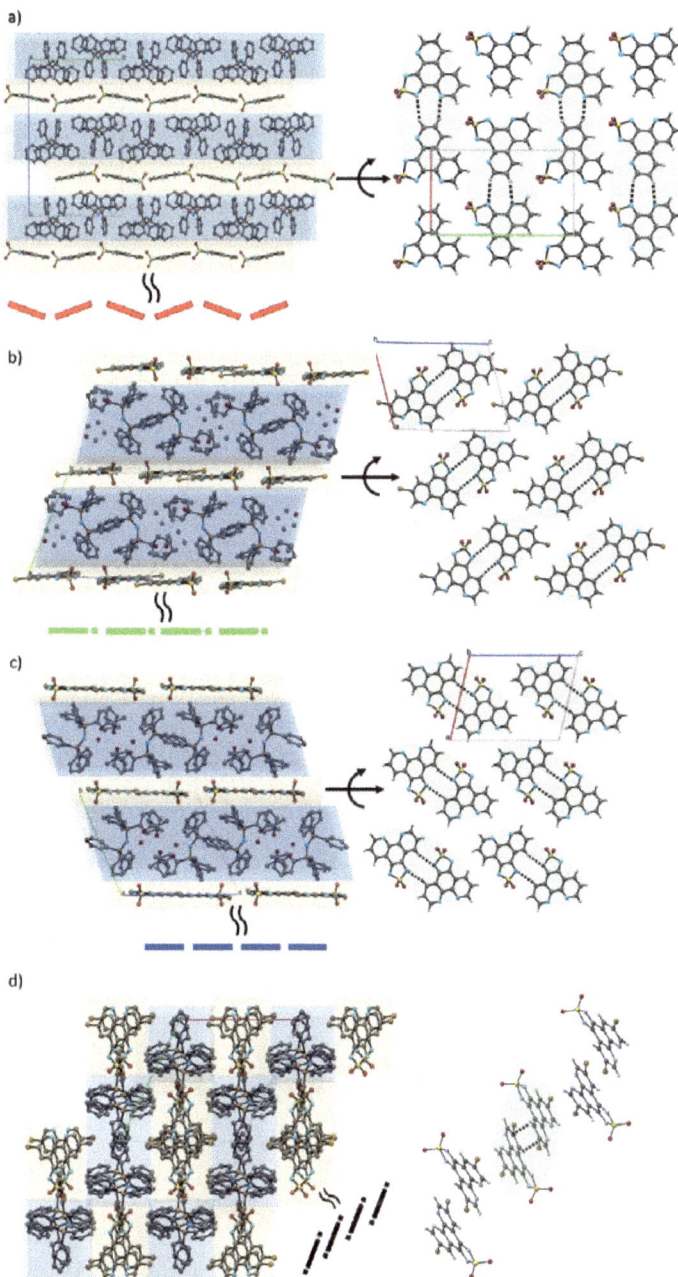

Figure 5. Packing diagrams of the layered structural models of (**a**) **PPN(4,7-L)**; (**b**) **PPN(BrL)**; (**c**) **PPN(L)**; and (**d**) **PPN(diBrL)**; Cationic PPN⁺ layers are marked with blue colour and anionic radical layers are marked with yellow for clarity. On the right side is the top view of a single radical anion layer. Note that the anionic layers of **PPN(4,7-L)** as well as **PPN(BrL)** are not completely flat, as the radical anions are slightly tilted. The green ovals indicate the shortest D···A (H-bond) or π-π contacts connecting the dioxothiadiazole groups: 3.403 Å (**a**), 3.520 Å (**b**), 3.455 Å (**c**), and 3.322 Å (**d**).

Table 2. Selected bond lengths (in Å) of BrL, diBrL vs. PPN(4,7-L), PPN(BrL), PPN(L), PPN(diBrL).

Compound	S=O	S-N	C=N	C-C
Br-L	1.423(4) 1.429(4)	1.693(5) 1.695(4)	1.284(7) 1.293(6)	1.505(7)
diBr-L	1.421(4) 1.426(4)	1.693(4) 1.696(5)	1.289(7) 1.284(7)	1.514(7)
av. in neutral molecules	1.425	1.694	1.288	1.510
PPN(4,7-L$^\bullet$)	1.439(2) 1.436(2)	1.648(2) 1.657(2)	1.333(2) 1.333(2)	1.443(2)
PPN(Br-L$^\bullet$)	1.437(3) 1.433(3)	1.649(3) 1.660(3)	1.333(5) 1.332(5)	1.441(5)
PPN(1,10-L$^\bullet$)	1.442(2) 1.448(2)	1.664(2) 1.664(3)	1.342(3) 1.340(3)	1.443(4)
PPN(diBr-L$^\bullet$) Molecule B	1.443(4) 1.443(4)	1.646(5) 1.660(5)	1.336(7) 1.333(8)	1.452(8)
PPN(diBr-L$^\bullet$) Molecule A	1.442(4) 1.444(4)	1.656(5) 1.660(5)	1.338(8) 1.333(8)	1.435(8)
av. in radical anions	1.441	1.656	1.335	1.443

3.5. PPN$^+$ Radical Salts—Magnetic Properties

The results of magnetic measurements are presented in both Figure 6 and Table 3. In all three salts that reveal layered structures (**PPN(4,7-L)**, **PPN(L)**, and **PPN(BrL)**) the $\chi T(T)$ curves show very similar behavior. This dependence is constant above ca. 60 K and takes the values which are close to the theoretical 0.375 cm^3 K mol^{-1} spin-only value assuming $S = \frac{1}{2}$ and $g = 2.0$ (Table 3 and Figure 6). Below ca. 60 K the $\chi T(T)$ decreases and plummets below 15 K achieving values close to zero due to antiferromagnetic interactions between the radical anions. $M(H)$ curves at 1.8 K differ slightly among these three compounds. The $M(H)$ are slowly, almost linearly increasing with field (Figure 6b), until at some point the increase becomes steep, which again is typical for weak local antiferromagnetic interactions between neighboring spins. For **PPN(4,7-L)** the inflection point is located around 4 T, while for **PPN(L)**—around 6 T and for **PPN(BrL)** well above 7 T. The $M(H)$ dependencies do not saturate at 7 T and the magnetization values at this field decreases along the series **PPN(4,7-L)**, **PPN(BrL)**, **PPN(L)**, suggesting that the strongest magnetic interactions operate within the **PPN(L)** salt.

Table 3. Values of χT at 300 K and 1.8 K, M(H) at 7T and the magnetic exchange constants *J* obtained from fitting of M(H) and $\chi T(T)$ in the whole temperature range using PHI software [55].

Compound	$\chi T(T)$ @300K/cm^3 K mol^{-1}	$\chi T(T)$ @80K/cm^3 K mol^{-1}	$\chi T(T)$ @1.8K/cm^3 K mol^{-1}	M(H) @7T/μ_B	J/cm^{-1}
PPN(4,7-L)	0.374	0.377	0.075	0.76	−2(1)
PPN(BrL)	0.374	0.363	0.024	0.20	−4(1)
PPN(L)	0.367	0.352	0.006	0.07	−5(1)
PPN(diBrL)	0.313 *	0.200	0.157 **	0.37	−116(10) −0.6(5)

* at 340 K; ** at 7 K.

$\chi T(T)$ curve for **PPN(diBrL)** (Figure 6 black dots), on the other hand, decreases in the whole 340–80 K temperature range from 0.313 cm^3 K mol^{-1} at 340 K to 0.196 cm^3 K mol^{-1} at 80 K with a plateau-like feature around 0.186 cm^3 K mol^{-1} below this temperature. The signal starts to decrease again below 30 K and reaches a minimum of 0.158 cm^3 K mol^{-1} at 7 K. Near 2.0 K additional small increase of the $\chi T(T)$ signal is observed which might be ascribed to very weak ferromagnetic interactions between the radical anions. The $M(H)$ curve increases in a Brillouin like fashion reaching the value of 0.37 μ_B (well below the expected 1.0 μ_B for $S = \frac{1}{2}$ and $g = 2.0$), but close to 0.5 μ_B which suggests the presence of strong antiferromagnetic interactions between half of the radical anions in the compound.

The magnetic data were fitted assuming local antiferromagnetic interactions between the pairs of anion radicals which in the case of **PPN(4,7-L)**, **PPN(L)**, and **PPN(BrL)** are transmitted through C-H···N hydrogen bonds (Figure 5) and for **PPN(diBrL)** through the π-π contacts with different interplane distances (two types of radical pairs with two types of magnetic interactions (one very strong and the other very weak). Figure S8 presents the magnetic coupling scheme for all four compounds. The results of the simultaneous fitting of $\chi T(T)$ and $M(H)$ (PHI program [55]) using the following Hamiltonians (Equation (1) for **PPN(4,7-L)**, **PPN(L)**, and **PPN(BrL)** and Equation (2) for **PPN(diBrL)**) are collected in Table 3 and presented as solid lines in Figure 6

$$\hat{H} = -2J_{12} \cdot S_1 \cdot S_2 + \mu_B \cdot g_1 \cdot S_1 \cdot B + \mu_B \cdot g_2 \cdot S_2 \cdot B \tag{1}$$

$$\hat{H} = -2J_{12} \cdot S_1 \cdot S_2 + -2J_{34} \cdot S_3 \cdot S_4 + \mu_B \cdot g_1 \cdot S_1 \cdot B + \mu_B \cdot g_2 \cdot S_2 \cdot B + \mu_B \cdot g_3 \cdot S_3 \cdot B + \mu_B \cdot g_4 \cdot S_4 \cdot B \tag{2}$$

where $S_1 = S_2 = S_3 = S_4 = \frac{1}{2}$ are the spin numbers of the radical anions, $g_1 = g_2 = g_3 = g_4 = 2.0$ is the g-factor, μ_B is the Bohr magneton, B is the magnetic field induction and J_{12} and J_{34} are the superexchange coupling constants—the fitting parameters with $J_{12} \gg J_{34}$.

The magnetic interaction pathways between pairs of radical anions are justified by the presence of hydrogen bonded supramolecular pairs highlighted in Figure 5b,c. In this simplified model, each radical anion interacts with only one neighbor utilizing two C-H···N hydrogen bonds. These hydrogen bonds are the strongest mediators of magnetic interactions in the structures of **PPN(4,7-L)**, **PPN(L)**, and **PPN(BrL)** justifying the use of single exchange coupling parameter. In case of **PPN(diBrL)** four spin carriers were taken into account operating with two different exchange coupling constants J_{12} and J_{34}, assuming that $J_{12} \gg J_{34}$. This is dictated by the presence of two step-like features in the $\chi T(T)$ dependence.

The $\chi T(T)$ fits correspond well with the experimental data above 10 K and the antiferromagnetic exchange coupling increases with the increasing temperature at which the decrease of the χT occurs. The weak match between the fitted and experimental $M(H)$ is a consequence of a simplified model employed in the analysis of the magnetic data and the presence of non-interacting $S = 1/2$ spins due to the defects in the crystal structure. However, the inflection of the $M(H)$ curves at 4 T for **PPN(4,7-L)**, 6 T for **PPN(L)**, and >7 T for **PPN(BrL)** is followed by the increase of the antiferromagnetic exchange coupling in this series.

The **PPN(4,7-L)**, **PPN(BrL)**, and **PPN(L)** belong to a structurally-related series where the anions and cations are arranged in layers. It appears that the magnetic interactions are strongly related to this arrangement. The strongest interactions are achieved in completely flat layers of **PPN(L)** with $J_{12} = -5(1)$ cm^{-1}. In **PPN(BrL)** the bulky bromine substituent increases the separation between the radicals and disrupts the C-H···hydrogen bonds resulting in slightly weaker magnetic interactions ($J_{12} = -4(1)$ cm^{-1}). Finally, in **PPN(4,7-L)** the layer is composed of tilted molecules with much weaker C-H···N H-bonds and the estimated magnetic interactions are even weaker ($J_{12} = -2(1)$ cm^{-1}). While this magneto-structural correlation is simplified, it clearly demonstrates how the derivatization of dioxothiadiazole-based radical anions enables fine-tuning of their magnetic behavior.

Magnetic behavior of **PPN(diBrL)** is also strongly correlated with the structural packing. The stacks of **diBr-L**$^{\bullet-}$ radical anions reveal four different π-contacts between them leading to two types of radical anion pairs within the infinite stack (Figure S9 in the Supplementary Materials). The most efficient one controls the magnetic behavior of one half of the radical anions and results in very strong antiferromagnetic coupling ($J_{12} = -116(10)$ cm^{-1}) that is comparable with the values reported for sodium salts of **4,7-L** [34] and **L** [35] and other types of molecular magnets ([56] and references therein). The antiferromagnetic interactions result in a χT value of 0.313 cm^3 K mol^{-1} at 340 K which is significantly lower than the expected 0.375 cm^3 K mol^{-1} for non-interaction $S = 1/2$ species. These interactions lead also to a plateau of 0.186 cm^3 K mol^{-1} below 70 K corresponding to half of the radical anions in the compound. The second exchange parameter J_{34} is much weaker than J_{12} but seems to be slightly underestimated as the fit does not reproduce the second step around 30 K where a further decrease of the $\chi T(T)$ to 0.157 cm^3 K mol^{-1} occurs. This weaker exchange controls the magnetic behavior of the remaining half of radical anions and is responsible for the observation of the 0.186 cm^3 K mol^{-1} plateau below 70 K and the Brillouin-like $M(H)$ curve reaching a saturation value approaching 0.5 μ_B expected for half of the radicals in **PPN(diBrL)** (Figure 6b).

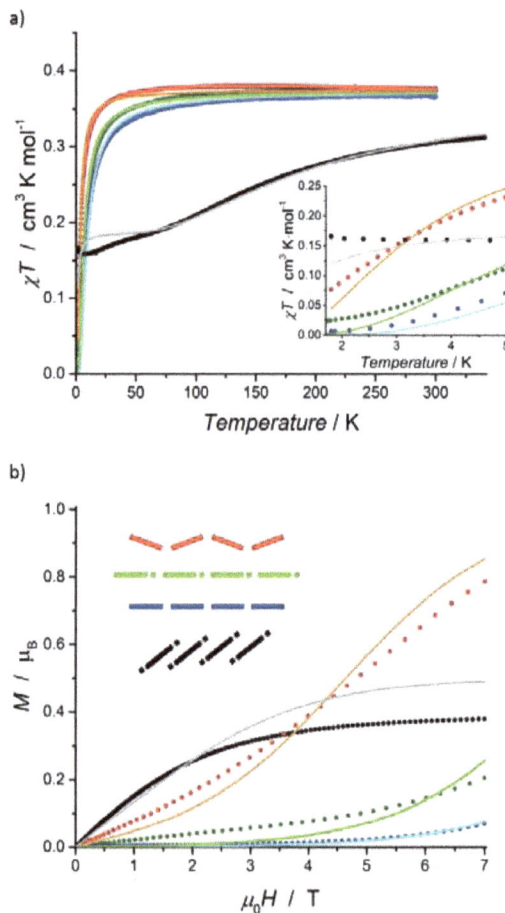

Figure 6. Experimental magnetic data (points) and best fits (solid lines) for **PPN(4,7-L)** (red), **PPN(L)** (blue), **PPN(BrL)** (green), and **PPN(diBrL)** (black): $\chi T(T)$ recorded at 0.1 T with an inset showing the low temperature window (**a**) and $M(H)$ recorded at 1.8 K with the schematic representation of the packing of radical anions (**b**). The χT and M values are calculated per one mole of radical anions.

4. Conclusions

Two new derivatives of [1,2,5]thiadiazole[3,4-f][1,10]phenanthroline 1,1-dioxide (**L**)—a redox active dioxothiadiazole—have been prepared and synthesized starting from 1,10-phenathroline: 5-bromo-[1,2,5]thiadiazolo[3,4-f][1,10]phenanthroline 2,2-dioxide (**BrL**), and 5,10-dibromo-thiadiazolo[3,4-f][1,10]phenanthroline 2,2-dioxide (**diBrL**). In the next step their organic paramagnetic salts with PPN$^+$ counter-cations have been prepared along with the previously unknown PPN$^+$ salts of **4,7-L** [34] and **L** [35]. The PPN salts show very good solubility in THF, acetonitrile, chloroform, and dichloromethane, which renders them suitable for the preparation of mixed-spin systems. All four radical anion-based compounds were characterized by means of single-crystal X-ray diffraction and magnetic measurements (SQUID magnetometry). **PPN(L)**, **PPN(4,7-L)**, and **PPN(BrL)** exhibit layered-type structures where the flat/weaved anionic layers are separated by layers of PPN⁺ cations. **PPN(diBrL)**, on the other hand, forms infinite chain-like π-π stacks of radical anions that are separated from each other by cations. The structures of the reported compounds directly influence the magnetic properties. The 'layered salts' show weak-to-moderate antiferromagnetic interactions despite slightly different substituents (bromine atoms) or the location of the nitrogen atoms, while the 'π-π-stacked-salt' exhibits very strong antiferromagnetic interactions transmitted through the direct overlap of the π orbitals of the radical anions. **PPN(L)**, **PPN(4,7-L)**, and **PPN(BrL)** constitute a rare example of a layered packing where the layers of radical anions are separated by the layers of cations (similar packing was observed for a few other PPN-based supramolecular systems [57]).

Supplementary Materials: The following are available online at http://www.mdpi.com/2073-4352/9/1/30/s1. Table S1. Details of single crystal X-ray data and structural refinement for **PPN(L)** (CCDC 1882329) **PPN(BrL)** (CCDC 1882331), **PPN(4,7-L)** (CCDC 1882328), **PPN(diBrL)** (CCDC 1882330), **diBrL** (CCDC 1882327), and **BrL** (CCDC 1882326). Figure S1. Illustration of molecular stacks in **BrL** with marked short σ-π contacts a) and fragment of a supramolecular layer with marked contacts that are shorter than the sum of the Van der Waals radii b) (CCDC Mercury program). Figure S2. Illustration of crystal packing of **diBrL**. The supramolecular chains of parallel hydrogen bonded dimers run through the structure interacting via π-orbitals and short contacts with bromine atoms. Figure S3. Illustration of **PPN(BrL)** supramolecular layers. Contacts between BrL anions that are shorter than the sum of the Van der Waals radii. Figure S4. NMR spectrum of 3-bromo-1,10-phenantroline-5,6-dione., Figure S5. NMR spectrum of 5-bromo-[1,2,5]thiadiazole[3,4-f] phenanthroline 2,2-dioxide. Figure S6. NMR spectrum of 3,8-dibromo-1,10-phenantroline-5,6-dione. Figure S7. NMR spectrum of 5,10-dibromo-[1,2,5]thiadiazole[3,4-f][1,10]phenanthroline 2,2-dioxide. Figure S8. Superexchange coupling scheme in **PPN(4,7-L)** (a), **PPN(BrL)** (b), **PPN(L)** (c), and **PPN(diBrL)** (d). The green ovals and dotted lines indicate the magnetic interaction pathways taken into account in the fitting of the magnetic data. In the case of **PPN(diBrL)** (d) two interaction pathways are considered: J_{12} and J_{34} with the assumption that $J_{12} \gg J_{34}$. Figure S9. Illustration of supramolecular stacks of diBrL anions in the crystal structure of **PPN(diBrL)**. The molecules are color coded to depict different intermolecular contacts between them. The asymmetric unit contains one green and one blue molecule. The most efficient π-π overlap is between the light blue and navy blue colored radical anions.

Author Contributions: Conceptualization: D.P., P.P., and M.A.; Methodology: P.P., M.A., and D.P.; Validation: D.P. and M.A.; Formal analysis: D.P. and M.A.; Investigation: P.P., M.A., and D.P.; Data curation: D.P. and M.A.; Writing—original draft preparation, M.A. and P.P.; Writing—review and editing: D.P.; Visualization: M.A.; Supervision: D.P.; Project administration: D.P.; Funding acquisition: M.A. and D.P.

Funding: This research was funded by the Polish National Science Centre within the Sonata Bis 6 (2016/22/E/ST5/00055) project. M.A. gratefully acknowledges the Polish Ministry of Science and Higher Education for the financial support within the Diamond Grant project (0041/DIA/2015/44).

Acknowledgments: This research was supported in part by PLGrid Infrastructure.

Conflicts of Interest: The authors declare no conflict of interest.

References

1. Miller, J.S. Organic Magnets—A History. *Adv. Mater.* **2002**, *14*, 1105–1110. [CrossRef]

2. Ferrando-Soria, J.; Vallejo, J.; Castellano, M.; Martínez-Lillo, J.; Pardo, E.; Cano, J.; Castro, I.; Lloret, F.; Ruiz-García, R.; Julve, M. Molecular magnetism, quo vadis? A historical perspective from a coordination chemist viewpoint. *Coord. Chem. Rev.* **2017**, *339*, 17–103. [CrossRef]

3. Sieklucka, B.; Pinkowicz, D. *Molecular Magnetic Materials: Concepts and Applications*; Sieklucka, B., Pinkowicz, D., Eds.; WILEY-VCH: Weinheim, Germany, 2016; ISBN 352769420X.

4. Ouahab, L. *Multifunctional Molecular Materials*; CRC Press Taylor & Francis Group: Boca Raton, FL, USA, 2012; ISBN 9789814364294.

5. Tezgerevska, T.; Alley, K.G.; Boskovic, C. Valence tautomerism in metal complexes: Stimulated and reversible intramolecular electron transfer between metal centers and organic ligands. *Coord. Chem. Rev.* **2014**, *268*, 23–40. [CrossRef]

6. Himmel, H.-J. Valence tautomerism in copper coordination chemistry. *Inorg. Chim. Acta* **2017**. [CrossRef]

7. Demir, S.; Jeon, I.-R.; Long, J.R.; Harris, T.D. Radical ligand-containing single-molecule magnets. *Coord. Chem. Rev.* **2014**, *289–290*, 149–176. [CrossRef]

8. Demir, S.; Gonzalez, M.I.; Darago, L.E.; Evans, W.J.; Long, J.R. Giant coercivity and high magnetic blocking temperatures for N23-radical-bridged dilanthanide complexes upon ligand dissociation /639/638/263/406 /639/638/911 /639/638/298/920 article. *Nat. Commun.* **2017**, *8*, 1–9. [CrossRef] [PubMed]

9. Dhers, S.; Feltham, H.L.C.; Brooker, S. A toolbox of building blocks, linkers and crystallisation methods used to generate single-chain magnets. *Coord. Chem. Rev.* **2015**, *296*, 24–44. [CrossRef]

10. Galán-Mascarós, J.R.; Coronado, E. Molecule-based ferromagnetic conductors: Strategy and design. *C. R. Chim.* **2008**, *11*, 1110–1116. [CrossRef]

11. Takahashi, K.; Cui, H.; Okano, Y.; Kobayashi, H.; Einaga, Y.; Sato, O. Electrical Conductivity Modulation Coupled to a High-Spin—Low-Spin Conversion in the Molecular System [FeIII(qsal)₂][Ni(dmit)₂]₃, CH_3CN, H_2O. *Inorg. Chem.* **2006**, *45*, 5739–5741. [CrossRef]

12. Bowman, A.C.; Tondreau, A.M.; Lobkovsky, E.; Margulieux, G.W.; Chirik, P.J. Synthesis and Electronic Structure Diversity of Pyridine(diimine)iron Tetrazene Complexes. *Inorg. Chem.* **2018**, *57*, 9634–9643. [CrossRef]

13. Le Gal, Y.; Roisnel, T.; Auban-Senzier, P.; Bellec, N.; Iñiguez, J.; Canadell, E.; Lorcy, D. Stable Metallic State of a Neutral Radical Single-Component Conductor at Ambient Pressure. *J. Am. Chem. Soc.* **2018**. [CrossRef] [PubMed]

14. Wang, Y.; Xie, Y.; Wei, P.; Blair, S.A.; Cui, D.; Johnson, M.K.; Schaefer, H.F.; Robinson, G.H. Stable Boron Dithiolene Radicals. *Angew. Chem. Int. Ed.* **2018**, 2–6. [CrossRef]

15. Wang, Y.; Hickox, H.P.; Xie, Y.; Wei, P.; Blair, S.A.; Johnson, M.K.; Schaefer, H.F.; Robinson, G.H. A Stable Anionic Dithiolene Radical. *J. Am. Chem. Soc.* **2017**, *139*, 6859–6862. [CrossRef]

16. Jeon, I.R.; Negru, B.; Van Duyne, R.P.; Harris, T.D. A 2D Semiquinone Radical-Containing Microporous Magnet with Solvent-Induced Switching from Tc = 26 to 80 K. *J. Am. Chem. Soc.* **2015**, *137*, 15699–15702. [CrossRef] [PubMed]

17. Rupp, F.; Chevalier, K.; Graf, M.; Schmitz, M.; Kelm, H.; Grün, A.; Zimmer, M.; Gerhards, M.; van Wüllen, C.; Krüger, H.J.; et al. Spectroscopic, Structural, and Kinetic Investigation of the Ultrafast Spin Crossover in an Unusual Cobalt(II) Semiquinonate Radical Complex. *Chem. Eur. J.* **2017**, *23*, 2119–2132. [CrossRef] [PubMed]

18. Yang, M.; Li, H.; Li, L. Unusual Ln-radical chains constructed from functionalized nitronyl nitroxides: Synthesis, structure and magnetic properties. *Inorg. Chem. Commun.* **2017**, *76*, 59–61. [CrossRef]

19. Sun, J.; Sun, Z.; Li, L.; Sutter, J.-P. Lanthanide-Nitronyl Nitroxide Chains Derived from Multidentate Nitronyl Nitroxides. *Inorg. Chem.* **2018**, *57*, 7507–7511. [CrossRef]

20. Calzado, C.J.; Rodríguez-García, B.; Galán Mascarós, J.R.; Hernández, N.C. Electronic Structure and Magnetic Interactions in the Radical Salt [BEDT-TTF]₂[CuCl₄]. *Inorg. Chem.* **2018**, *57*, 7077–7089. [CrossRef]

21. Ke, X.; Hong, Y.; Lynch, V.M.; Kim, D.; Sessler, J.L. Metal-Stabilized Quinoidal Dibenzo[g, p]chrysene-Fused Bis- dicarbacorrole System. *J. Am. Chem. Soc.* **2018**, *140*, 7579–7586. [CrossRef]

22. M, M.J.; Yee, G.T.; Mclean, R.S.; Epstein, A.J.; Miller, J.S. A ROOM-TEMPERATURE MOLECULAR ORGANIC BASED MAGNET. *Science* **1991**, *252*, 1415–1417. [CrossRef]

23. Pokhodnya, K.I.; Bonner, M.; Her, J.H.; Stephens, P.W.; Miller, J.S. Magnetic ordering (Tc= 90 K) observed for layered [FeII(TCNE.-)(NCMe)2]+[FeIIICl4]-(TCNE = tetracyanoethylene). *J. Am. Chem. Soc.* **2006**, *128*, 15592–15593. [CrossRef] [PubMed]

24. Abrahams, B.F.; Elliott, R.W.; Hudson, T.A.; Robson, R.; Sutton, A.L. X4TCNQ2-dianions: Versatile building blocks for supramolecular systems. *CrystEngComm* **2018**, *20*, 3131–3152. [CrossRef]

25. Miyasaka, H.; Motokawa, N.; Matsunaga, S.; Yamashita, M.; Sugimoto, K.; Mori, T.; Toyota, N.; Dunbar, K.R. Control of charge transfer in a series of Ru2 II,II/TCNQ two-dimensional networks by tuning the electron affinity of TCNQ units: A route to synergistic magnetic/conducting materials. *J. Am. Chem. Soc.* **2010**, *132*, 1532–1544. [CrossRef] [PubMed]

26. Novitchi, G.; Shova, S.; Lan, Y.; Wernsdorfer, W.; Train, C. Verdazyl Radical, a Building Block for a Six-Spin-Center 2p–3d–4f Single-Molecule Magnet. *Inorg. Chem.* **2016**, *55*, 12122–12125. [CrossRef] [PubMed]

27. Xie, Y.; Shuku, Y.; Matsushita, M.M.; Awaga, K. Thiadiazole dioxide-fused picene: Acceptor ability, anion radical formation, and n-type charge transport characteristics. *Chem. Commun.* **2014**, *50*, 4178–4180. [CrossRef] [PubMed]

28. Mirífico, M.V.; Caram, J.A.; Gennaro, A.M.; Cobos, C.J.; Vasini, E.J. Radical anions containing the dioxidated 1,2,5-thiadiazole heterocycle. Part II. *J. Phys. Org. Chem.* **2011**, *24*, 1039–1044. [CrossRef]

29. Keller, S.N.; Bromby, A.D.; Sutherland, T.C. Optical Effect of Varying Acceptors in Pyrene Donor–Acceptor–Donor Chromophores. *Eur. J. Org. Chem.* **2017**, *2017*, 3980–3985. [CrossRef]

30. Schüttler, C.; Li-Böhmer, Z.; Harms, K.; Von Zezschwitz, P. Enantioselective synthesis of 3,4-disubstituted cis- and trans-1,2,5-thiadiazolidine-1,1-dioxides as precursors for chiral 1,2-diamines. *Org. Lett.* **2013**, *15*, 800–803. [CrossRef]

31. Arroyo, N.R.; Rozas, M.F.; Vázquez, P.; Romanelli, G.P.; Mirífico, M.V. Solvent-Free Condensation Reactions to Synthesize Five-Membered Heterocycles Containing the Sulfamide Fragment. *Synth* **2016**, *48*, 1344–1352. [CrossRef]

32. Linder, T.; Badiola, E.; Baumgartner, T.; Sutherland, T.C. Synthesis of π-extended thiadiazole (oxides) and their electronic properties. *Org. Lett.* **2010**, *12*, 4520–4523. [CrossRef]

33. Arslan, N.B.; Ertürk, A.G.; Kazak, C.; Bekdemir, Y. 3-Amino-4-[4-(dimethylamino)phenyl]-4,5-dihydro-1,2,5-thiadiazole 1,1-dioxide. *Acta Crystallogr. Sect. E Struct. Rep. Online* **2011**, *67*, o1736. [CrossRef] [PubMed]

34. Pinkowicz, D.; Li, Z.; Pietrzyk, P.; Rams, M. New thiadiazole dioxide bridging ligand with a stable radical form for the construction of magnetic coordination Chains. *Cryst. Growth Des.* **2014**, *14*, 4878–4881. [CrossRef]

35. Shuku, Y.; Suizu, R.; Awaga, K. Monovalent and mixed-valent potassium salts of Thiadiazolo phenanthroline 1, 1-Dioxide: A radical anion for multidimensional network structures. *Inorg. Chem.* **2011**, *50*, 11859–11861. [CrossRef] [PubMed]

36. Shuku, Y.; Awaga, K. Transition metal complexes and radical anion salts of 1,10-Phenanthroline derivatives annulated with a 1,2,5-Tiadiazole and 1,2,5-Tiadiazole 1,1-Dioxide Moiety: Multidimensional crystal structures and various magnetic properties. *Molecules* **2014**, *19*, 609–640. [CrossRef] [PubMed]

37. Shuku, Y.; Suizu, R.; Domingo, A.; Calzado, C.J.; Robert, V.; Awaga, K. Multidimensional network structures and versatile magnetic properties of intermolecular compounds of a radical-anion ligand, [1,2,5]thiadiazolo[3,4-f][1,10]phenanthroline 1,1-dioxide. *Inorg. Chem.* **2013**, *52*, 9921–9930. [CrossRef] [PubMed]

38. Mercuri, M.L.; Congiu, F.; Concas, G.; Sahadevan, S.A. Recent Advances on Anilato-Based Molecular Materials with Magnetic and/or Conducting Properties. *Magnetochemistry* **2017**, *3*, 17. [CrossRef]

39. Paw, W.; Eisenberg, R. Synthesis, Characterization, and Spectroscopy of Dipyridocatecholate Complexes of Platinum. *Inorg. Chem.* **1997**, *36*, 2287–2293. [CrossRef]

40. Zhao, J.F.; Chen, L.; Sun, P.J.; Hou, X.Y.; Zhao, X.H.; Li, W.J.; Xie, L.H.; Qian, Y.; Shi, N.E.; Lai, W.Y.; et al. One-pot synthesis of 2-bromo-4,5-diazafluoren-9-one via a tandem oxidation-bromination-rearrangement of phenanthroline and its hammer-shaped donor-acceptor organic semiconductors. *Tetrahedron* **2011**, *67*, 1977–1982. [CrossRef]

41. Bain, G.; Berry, J.F. Diamagnetic Corrections and Pascal's Constants. *J. Chem. Educ.* **2008**, *85*, 532–536. [CrossRef]

42. Bruker. *SAINT*; Bruker AXS Inc.: Madison, WI, USA, 2012.

43. Bruker. *SADABS*; Bruker AXS Inc.: Madison, WI, USA, 2001.

44. Bruker. *TWINABS*; Bruker AXS Inc.: Madison, WI, USA, 2001.

45. Bruker. *APEX3*; Bruker AXS Inc.: Madison, WI, USA, 2012.

46. Sheldrick, G.M. SHELXT—Integrated space-group and crystal-structure determination. *Acta Crystallogr. Sect. A Found. Crystallogr.* **2015**, *71*, 3–8. [CrossRef]

47. Sheldrick, G.M. Crystal structure refinement with SHELXL. *Acta Crystallogr. Sect. C Struct. Chem.* **2015**, *71*, 3–8. [CrossRef] [PubMed]

48. Sheldrick, G.M. A short history of SHELX. *Acta Crystallogr. Sect. A Found. Crystallogr.* **2008**, *64*, 112–122. [CrossRef]

49. Dolomanov, O.V.; Bourhis, L.J.; Gildea, R.J.; Howard, J.A.K.; Puschmann, H. OLEX2: A complete structure solution, refinement and analysis program. *J. Appl. Crystallogr.* **2009**, *42*, 339–341. [CrossRef]

50. Guzei, I.A. An idealized molecular geometry library for refinement of poorly behaved molecular fragments with constraints. *J. Appl. Crystallogr.* **2014**, *47*, 806–809. [CrossRef]

51. Frisch, M.J.; Trucks, G.W.; Schlegel, H.B.; Scuseria, G.E.; Robb, M.A.; Cheeseman, J.R.; Scalmani, G.; Barone, V.; Petersson, G.A.; Nakatsuji, H.; et al. *Gaussian09 Program*; Gaussian, Inc.: Wallingford, CT, USA, 2016.

52. Becke, A.D. Density-functional thermochemistry. III. The role of exact exchange. *J. Chem. Phys.* **1993**, *98*, 5648–5652. [CrossRef]

53. Lee, C.; Yang, W.; Parr, R.G. Development of the Colle-Salvetti correlation-energy formula into a functional of the electron density. *Phys. Rev. B* **1988**, *37*, 785–789. [CrossRef]

54. Petersson, G.A.; Bennett, A.; Tensfeldt, T.G.; Al-Laham, M.A.; Shirley, W.A.; Mantzaris, J. A complete basis set model chemistry. I. The total energies of closed-shell atoms and hydrides of the first-row elements. *J. Chem. Phys.* **1988**, *89*, 2193–2218. [CrossRef]

55. Chilton, N.F.; Anderson, R.P.; Turner, L.D.; Soncini, A.; Murray, K.S. PHI: A powerful new program for the analysis of anisotropic monomeric and exchange-coupled polynuclear d- and f-block complexes. *J. Comput. Chem.* **2013**, *34*, 1164–1175. [CrossRef]

56. Pinkowicz, D.; Southerland, H.; Wang, X.-Y.; Dunbar, K. Record Antiferromagnetic Coupling for a 3d/4d Cyanide-Bridged. *Compd. J. Am. Chem. Soc.* **2014**, *136*, 9922–9924. [CrossRef]

57. Pinkowicz, D.; Southerland, H.; Avendano, C.; Prosvirin, A.; Sanders, C.; Wernsdorfer, W.; Pederesen, K.S.; Dreiser, J.; Clérac, R.; Nehrkorn, J.; et al. Cyanide Single-Molecule Magnets Exhibiting Solvent Dependent Reversible "On" and "Off" Exchange Bias Behavior. *J. Am. Chem. Soc.* **2015**, *137*, 14406–14422. [CrossRef]

crystals

MDPI

Review

Magnetic and Electronic Properties of π-d Interacting Molecular Magnetic Superconductor κ-(BETS)$_2$FeX$_4$ (X = Cl, Br) Studied by Angle-Resolved Heat Capacity Measurements

Shuhei Fukuoka [1,*]**, Sotarou Fukuchi** [2]**, Hiroki Akutsu** [2]**, Atsushi Kawamoto** [1] **and Yasuhiro Nakazawa** [2]

[1] Department of Condensed Matter Physics, Graduate School of Science, Hokkaido University, Kita-ku Sapporo, Hokkaido 060-0810, Japan; atkawa@phys.sci.hokudai.ac.jp
[2] Department of Chemistry, Graduate School of Science, Osaka University, Machikaneyama 1-1, Toyonaka, Osaka 560-0043, Japan; fukuchis17@chem.sci.osaka-u.ac.jp (S.F.); akutsu@chem.sci.osaka-u.ac.jp (H.A.); nakazawa@chem.sci.osaka-u.ac.jp (Y.N.)
* Correspondence: fukuoka@phys.sci.hokudai.ac.jp; Tel.: +81-01-706-4424

Received: 8 December 2018; Accepted: 19 January 2019; Published: 26 January 2019

Abstract: Thermodynamic picture induced by π-d interaction in a molecular magnetic superconductor κ-(BETS)$_2$FeX$_4$ (X = Cl, Br), where BETS is bis(ethylenedithio)tetraselenafulvalene, studied by single crystal calorimetry is reviewed. Although the $S = 5/2$ spins of Fe^{3+} in the anion layers form a three-dimensional long-range ordering with nearly full entropy of $R\ln 6$, a broad hump structure appears in the temperature dependence of the magnetic heat capacity only when the magnetic field is applied parallel to the a axis, which is considered as the magnetic easy axis. The scaling of the temperature dependence of the magnetic heat capacity of the two salts is possible using the parameter of $|J_{dd}|/k_B$ and therefore the origin of the hump structure is related to the direct magnetic interaction, J_{dd}, that is dominant in the system. Quite unusual crossover from a three-dimensional ordering to a one-dimensional magnet occurs when magnetic fields are applied parallel to the a axis. A notable anisotropic field-direction dependence against the in-plane magnetic field was also observed in the transition temperature of the bulk superconductivity by the angle-resolved heat capacity measurements. We discuss the origin of this in-plane anisotropy in terms of the $3d$ electron spin configuration change induced by magnetic fields.

Keywords: π-d system; thermodynamic measurement; superconductivity; antiferromagnetism; single crystal heat capacity measurement; magnetic conductor

---

1. Introduction

There are increasing interests in studying magnetic properties of molecule-based materials from fundamental science and application, since they show a variety of functionalities related to spin degrees of freedom, which have different appearance manners from those of intermetallic compounds, such as transition metal oxides and pnictide, etc. Not only the development of new materials, which have ferromagnetic or ferrimagnetic ordering of molecular spins, but also the synthesis of new compounds with large cluster spins, such as single-molecule magnets (SMM), single-chain magnets (SCM), and those with spin crossover features, have been performed [1,2]. Some molecular magnets show a large magnetocaloric effect (MCE) derived from the large magnetic entropy change, which is applicable to cryogenic coolants [3,4]. The fabrication of devices applicable to spintronics and topological phenomena are becoming challenging subjects both for synthetic chemists and physicists [5–7]. The molecule-based magnets studied up to now contain organic radical compounds, assembled magnetic metal complexes with open shell metal cations, and coordination polymers of them, etc. [8–11]. The unpaired π electrons in the singly occupied molecular orbital (SOMO), the highest occupied molecular orbital (HOMO), and the lowest unoccupied molecular orbital (LUMO) of organic molecules in the radical compounds possess delocalized character over the molecules and have strong quantum mechanical features [12–14]. The magnetic metal complexes have potential for showing various net magnetic properties of *d*- or *f*- electrons in the inorganic ions in various coordination fields produced by surrounding organic ligands [15,16]. The design and synthesis of new molecules and the provision of new functionalities are extensively performed as interesting challenges in chemistry. The rich variety of structures and relatively sensitive responses against external stimuli are advantages for applications of them [17,18]. Due to these characteristic natures, molecule-based magnets are considered as promising materials for studying important topics, such as spin crossover, spin frustration, light irradiation-induced magnetic structures, and single molecule magnets [19–22].

In addition to these materials, the magnetic properties of charge transfer complexes are also attracting attention, since they show various conducting and magnetic properties inherent in the multi-composition of molecules with different electronic structures and functionalities [23,24]. The π-*d* interacting systems, which consist of organic donor molecules and counter anions containing magnetic ions, such as Cu^{2+}, Fe^{3+}, Mn^{2+}, etc., with localized 3*d* electron spin moments have been studied in terms of the developing cooperative phenomena between conducting electrons and localized spins. They are recognized as organic-inorganic hybrid molecular magnets. Until now, several π-*d* interacting compounds with D_2A composition, where *D* denotes a donor molecule, such as BETS and BEDT-TTF (bis(ethylenedithio)tetrathiafulvalene), and *A* denotes monovalent magnetic counter anions, such as FeX_4^- (*X* = Cl, Br), have been studied [25–30]. In such π-*d* interacting systems, the charge transfer from donor molecules to counter anion molecules occurs and unpaired π electrons and 3*d* electron spins coexist in a crystal. Since the donor molecules and the magnetic counter anion molecules form respective layers, the π electrons show two-dimensional conducting and magnetic properties with strong electron correlations. In addition, magnetic interaction between the π electrons and the 3*d* electrons, which is called π-*d* interaction, appears in the systems [31]. By combining the π-*d* interacting effect and the electron correlation, these π-*d* systems show various unique magnetic and conducting properties. For example, the metal-insulator transition and the coexistent state of a magnetic long-range ordering and superconductivity are observed in them [32–36]. Furthermore, a rich variety of electronic and magnetic phases appear with the tuning of external parameters, such as temperature, pressure, and magnetic and electric fields, etc., in these compounds. A magnetic-field-induced superconducting (FISC) state observed in λ-$(BETS)_2FeCl_4$ under extremely large magnetic fields higher than 17 T is a representative phenomenon that the electronic state shows drastic change by controlling external parameters [37].

In this article, we review the results of the thermodynamic experiments for π-*d* interacting systems of κ-$(BETS)_2FeX_4$ (*X* = Cl, Br) performed by single crystal calorimetry as well as the development of the calorimetry system using the measurements and discuss the physical properties of these compounds in relation to the π-*d* interacting effects.

2. Electronic Structure of the κ-$(BETS)_2FeX_4$ Systems

The crystal structures of κ-$(BETS)_2FeX_4$ are shown in Figure 1. The BETS molecules and the FeX_4^- anions form conducting π electron layers and insulating FeX_4^- layers, respectively, in the *ac* plane and these layers are stacked alternately in the direction parallel to the *b* axis [34]. In the donor layers, these molecules form dimers with face to face contact in the layers. The dimer units are arranged in a nearly orthogonally tilted structure to form a zig-zag lattice. This structure is called a κ-type structure and various superconductors with relatively high T_c have this structure. Since one electron is removed from two BETS molecules, namely one dimer, to form a charge transfer complex with FeX_4^- anions, the 3/4-filled band is expected in BETS layers as is usual for the compounds with a 2:1 concentration. However, since the degree of dimerization in the κ-type structure is higher than the other packing, the 3/4-filled band splits into bonding and antibonding bands, changing the 3/4-filled band into an effective half-filled band system with strong electron correlation. In the FeX_4^- layers, Fe^{3+} ions have localized $3d$ electrons with $S = 5/2$ spins. The localized $3d$ electrons and the conducting π electrons are strongly coupled through the π-d interaction, and the conducting and magnetic properties of the π electrons and the $3d$ electrons influence each other.

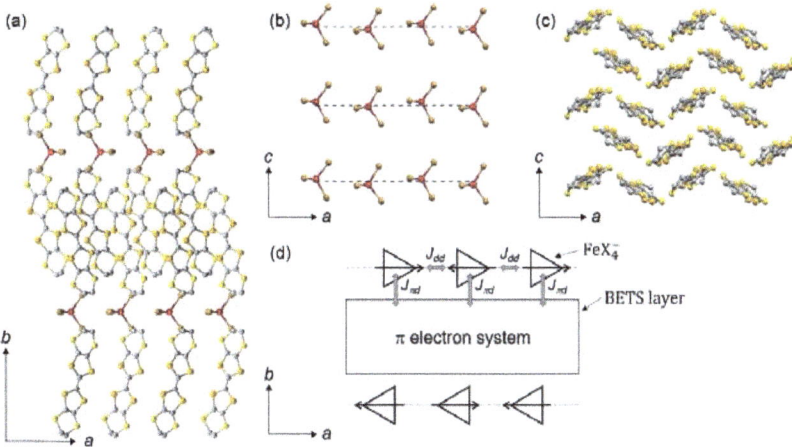

Figure 1. (**a**) Crystal structure of κ-$(BETS)_2FeX_4$ projected along the c axis. Molecular arrangement of the (**b**) anion (FeX_4^-) layer and (**c**) donor (BETS) layer in the ac plane. BETS and FeX_4^- are drawn in ball-and-stick representation. Carbon, sulfur, selenium, iron, and halogen atoms are shown in gray, yellow, orange, red, and brown, respectively. The dashed lines represent the one-dimensional direct magnetic interaction network. (**d**) Schematic view of the magnetic interaction network in the κ-$(BETS)_2FeX_4$ system.

3. Transport and Magnetic Properties

The mechanism of magnetic interaction between π electron spins and $3d$ electron spins and the possibility of π-d hybridization in electronic bands have been discussed theoretically and experimentally. The conducting and magnetic properties of κ-$(BETS)_2FeX_4$ have been studied in the previous works by Fujiwara and Otsuka et al. and they reported that both κ-$(BETS)_2FeBr_4$ and κ-$(BETS)_2FeCl_4$ show superconductivity coexisting with an antiferromagnetic long-range ordering [34–36]. The $3d$ electrons in the Fe^{3+} show an antiferromagnetic transition at 2.47 K in κ-$(BETS)_2FeBr_4$ and 0.47 K in κ-$(BETS)_2FeCl_4$, respectively, while the π electrons are metallic in a wide temperature range. At extremely low temperatures, they show superconducting transitions. The transition temperature is 1.5 K in κ-$(BETS)_2FeBr_4$ and 0.1 K in κ-$(BETS)_2FeCl_4$, respectively. The superconducting ordering and the antiferromagnetic ordering occur independently at a glance, since only a small kink is observed around T_N in the temperature dependence of resistivities [34,35]. Furthermore, the superconducting transition occurs at almost the same temperature in κ-$(BETS)_2GaX_4$ (X = Cl, Br), which have non-magnetic counter anions [38]. However, the electronic phase diagram under magnetic fields indicates the anisotropy against magnetic field direction, which demonstrates that the internal field produced by the $3d$ electron spin ordering affects the superconductivity of the π electrons through the π-d interaction [34,39]. The electron correlation of the π electrons which not only induces antiferromagnetic fluctuations but also produces charge fluctuations is also considered as an important factor to determine the magnetic phase diagram of these compounds. Moreover, it is reported that κ-$(BETS)_2FeBr_4$ shows a FISC state similar to the case of well-known λ-$(BETS)_2FeCl_4$ [39]. Since the mechanism of the FISC state is explained by the Jaccarino-Peter compensation effect realized by the internal field produced by the aligned $3d$ electron spins, the existence of the π-d interactions should be taken into account [40,41]. These results demonstrate that the coupling of the π electrons and the $3d$ electrons is crucial to characterize the conducting and magnetic properties of these systems and this coupling can give unique features as molecular magnets.

4. Calorimetry System Applicable to Tiny Single Crystals of Molecular Magnetic Materials

The heat capacity measurements were performed by the thermal relaxation method, which is suitable for single crystal measurements at a low temperature region. Since the crystals of the κ-(BETS)$_2$FeX$_4$ system are tiny thin plates, we used custom-made calorimetry cells of which details were already reported in [42]. The sample stage consists of a RuOx thermometer of which the room temperature resistance is 1 kΩ, and a strain gauge (EFLK-1000) heater with 1 kΩ in resistance. By using ϕ13 μm constantan wires as the electric leads for the heater and the thermometer, it is possible to adjust the thermal relaxation time between 0.1 s to 100 s depending on the sample size and the experimental temperature region. The temperature of the sample was monitored by an ac resistance bridge (LakeShore model 370N) with a scanner system and a pre-amplifier. The calorimetry cells are also designed for conducting angle-resolved heat capacity measurements under in-plane magnetic fields. The sample stage was suspended by thin stainless wires to prevent the sample from tilting by the magnetic torque produced by a magnetic field. We confirmed that the in-plane direction is kept just parallel to the magnetic field direction by monitoring the resistance of Hall sensor. We succeeded in reducing the misalignment of the field-angle direction within $\pm 1°$ in all magnetic fields, which is satisfactory for the present experiments. Therefore, investigations of the anisotropy of the superconducting transition against the in-plane magnetic field were possible with high accuracy by using the calorimetry cells. The angle-resolved system was subsequently modified to reduce the blank heat capacity by Imajo et al. of which details were reported in [43].

In the experiments, the calorimetry cells were mounted on a top loading type ^{3}He cryostat and a dilution refrigerator (TS-3H100 Taiyo Nissan), which are available in the variable temperature insert (VTI) system with superconducting magnets. The minimum temperature of the former is 0.6 K and that of the latter is 100 mK.

In the heat capacity measurements in this study, we used a single crystal of κ-(BETS)$_2$FeBr$_4$ with 94 μg and that of κ-(BETS)$_2$FeCl$_4$ with 45 μg. The BETS molecules were solved in the solvent of 1,1,2-trichloroethane with tetrabutylammonium salts of FeX$_4{}^-$. The electrochemical oxidation technique was used to grow the single crystals. The sample was adhered on the sample stage with a small amount of Apiezon N grease to attain good thermal contact. We confirmed that the thermal relaxation curves of the measurements obey the simple single exponential function in all temperature range between 100 mK and 10 K, which means that the thermal contact between the sample and the stage is adequate to attain absolute values of the heat capacity. As a matter of fact, the absolute values of the heat capacity, including the height of the peak, coincide well with the data of previous works [34,35].

5. Thermodynamic Properties of Antiferromagnetic Ordered State of κ-(BETS)$_2$FeX$_4$

Here, we review the magnetic nature of the $3d$ electron systems investigated by heat capacity measurements. Temperature dependences of the magnetic heat capacity of κ-(BETS)$_2$FeBr$_4$ and κ-(BETS)$_2$FeCl$_4$ are shown in Figure 2 by the $C_{mag}T^{-1}$ vs. T plot [44]. The contribution of the $3d$ electron spins were evaluated by subtracting the lattice heat capacity. Although the π electrons contribute to the total heat capacity (C_p), the entropic contribution of them is almost negligible in this plot. Their electronic state can be explained by the band state, and the electronic heat capacity should give a simple formula of $C_{el} = \gamma T$. The value of the γ term is expected to be about 10–30 mJK^{-2}mol^{-1} if the π electron band gives a similar band width, W, as for the case of metallic compounds of κ-(BEDT-TTF)$_2X$ systems [45–49]. This indicates that the contribution of the π electrons should be two orders of magnitude smaller than that of the $3d$ electrons. The sharp peaks at 2.47 K and 0.47 K are attributed to the formation of the long-range ordering of the $3d$ electron spins in the anion layers. The temperature of the magnetic transition coincides well with the results of magnetic susceptibility measurements by Fujiwara and Otsuka et al. [34–36]. As a matter of fact, the evaluated magnetic entropy reaches S_{mag} = 14.9 JK^{-1}mol^{-1} at 8 K for κ-(BETS)$_2$FeBr$_4$ and 2 K for κ-(BETS)$_2$FeCl$_4$. These values are consistent with the full entropy of the 3d electron spins in FeX$_4^-$ possessing the spin multiplicity of $S = 5/2$ spins, namely, $S_{mag} = R\ln6$. This result means that the magnetic orders occur with a bulk feature in both compounds in which all the 3d electron spins form an antiferromagnetic structure.

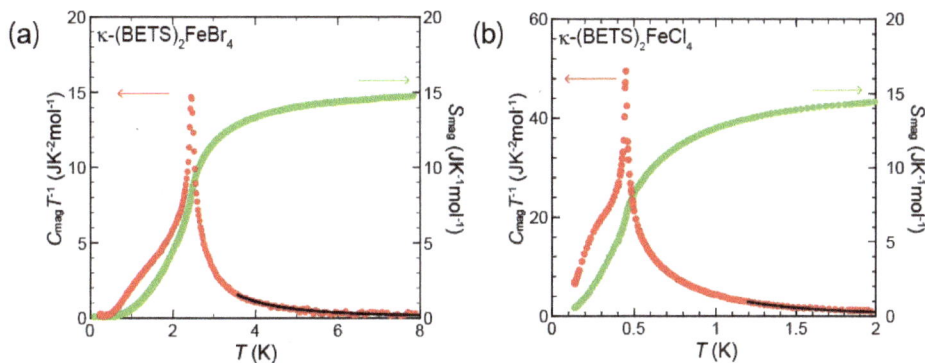

Figure 2. Temperature dependence of the magnetic heat capacity (red) and the magnetic entropy (green) of (**a**) κ-(BETS)$_2$FeBr$_4$ and (**b**) κ-(BETS)$_2$FeCl$_4$. The solid lines represent the fitting curves of high-temperature magnetic heat capacity data using the AT^{-2} term. Reproduced with permission from [44].

The temperature dependence of the magnetic entropy shown in Figure 2 indicates that most of the magnetic entropy is distributed around T_N. However, a nearly symmetric peak shape of C_p against temperature and the existence of the higher-temperature tail of the magnetic heat capacity prompted us to consider a kind of low dimensional fluctuation effect. From the data in Figure 2, it is notable that nearly 40% of the contribution for the magnetic entropy is distributed above T_N. There are two types of magnetic interactions between the 3d electron spins in the FeX$_4^-$ sites in these two compounds [31,50]. One is the direct magnetic interaction expressed as J_{dd}, and the other is the indirect magnetic interaction mediated by the coupling between the 3d electron spins and the π electron spins on the BETS layers, which is expressed as $J_{\pi d}$. The schematic view of the interaction is shown in Figure 1d. Although both interactions are antiferromagnetic, the contribution for the antiferromagnetic ordering of the direct magnetic interaction is expected to be relatively larger than that of the indirect magnetic interaction. Mori et al. estimated the contribution of the direct magnetic interaction and the indirect magnetic interaction to T_N by theoretical calculation and revealed the dominant contribution of the direct magnetic interaction [50]. Since the counter anions of FeX$_4^-$ form one-dimensional chain like structures along the a axis, as is shown in the crystal structure in Figure 1b, the J_{dd} forms a one-dimensional interaction network. The direct magnetic interactions along the inter-chain (parallel to the c axis) and the inter-layer (parallel to the b axis) directions are one order of magnitude smaller and are almost negligible. However, the indirect magnetic interaction between the 3d electron spins works in all directions to form a three-dimensional magnetic interaction network. In the inter-chain and the inter-layer directions, magnetic interactions are dominated by this indirect magnetic interaction. In the case of the a axis direction, the direct magnetic interaction and the indirect interaction coexist. Therefore, the short-range ordering of the 3d electron spins due to the dominant direct magnetic interaction with one-dimensional character should develop from the higher-temperature region above T_N, leading to the distribution of the magnetic entropy above T_N. The J_{dd} values can be evaluated by fitting the high-temperature magnetic heat capacity data using the AT^{-2} term [51,52]. The fitting results give the value of $|J_{dd}|/k_B$ = 0.27 K for κ-(BETS)$_2$FeBr$_4$ and $|J_{dd}|/k_B$ = 0.081 K for κ-(BETS)$_2$FeCl$_4$, respectively. The fitting curves of the high-temperature magnetic heat capacity using the AT^{-2} term is shown by black solid lines in Figure 2. Here, we must mention that the single-ion anisotropy of Fe^{3+} is not taken into account and these values contain ambiguity. Although the one-dimensional fluctuations appear at high temperatures, the indirect magnetic interaction through the π electron layers works cooperatively and forms three-dimensional ordering.

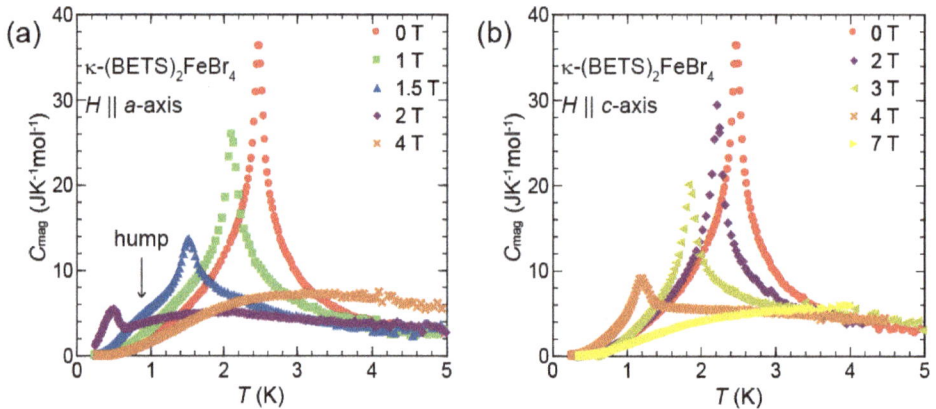

Figure 3. Temperature dependences of the magnetic heat capacity of κ-(BETS)$_2$FeBr$_4$ under magnetic fields parallel to (**a**) the a axis and (**b**) the c axis. Reproduced with permission from [44].

The magnetic field dependences of the magnetic heat capacity of the $3d$ electron spins of the two compounds were also investigated. By applying magnetic fields, the transition temperature decreases with the increase of magnetic fields, which is a typical behavior of conventional antiferromagnetic compounds. The temperature dependence of the magnetic heat capacity of κ-(BETS)$_2$FeBr$_4$ obtained under magnetic fields is shown in Figure 3. The magnetic fields are applied to the a axis and the c axis directions in the plane. The a axis is the magnetic easy axis of the $3d$ electron spins, which is confirmed by the single crystal magnetic susceptibility measurements [34]. The suppression of the T_N is largest in the a axis direction and T_N decreases down to about 2.07 K at 1 T and 0.45 K at 2 T, while that of the $H \parallel c$ axis direction is 2.21 K at 2 T and 1.18 K at 4 T, respectively.

Figure 4a shows the temperature dependences of the heat capacity under magnetic field at 1 T applied in several directions from the b axis (out-of-plane direction) to the a axis (in-plane direction) in the ab plane measured by the long relaxation method using the same calorimetry cell. By tilting the field direction from the b axis to the a axis, T_N shifts to the lower temperature region. The field-direction dependence of the T_N at 1 T and 2 T are summarized in Figure 4b. The T_N is 2.37 K at 1 T and 2.24 K at 2 T, respectively, in the b axis direction. The degree of the suppression of T_N by the magnetic field is almost the same between the b axis direction and the c axis direction, which is also consistent with the previous magnetic susceptibility measurements [34].

The curious features appear in the temperature dependence of the magnetic heat capacity when the magnetic fields are applied to the a axis direction. As confirmed in Figure 3, the magnetic field works to suppress the peak gradually and a kind of hump structure appears. In the data of Figure 3a, the hump structure is observed clearly at 1.5 T and 2 T. It is important to mention that the magnetic entropy at 8 K is retained as $R\ln6$ even though the temperature dependence of the magnetic heat capacity shows such a drastic change. This result indicates that the hump structure is derived from the spin degrees of freedom of the $3d$ electron spins and the magnetic nature is gradually changed inside the antiferromagnetic phase. Note that the temperature dependences of the magnetic heat capacity under the magnetic field parallel to the b axis and the c axis are almost the same, and the hump structure is not observed when the magnetic field is applied in both directions.

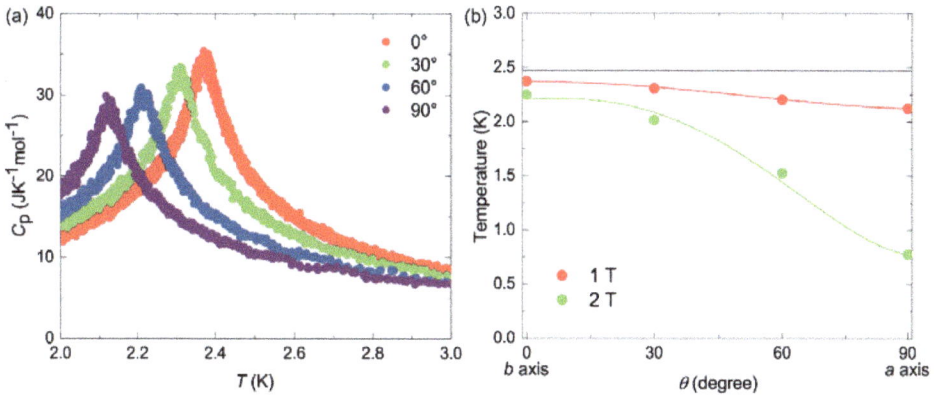

Figure 4. (**a**) Temperature dependences of the heat capacity of κ-(BETS)$_2$FeBr$_4$. The magnetic field of 1 T is applied in several directions from the *b* axis (0°) to the *a* axis (90°). (**b**) Field-direction dependence of the antiferromagnetic transition temperature at 1 T (red) and 2 T (green). The black line represents the antiferromagnetic transition temperature of 2.47 K at 0 T.

A similar tendency is observed in κ-(BETS)$_2$FeCl$_4$, though the magnetic features are shifted to the lower energy scale. Figure 5 shows the temperature dependences of the magnetic heat capacity under magnetic fields applied parallel to the *a* axis and the *c* axis. The hump structure is also observed in κ-(BETS)$_2$FeCl$_4$ at 0.5 T as is indicated in Figure 5a. These results certainly suggest that the same magnetic nature in κ-(BETS)$_2$FeBr$_4$ also appears in κ-(BETS)$_2$FeCl$_4$.

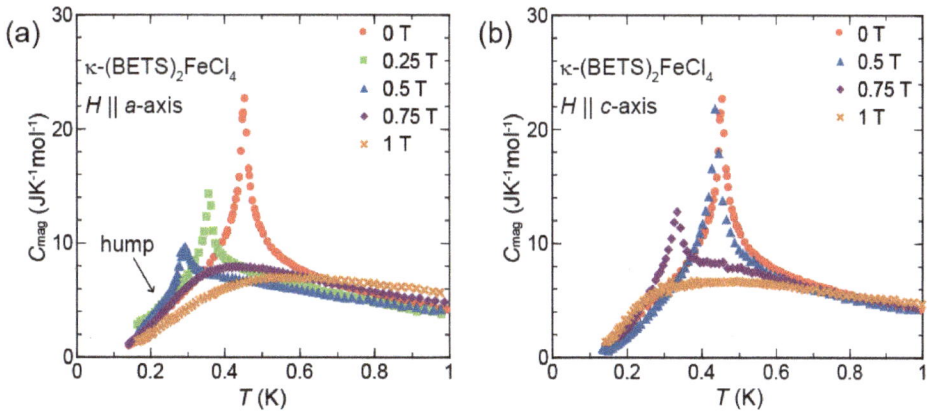

Figure 5. Temperature dependences of the magnetic heat capacity of κ-(BETS)$_2$FeCl$_4$ under magnetic fields parallel to (**a**) the *a* axis and (**b**) the *c* axis. Reproduced with permission from [44].

In order to compare the data of the two compounds in the same framework, the magnetic heat capacity data obtained in the configuration of the $H \parallel a$ axis of the two compounds are displayed in the same figure in Figure 6. In this figure, the temperatures are scaled by the dominant direct magnetic interaction, $|J_{dd}|/k_B$, of each compound. It is worthy of note that not only the high temperature tails derived from the short-range ordering, but also the hump structures scale well between the two compounds, suggesting that the origin of the hump structure is related to the energy scale of the direct magnetic interaction. The temperature dependence of the magnetic heat capacity of the κ-(BETS)$_2$FeCl$_4$ at 0 T coincides with that of κ-(BETS)$_2$FeBr$_4$ at 1.5 T. The magnetic field of 1.5 T corresponds to the difference of the magnitude of the direct magnetic interactions between κ-(BETS)$_2$FeBr$_4$ and

κ-(BETS)$_2$FeCl$_4$ compounds. From this scaling result and quantitative evaluation of the magnetic interactions, we can claim that even though the 3*d* electron spin systems undergo the antiferromagnetic ordered state by the indirect magnetic interaction, the characteristics of the direct magnetic interactions, J_{dd}, still remains as internal degrees of freedom below T_N, which is observed as a hump structure in the magnetic heat capacity.

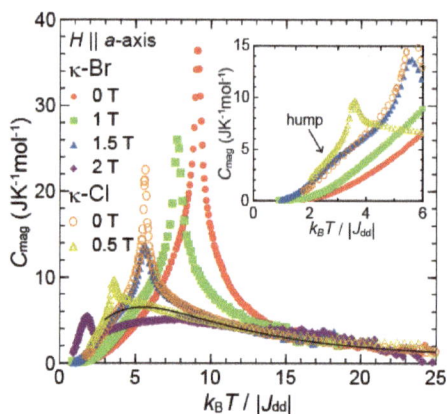

Figure 6. Temperature dependences of the magnetic heat capacities of κ-(BETS)$_2$FeBr$_4$ (κ-Br) and κ-(BETS)$_2$FeCl$_4$ (κ-Cl). The solid line represents the theoretical curve of the $S = 5/2$ one-dimensional Heisenberg chain [53]. Reproduced with permission from [44].

It is generally recognized that the compounds possessing a low dimensional structure of the magnetic ions or molecules usually show short-range fluctuations in their thermodynamic and magnetic properties. In the case of κ-(BETS)$_2$FeX$_4$ systems, the tail structure in the magnetic heat capacity observed at higher temperatures is due to the short-range ordering derived from the dominant direct magnetic interaction, and the long-range ordering is induced by the indirect magnetic interaction. Numerous works to investigate the thermodynamic nature of such low dimensional magnetic systems by heat capacity measurements have been performed and comprehensive discussion is given in several review papers and textbooks [52,54,55]. However, the situation of the present compounds is quite different from the conventional low dimensional magnetic systems. The inter-layer magnetic interactions between the π electrons and the 3*d* electrons are relatively large and in the same order with the direct magnetic interactions in the *a* axis direction. As is shown schematically in Figure 1d, the indirect magnetic interaction between the 3*d* electron spins through the π-*d* interaction exists in all directions and they can form a three-dimensional ordering at rather high temperatures. However, only in the *a* axis direction, the direct magnetic interaction of J_{dd} exists with a similar order as the indirect magnetic interaction. This interaction is quite anisotropic, like one-dimensional magnetic systems. Although the three-dimensional ordering occurs as a cooperative effect of the direct and the indirect magnetic interactions, the effect of the direct magnetic interaction remains as internal degrees of freedom even though the three-dimensional ordering is established at T_N. The low dimensional magnetic system, which has direct and indirect magnetic interactions with a similar order, is quite rare and probably the unique point for this material. Moreover, the strong electron correlation among the π electrons also influences the magnetic nature of the 3*d* electron spin system. Such situation characteristic in the π-*d* interacting system changes the three-dimensional ordered state to the unconventional magnetic state such that the nature of the one-dimensional direct magnetic interaction remains even below T_N. Interestingly, the similar hump structure is also observed more clearly in the λ-(BETS)$_2$FeCl$_4$, which is another π-*d* interacting system, which is reviewed in [56]. Recently, some theoretical studies on the emergence of the hump structure in the π-*d* interacting system have been reported [57]. These results claim that the curious magnetic behaviors reviewed in this section are characteristic behavior of the π-d interacting systems and shed light on a new aspect of molecular magnets.

6. Anisotropic Magnetic Field Dependence of the Superconducting Transition of κ-(BETS)$_2$FeBr$_4$

The coupling of the $3d$ electrons and the π electrons also influences the superconducting nature. Below the antiferromagnetic transition temperature, κ-(BETS)$_2$FeBr$_4$ and κ-(BETS)$_2$FeCl$_4$ show a superconducting transition at 1.5 K and 0.1 K, respectively [34–36]. The superconductivity of the κ-type compounds has two-dimensional characters with line-nodes in the cylindrical Fermi surface and the pairing state is considered as the anisotropic d_{x2-y2} or d_{xy} type depending on the magnitude of dimerization and frustration factor inherent in the triangularity of the κ-type packing [58,59]. To discuss the superconducting nature, it is necessary to extract the electronic heat capacity of the π electrons. In the case of κ-(BETS)$_2$FeBr$_4$, the magnetic heat capacity originating from the $3d$ electron spins gives a dominant contribution to the total heat capacity, and this situation makes it difficult to analyze the accurate electronic heat capacity of the π electrons. However, by subtracting the appropriate magnetic heat capacity by the procedure explained in [60], the thermal anomaly due to the superconducting transition was evaluated. Figure 7 shows the temperature dependence of the heat capacity of κ-(BETS)$_2$FeBr$_4$ around the superconducting transition temperature and the thermal anomaly due to the superconductivity. The magnitude of the heat capacity jump at T_c = 1.5 K is about ΔC_p ~ 50 mJK^{-1}mol^{-1}, which is a typical value for organic superconductors [45–49]. The deviation from the Bardeen-Cooper-Schrieffer (BCS) curves may imply a possible nodal superconductor. However, it should be emphasized that the ambiguity of the background evaluation exists as a serious factor for further discussion and it is difficult to discuss the temperature dependence of ΔC_p and pair symmetry using BCS theory or other models. We cannot discuss the origin of the small anomaly around 0.8 K at present, since we cannot exclude an extrinsic origin due to the slight change of the thermal conductivity of the wires used as a heat leak in the calorimetry cell.

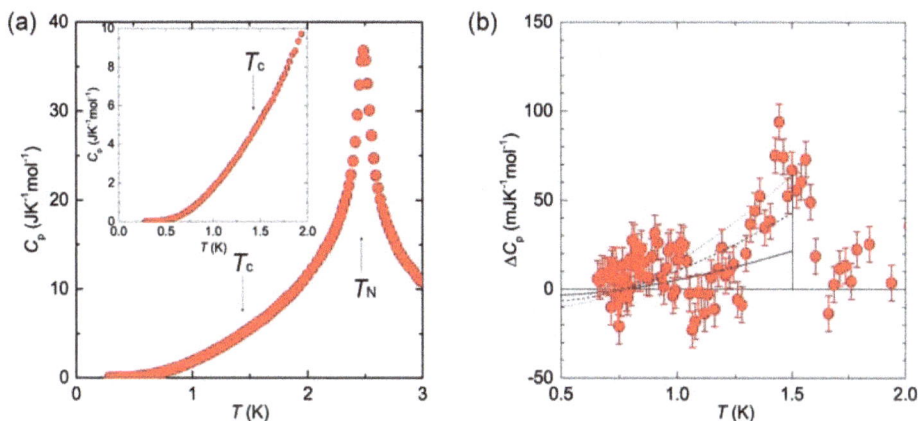

Figure 7. (a) C_p vs. T curve of κ-(BETS)$_2$FeBr$_4$ at 0 T. The inset shows the enlarged view around the superconducting transition temperature. (b) ΔC_p vs. T curve of κ-(BETS)$_2$FeBr$_4$ at 0 T. The solid, dashed, and dotted lines represent the BCS curves assuming that the γ values are 10, 20, 30 mJK^{-2}mol^{-1}, respectively [61]. Reproduced with permission from [60].

The magnetic field dependence of the peak temperature due to the superconductivity shows unusual in-plane anisotropy. Although the exact estimation of the electronic heat capacity under the magnetic field is difficult, the relative field-dependent change of the thermal anomaly and the superconducting transition temperature can be traced by comparing the heat capacity data under magnetic fields parallel to the b axis, which is the perpendicular direction of the conducting layers as was reported in [60]. Figure 8 shows the ΔC_p vs. T curves at several magnetic fields applied parallel to the c axis. The anomaly shows no significant field dependence up to 2 T. Above 2 T, the superconducting transition temperature is gradually decreased and disappears around 2.5 T.

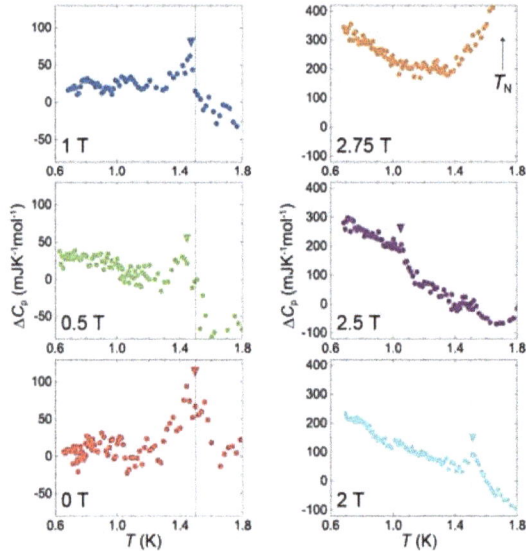

Figure 8. ΔC_p vs. T curves of κ-(BETS)$_2$FeBr$_4$ at several magnetic fields applied parallel to the c axis. The dashed lines represent the superconducting transition temperature of 1.5 K at 0 T. Reproduced with permission from [60].

Figure 9 shows the field-direction dependence of the ΔC_p vs. T curves at 1 T and 2 T. Although several ΔC_p curves contain large offset inherent in the ambiguity of the estimation of the magnetic contribution derived from the $3d$ electron spins, the anomaly associated with the superconducting transition can be traced. Although the superconducting transition shows almost isotropic field-direction dependence at 1 T, it shows significant anisotropic field-direction dependence at 2 T. When the magnetic field-direction is tilted from the c axis to the a axis, the superconducting transition shifts to the lower temperature region drastically, and it is no longer observed above 30° from the c axis in the experimentally available temperature range down to 0.65 K. Such field-direction dependence is not explained by the anisotropy of the Fermi surfaces and the symmetry of the superconducting gap structure [62–64].

Figure 10 shows a magnetic field vs. temperature (B-T) phase diagram determined by the heat capacity measurements. This phase diagram is almost the same with that determined by the transport and magnetic susceptibility measurements by Fujiwara et al. [26,34]. Since the magnetic fields are applied parallel to the in-plane direction of the donor layers, the pair breaking by the orbital effect is not so large. Therefore, in this configuration, the suppression mechanism of electron pairs is mainly determined by the Zeeman effect [65]. The Pauli limit of the weak coupling superconductor is given as $H_P = 1.84T_c$, which corresponds to the 2.6 T for κ-(BETS)$_2$FeBr$_4$. This field is close to the field where the thermal anomaly due to the superconducting transition disappears when the magnetic field is applied parallel to the c axis [66]. The origin of the anisotropy produced above 1 T is attributed to the change of internal magnetic fields induced by the change of antiferromagnetic spin structures. From the magnetization measurement, it is confirmed that when the magnetic field is applied parallel to the a axis, corresponding to the magnetic easy axis of the $3d$ electron spins, the $3d$ electron spins show a metamagnetic transition around 2 T, leading to the drastic change of the internal field [34]. Fujiwara et al. calculated the field dependence of the effective magnetic field for several directions and suggested that the drastic increase of the internal field occurs at the metamagnetic transition field [67]. From these results, it is considered that the increase of the effective magnetic field induced by the change of the antiferromagnetic spin structure destabilizes the superconducting state even below the Pauli limit. In contrast to this, only a slight change is expected for the antiferromagnetic spin arrangement below the metamagnetic transition field. Therefore, the suppression of the superconducting transition temperature is moderate at the weak magnetic field region. On the other hand, only a gradual change of the internal field occurs up to the Pauli-limit value when the magnetic field is applied parallel to the *c* axis. In this direction, the magnetic field is perpendicular to the magnetic easy axis and the magnetic field gradually cants the direction of the $3d$ electron spins while keeping the antiferromagnetically ordered structure. Therefore, the change of the internal field is gradual and the drastic suppression of the superconducting transition temperature is not observed up to the Pauli-limit value. These results suggest that the magnetic state of the $3d$ electron spins influences the superconducting state of the π electron system through the π-d interaction as the change of the effective magnetic field.

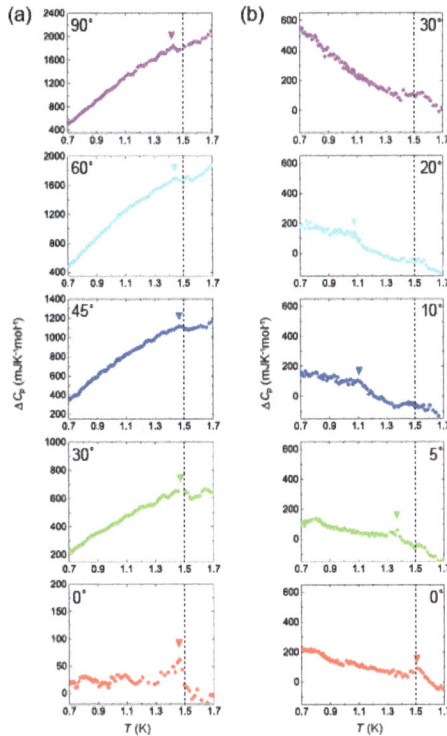

Figure 9. Field-direction dependences of the ΔC_p vs. *T* curves of κ-(BETS)$_2$FeBr$_4$ under the in-plane magnetic field at (**a**) 1 T and (**b**) 2 T. 0° and 90° correspond to the *c* axis and the *a* axis, respectively. Dashed lines represent the superconducting transition temperature of 1.5 K at 0 T. Reproduced with permission from [60].

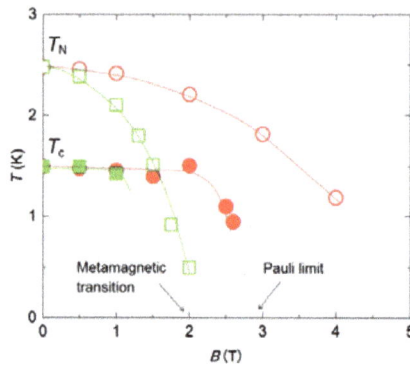

Figure 10. B-T phase diagram of κ-(BETS)$_2$FeBr$_4$ determined by heat capacity measurements. The closed and open symbols represent the superconducting transition temperature and the antiferromagnetic transition temperature. The magnetic field is applied parallel to the a axis (green) and the c axis (red). Reproduced with permission from [60].

7. Summary

In this article, the unique magnetic and superconducting properties of κ-(BETS)$_2$FeX$_4$ (X = Cl, Br) investigated by single crystal heat capacity measurements were reviewed. In the former part, we showed the results of the magnetic heat capacity of the $3d$ electron spin system. The heat capacity measurements revealed that both κ-(BETS)$_2$FeBr$_4$ and κ-(BETS)$_2$FeCl$_4$ showed hump structures only when the external magnetic fields were applied parallel to the a axis, which is the easy axis of the $3d$ electron spins. The hump structures scale well between two compounds by using the value of dominant direct magnetic interaction, $|J_{dd}|/k_B$. Such anisotropic magnetic properties were produced by the coexistence of the direct and the indirect magnetic interactions in this direction. In the latter part, we showed the results of the electronic heat capacity of the π electron system. The superconducting transition anomaly in the electronic heat capacity was observed in κ-(BETS)$_2$FeBr$_4$ at 1.5 K, although the ambiguity of the subtraction of the large magnetic heat capacity of the $3d$ electron spins remains. When the magnetic field was applied parallel to the conducting layers, the superconducting transition temperature also showed anisotropic field dependence even though the superconductivity itself has two-dimensional characters. Although the superconducting transition could be observed near the Pauli limit if the magnetic field was applied parallel to the c axis, it was drastically suppressed if the magnetic field was applied parallel to the a axis similar to the case of T_N. The anisotropic field-direction dependence in the in-plane configuration can be understood by considering the change of the internal magnetic field produced by the $3d$ electron spin system. It should be emphasized that the coexistence of the direct magnetic interaction and the indirect magnetic interaction through the π-d interaction plays a crucial role for both the origin of the unconventional magnetic properties of the $3d$ electron spins and that of the anisotropic field-direction dependence of the superconducting transition temperature.

Author Contributions: Conceptualization, S.F. (Shuhei Fukuoka) and Y.N.; Investigation, S.F. (Shuhei Fukuoka) and S.F. (Sotarou Fukuchi); Writing—original draft preparation, S.F. (Shuhei Fukuoka) and Y.N.; Writing—review and editing, H.A. and A.K.; Funding acquisition, S.F. (Shuhei Fukuoka) and Y.N.

Funding: This work was partly supported by a Grant-in-Aid for JSPS Fellow.

Acknowledgments: Authors are grateful to S. Yamashita and T. Yamamoto for fruitful discussion, and H. Fujiwara, T. Shirahata, and K. Takahashi for the synthesis of large high quality κ-(BETS)$_2$FeX$_4$ (X = Cl, Br) crystals.

Conflicts of Interest: The authors declare no conflict of interest.

References

1. Miyasaka, H.; Nakata, K.; Lecren, L.; Coulon, C.; Nakazawa, Y.; Fujisaki, T.; Sugiura, K.; Yamashita, M.; Clérac, R. Two-dimensional networks based on Mn_4 complex linked by dicyanamide anion: From single-molecule magnet to classical magnet behavior. *J. Am. Chem. Soc.* **2006**, *128*, 3770–3783. [CrossRef] [PubMed]

2. Miyasaka, H.; Madanbashi, T.; Sugimoto, K.; Nakazawa, Y.; Wernsdorfer, W.; Sugiura, K.; Yamashita, M.; Coulon, C.; Clérac, R. Single-chain magnet behavior in an alternated one-dimensional assembly of a Mn^{III} Schiff-base complex and a TNCQ radical. *Chem. Eur. J.* **2006**, *12*, 7028–7040. [CrossRef] [PubMed]

3. Fitta, M.; Pełka, R.; Konieczny, P.; Bałanda, M. Multifunctional molecular magnets: Magnetocaloric effect in octacyanometallates. *Crystals* **2019**, *9*, 9. [CrossRef]

4. Liu, J.-L.; Chen, Y.-C.; Guo, F.-S.; Tong, M.-L. Recent advances in the design of magnetic molecules for use as cryogenic magnetic coolants. *Coord. Chem. Rev.* **2014**, *281*, 26–49. [CrossRef]

5. Miller, J.S.; Epstein, A.J.; Reiff, W.M. Ferromagnetic molecular charge-transfer complexes. *Chem. Rev.* **1988**, *88*, 201–220. [CrossRef]

6. Qiu, Z.; Uruichi, M.; Hou, D.; Uchida, K.; Yamamoto, H.M.; Saitoh, E. Spin-current injection and detection in κ-$(BEDT-TTF)_2Cu[N(CN)_2]Br$. *AIP Adv.* **2015**, *5*, 057167. [CrossRef]

7. Liu, M.; Chuang, Z.; Malissa, H.; Groesbeck, M.; Kavand, M.; McLaughlin, R.; Jamali, S.; Hao, J.; Sun, D.; Davidson, R.A.; et al. Organic-based magnon spintronics. *Nat. Mater.* **2018**, *17*, 308–312. [CrossRef] [PubMed]

8. Kinoshita, M.; Turek, P.; Tamura, M.; Nozawa, K.; Shiomi, D.; Nakazawa, Y.; Ishikawa, M.; Takahashi, M.; Awaga, K.; Inabe, T.; et al. An organic radical ferromagnet. *Chem. Lett.* **1991**, *20*, 1225–1228. [CrossRef]

9. Tamura, M.; Nakazawa, Y.; Shiomi, D.; Nozawa, K.; Hosokoshi, Y.; Ishikawa, M.; Takahashi, M.; Kinoshita, M. Bulk ferromagnetism in the β-phase crystal of the *p*-nitrophenyl nitroxide. *Chem. Phys. Lett.* **1991**, *186*, 401–404. [CrossRef]

10. Korshak, Y.V.; Medvedeva, T.V.; Ovchinnikov, A.A.; Spector, V.N. Organic polymer ferromagnet. *Nature* **1987**, *326*, 370–372. [CrossRef]

11. Miller, J.S.; Epstein, A.J. Organic and organometallic molecular magnetic materials—Designer magnets. *Angew. Chem. Int. Ed. Engl.* **1994**, *33*, 385–415. [CrossRef]

12. Tamura, M.; Hosokoshi, Y.; Shiomi, D.; Kinoshita, M.; Nakazawa, Y.; Ishikawa, M.; Sawa, H.; Kitazawa, T.; Eguchi, A.; Nishio, Y.; et al. Magnetic properties and structures of the α- and δ-phases of *p*-NPNN. *J. Phys. Soc. Jpn.* **2003**, *72*, 1735–1744. [CrossRef]

13. Okabe, T.; Yamaguchi, H.; Kittaka, S.; Sakakibara, T.; Ono, T.; Hosokoshi, Y. Magnetic properties of the $S = 1/2$ honeycomb lattice antiferromagnet 2-Cl-3,6-F_2-V. *Phys. Rev. B* **2017**, *95*, 075120. [CrossRef]

14. Kono, Y.; Kittaka, S.; Yamaguchi, H.; Hosokoshi, Y.; Sakakibara, T. Quasi-one-dimensional Bose-Einstein condensation in the spin-1/2 ferromagnetic-leg ladder 3-I-V. *Phys. Rev. B* **2018**, *97*, 100406. [CrossRef]

15. Caneschi, A.; Gatteschi, D.; Ray, P.; Sessoli, R. Structure and magnetic ordering of a ferrimagnetic helix formed by manganese (II) and a nitronyl nitroxide radical. *Inorg. Chem.* **1991**, *30*, 3936–3941. [CrossRef]

16. Kumagai, H.; Inoue, K. A chiral molecular based metamagnet prepared from manganese ions and a chiral triplet organic radical as a bridging ligand. *Angew. Chem. Int. Ed.* **1999**, *38*, 1601–1603. [CrossRef]

17. Gatteschi, D.; Sessoli, R.; Villain, J. *Molecular Nanomagnets*; Oxford University Press: Oxford, UK, 2010.

18. Sato, O. Dynamic molecular crystals with switchable physical properties. *Nat. Chem.* **2016**, *8*, 644–656. [CrossRef] [PubMed]

19. Nihei, M.; Shiga, T.; Maeda, Y.; Oshio, H. Spin crossover iron(III) complexes. *Coord. Chem. Rev.* **2007**, *251*, 2606–2621. [CrossRef]

20. Shimizu, Y.; Miyagawa, K.; Kanoda, K.; Maesato, M.; Saito, G. Spin liquid state in an organic Mott insulator with a triangular lattice. *Phys. Rev. Lett.* **2003**, *91*, 107001. [CrossRef]

21. Christou, G.; Gatteschi, D.; Hendrickson, D.N.; Sessoli, R. Single-molecule magnets. *Mrs Bull.* **2000**, *25*, 66–71. [CrossRef]

22. Ohkoshi, S.; Imoto, K.; Tsunobuchi, Y.; Takano, S.; Tokoro, H. Light-induced spin-crossover magnet. *Nat. Chem.* **2011**, *3*, 564–569. [CrossRef] [PubMed]

23. Enoki, T.; Miyazaki, A. Magnetic TTF-based charge-transfer complexes. *Chem. Rev.* **2004**, *104*, 5449–5477. [CrossRef] [PubMed]

24. Coronado, E.; Day, P. Magnetic molecular conductors. *Chem. Rev.* **2004**, *104*, 5419–5448. [CrossRef] [PubMed]
25. Kobayashi, H.; Kobayashi, A.; Cassoux, P. BETS as a source of molecular magnetic superconductors (BETS = bis(ethylenedithio)tetraselenafulvalene. *Chem. Soc. Rev.* **2000**, *29*, 325–333. [CrossRef]
26. Kobayashi, H.; Cui, H.B. Organic metal and superconductors based on BETS (BETS = Bis (ethylenedithio)tetraselenafulvalene). *Chem. Rev.* **2004**, *104*, 5265–5288. [CrossRef] [PubMed]
27. Kikuchi, K.; Nishikawa, H.; Ikemoto, I.; Toita, T.; Akutsu, H.; Nakatsuji, S.; Yamada, J. Tetrachloroferrate (III) salts of BDH-TTP [2,5-Bis(1,3-dithiolan-2-ylidene)-1,3,4,6-tetrathiapentalene] and BDA-TTP [2,5-Bis(1,3-dithian-2-ylidene)-1,3,4,6-tetrathiapentalene]: Crystal structures and physical properties. *J. Solid State Chem.* **2002**, *168*, 503–508. [CrossRef]
28. Fujiwara, H.; Wada, K.; Hiraoka, T.; Hayashi, T.; Sugimoto, T.; Nakazumi, H.; Yokogawa, K.; Teramura, M.; Yasuzuka, S.; Murata, K.; et al. Stable metallic behavior and antiferromagnetic ordering of Fe(III) *d* spins in (EDO-TTFVO)$_2$·FeCl$_4$. *J. Am. Chem. Soc.* **2005**, *127*, 14166–14167. [CrossRef] [PubMed]
29. Xiao, X.; Hayashi, T.; Fujiwara, H.; Sugimoto, T.; Noguchi, S.; Weng, Y.; Yoshino, H.; Murata, K.; Aruga-Katori, H. An antiferromagnetic molecular metal based on a new bent-donor molecule. *J. Am. Chem. Soc.* **2007**, *129*, 12618–12619. [CrossRef] [PubMed]
30. Coronado, E.; Galán-Mascarós, J.R.; Gómez-Garća, C.J.; Laukhin, V. Coexistence of ferromagnetism and metallic conductivity in a molecule-based layered compound. *Nature* **2000**, *408*, 447–449. [CrossRef]
31. Hotta, C.; Fukuyama, H. Effects of localized spins in quasi-two dimensional organic conductor. *J. Phys. Soc. Jpn.* **2000**, *69*, 2577–2596. [CrossRef]
32. Kobayashi, A.; Udagawa, T.; Tomita, H.; Naito, T.; Kobayashi, H. New organic metals based on BETS compounds with MX$_4^-$ anions (BETS =bis(ethylenedithio)tetraselenafulvalene; M = Ga, Fe, In; X = Cl, Br). *Chem. Lett.* **1993**, *22*, 2179–2182. [CrossRef]
33. Kobayashi, H.; Tomita, H.; Naito, T.; Kobayashi, A.; Sakai, F.; Watanabe, T.; Cassoux, P. New BETS conductors with magnetic anions (BETS = bis(ethylenedithio)tetraselenafulvalene). *J. Am. Chem. Soc.* **1996**, *118*, 368–377. [CrossRef]
34. Fujiwara, H.; Fujiwara, E.; Nakazawa, Y.; Narymbetov, B.Z.; Kato, K.; Kobayashi, H.; Kobayashi, A.; Tokumoto, M.; Cassoux, P. A novel antiferromagnetic organic superconductor κ-(BETS)$_2$FeBr$_4$ [Where BETS = Bis(ethylenedithio)tetraselenafulvalene]. *J. Am. Chem. Soc.* **2001**, *123*, 306–314. [CrossRef] [PubMed]
35. Otsuka, T.; Kobayashi, A.; Miyamoto, Y.; Kiuchi, J.; Nakamura, S.; Wada, N.; Fujiwara, E.; Fujiwara, H.; Kobayashi, H. Organic antiferromagnetic metals exhibiting superconducting transitions κ-(BETS)$_2$FeX$_4$ (X = Cl, Br); Drastic effect of halogen substitution on the successive phase transitions. *J. Solid State Chem.* **2001**, *159*, 407–412. [CrossRef]
36. Otsuka, T.; Kobayashi, A.; Miyamoto, Y.; Kikuchi, J.; Wada, N.; Ojima, E.; Fujiwara, H.; Kobayashi, H. Successive antiferromagnetic and superconducting transitions in an organic metal, κ-(BETS)$_2$FeCl$_4$. *Chem. Lett.* **2000**, *29*, 732–733. [CrossRef]
37. Uji, S.; Shinagawa, H.; Terashima, T.; Yakabe, T.; Terai, Y.; Tokumoto, M.; Kobayashi, A.; Tanaka, H.; Kobayashi, H. Magnetic-field-induced superconductivity in a two-dimensional organic conductor. *Nature* **2001**, *410*, 908–910. [CrossRef]
38. Tanaka, H.; Ojima, E.; Fujiwara, H.; Nakazawa, Y.; Kobayashi, H.; Kobayashi, A. A new κ-type organic superconductor based on BETS molecules, κ-(BETS)$_2$GaBr$_4$ [BETS = bis(ethylenedithio)tetraselenafulvalene]. *J. Mater. Chem.* **2000**, *10*, 245–247. [CrossRef]
39. Konoike, T.; Uji, S.; Terashima, T.; Nishimura, M.; Yasuzuka, S.; Enomoto, K.; Fujiwara, H.; Zhang, B.; Kobayashi, H. Magnetic-field-induced superconductivity in the antiferromagnetic organic superconductor κ-(BETS)$_2$FeBr$_4$. *Phys. Rev. B* **2004**, *70*, 094514. [CrossRef]
40. Jaccarino, V.; Peter, M. Ultra-high-field superconductivity. *Phys. Rev. Lett.* **1962**, *9*, 290–292. [CrossRef]
41. Fujiyama, S.; Takigawa, M.; Kikuchi, J.; Cui, H.-B.; Fujiwara, H.; Kobayashi, H. Compensation of effective field in the field-induced superconductor κ-(BETS)$_2$FeBr$_4$ observed by ^{77}Se NMR. *Phys. Rev. Lett.* **2006**, *96*, 217001. [CrossRef]
42. Fukuoka, S.; Horie, Y.; Yamashita, S.; Nakazawa, Y. Development of heat capacity measurement system for single crystals of molecule-based compounds. *J. Therm. Anal. Calorim.* **2013**, *113*, 1303–1308. [CrossRef]

43. Imajo, S.; Fukuoka, S.; Yamashita, S.; Nakazawa, Y. Construction of relaxation calorimetry for 10^{1-2} µg samples and heat capacity measurements of organic complexes. *J. Therm. Anal. Calorim.* **2016**, *123*, 1871–1876. [CrossRef]

44. Fukuoka, S.; Yamashita, S.; Nakazawa, Y.; Yamamoto, T.; Fujiwara, H.; Shirahata, T.; Takahashi, K. Thermodynamic properties of antiferromagnetic ordered states of π-d interacting systems of κ-(BETS)$_2$FeX$_4$ (X = Br, Cl). *Phys. Rev. B* **2016**, *93*, 245136. [CrossRef]

45. Andraka, B.; Kim, J.S.; Stewart, G.R.; Carlson, K.D.; Wang, H.H.; Williams, J.M. Specific heat in high magnetic field of κ-di[Bis(ethylenedithio)tetrathiafulvalene]-di(thiocynano)Cuprate [κ-(ET)$_2$Cu(NCS)$_2$]; Evidence for strong-coupling superconductivity. *Phys. Rev. B* **1989**, *40*, 11345. [CrossRef]

46. Andraka, B.; Jee, C.S.; Kim, J.S.; Stewart, G.R.; Carlson, K.D.; Wang, H.H.; Crouch, A.V.S.; Kini, A.M.; Williams, J.M. Specific heat of the high T_c organic superconductor κ-(ET)$_2$Cu[N(CN)$_2$]Br. *Solid State Commun.* **1991**, *79*, 57–59. [CrossRef]

47. Wosnitza, J.; Liu, X.; Schweitzer, D.; Keller, H.J. Specific heat of the organic superconductor κ-(BEDT-TTF)$_2$I$_3$. *Phys. Rev. B* **1994**, *59*, 12747. [CrossRef]

48. Nakazawa, Y.; Yamashita, S. Thermodynamic properties of κ-(BEDT-TTF)$_2$$X$ salts: Electron correlation and superconductivity. *Crystals* **2012**, *2*, 741–761. [CrossRef]

49. Nakazawa, Y.; Imajo, S.; Matsuoka, Y.; Yamashita, S.; Akutsu, H. Thermodynamic picture of dimer-Mott organic superconductors revealed by heat capacity measurements with external and chemical pressure control. *Crystals* **2018**, *8*, 143. [CrossRef]

50. Mori, T.; Katsuhara, M. Estimation of πd-interaction in organic conductors including magnetic anions. *J. Phys. Soc. Jpn.* **2002**, *71*, 826–844. [CrossRef]

51. Gopal, E.S.R. *Specific Heats at Low Temperature*; Heywood Books: London, UK, 1966.

52. De Jongh, L.J.; Miedema, A.R. Experiments on simple magnetic model systems. *Adv. Phys.* **1974**, *23*, 1–260. [CrossRef]

53. Blöte, H.W. The heat capacities of linear Heisenberg chains. *Physica* **1974**, *78*, 302–307. [CrossRef]

54. Sorai, M.; Nakazawa, Y.; Nakano, M.; Miyazaki, Y. Calorimetric investigation of phase transitions occurring in molecule-based magnets. *Chem. Rev.* **2013**, *113*, PR41–PR122. [CrossRef] [PubMed]

55. Blundell, S. *Magnetism in Condensed Matter*; Oxford University Press: Oxford, UK, 2001.

56. Akiba, H.; Shimada, K.; Tajima, N.; Kajita, K.; Nishio, Y. Paramagnetic metal-antiferromagnetic insulator transition in π-d system λ-BETS$_2$FeCl$_4$, BETS = Bis(ethylenedithio)tetraselenafulvalene. *Crystals* **2012**, *2*, 984–995. [CrossRef]

57. Ito, K.; Shimahara, H. Mean field theory of a coupled Heisenberg model and its application to an organic antiferromagnet with magnetic anions. *J. Phys. Soc. Jpn.* **2016**, *85*, 024704. [CrossRef]

58. Nakazawa, Y.; Kanoda, K. Low-temperature specific heat of κ-(BEDT-TTF)$_2$Cu[N(CN)$_2$]Br in the superconducting state. *Phys. Rev. B* **1997**, *55*, R8670–R8673. [CrossRef]

59. Taylor, O.J.; Carrington, A.; Schlueter, J.A. Specific-heat measurements of the gap structure of the organic superconductors κ-(ET)$_2$Cu[N(CN)$_2$]Br and κ-(ET)$_2$Cu(NCS)$_2$. *Phys. Rev. Lett.* **2007**, *99*, 057001. [CrossRef] [PubMed]

60. Fukuoka, S.; Yamashita, S.; Nakazawa, Y.; Yamamoto, T.; Fujiwara, H. Anisotropic field dependence of the superconducting transition in the magnetic molecular superconductor κ-(BETS)$_2$FeBr$_4$. *J. Phys. Soc. Jpn.* **2017**, *86*, 014706. [CrossRef]

61. Muhlschlegel, B. Die thermodynamischen Funktionen des Supraleiters. *Z. Phys.* **1959**, *155*, 313–327. [CrossRef]

62. Sasaki, T.; Toyota, N. Anisotropic galvanomagnetic effect in the quasi-two-dimensional organic conductor α-(BEDT-TTF)$_2$KHg(SCN)$_4$, where BEDT-TTF is bis(ethylenedithio)tetrathiafulvalene. *Phys. Rev. B* **1994**, *49*, 10120–10130. [CrossRef]

63. Wang, H.H.; Vanzile, M.L.; Schlueter, J.A.; Kini, A.M.; Sche, P.P. In-plane ESR microwave conductivity measurements and electronic band structure studies of the organic superconductor β''-(BEDT-TTF)$_2$SF$_5$CH$_2$CF$_2$SO$_3$. *J. Phys. Chem. B* **1999**, *103*, 5493–5499. [CrossRef]

64. Ishiguro, T.; Kajimura, K.; Bando, H.; Murata, K.; Anzai, H. Electronic state of (TMTSF)$_2$ClO$_4$ in metallic region. *Mol. Cryst. Liq. Cryst.* **1985**, *119*, 19–26. [CrossRef]

65. Tinkham, M. *Introduction to Superconductivity*, 2nd ed.; McGraw-Hill: New York, NY, USA, 1996.

66. Clogston, A.M. Upper limit for the critical field in hard superconductors. *Phys. Rev. Lett.* **1962**, *9*, 266–267. [CrossRef]

67. Fujiwara, H.; Kobayashi, H.; Fujiwara, E.; Kobayashi, A. An indication of magnetic-field-induced superconductivity in a bifunctional layered organic conductor, κ-(BETS)$_2$FeBr$_4$. *J. Am. Chem. Soc.* **2002**, *124*, 6816–6817. [CrossRef] [PubMed]

MDPI

St. Alban-Anlage 66

4052 Basel

Switzerland

Tel. +41 61 683 77 34

Fax +41 61 302 89 18

www.mdpi.com

Crystals Editorial Office

E-mail: crystals@mdpi.com

www.mdpi.com/journal/crystals

www.ingramcontent.com/pod-product-compliance
Lightning Source LLC
Chambersburg PA
CBHW041217220326
41597CB00033BA/6003